DERIVED SI UNITS AND CONVERSION FACTORS

$1\ N = 1\ kg\ m\ s^{-2}$ Force (mass $\cdot$ acceleration)

$1\ Pa = 1\ N\ m^{-2} = 1\ kg\ m^{-1}\ s^{-2}$ Stress

$1\ J = 1\ N\ m = 1\ kg\ m^2\ s^{-2}$ Work or energy

$1\ W = 1\ J\ s^{-1}$ Power

$1\ bar = 0.1\ MN\ m^{-2} = 0.1\ MPa = 0.9868\ atm$

$1\ cal = 4.18\ J$

$1\ a = 3.15569 \times 10^7\ s$

$0°C = 273.15\ K$

Principles
of
Glacier Mechanics

Roger LeB. Hooke

Adjunct Professor
Department of Geology and Institute for Quaternary Studies
University of Maine, Orono

Professor
Department of Geology and Geophysics
University of Minnesota, Minneapolis

A textbook for upper division and graduate students with an interest in the flow of glaciers and the origins of landforms produced by glaciers

Prentice Hall, Upper Saddle River, New Jersey 07458

Library of Congress Cataloging-in-Publication Data

Hooke, Roger LeB
    Principles of glacier mechanics / Roger LeB. Hooke.
        p.   cm.
    "A textbook for upper division and graduate students with an
interest in the flow of glaciers and the orgins of landforms
produced by glaciers."
    Includes bibliographical references and index.
    ISBN 0-13-243312-5
    1. Glaciers.   2. Glacial landforms.   I. Title.
    GB2403.2.H66   1998
    551.31'3—dc21   97–5639
                                                                 CIP

Executive Editor: Robert McConnin
Manufacturing Manager: Trudy Pisciotti
Editorial production supervision
  and interior design: Innodata
Cover design: Karen Salzbach

©1998 by Prentice Hall
Simon and Schuster/A Viacom Company
Upper Saddle River, New Jersey 07458

Printed in the United States of America
10 9 8 7 6 5 4 3 2 1

ISBN 0-13-243312-5

Prentice-Hall International (UK) Limited, *London*
Prentice-Hall of Australia Pty. Limited, *Sydney*
Prentice-Hall Canada Inc. *Toronto*
Prentice-Hall Hispanoamericana, S.A., *Mexico*
Prentice-Hall of India Private Limited, *New Delhi*
Prentice-Hall of Japan, Inc., *Tokyo*
Simon & Schuster Asia Pte. Ltd., *Singapore*
Editora Prentice-Hall do Brasil, Ltda., *Rio de Janeiro*

Contents

It is with a deep sense of gratitude that I dedicate this book to those who, at various times through the formative stages of my life, guided me into the most exciting and rewarding career I can imagine: the study of our Earth.

> *To my parents, who opened many doors for me;*
>
> *To my older brother, Richard, who led me through a door leading to the wilderness;*
>
> *To John Muir, who opened my eyes to the spirituality in wilderness;*
>
> *To my wife, Ann, who introduced me to geology;*
>
> *To John P. Miller, who focused my attention on processes at the Earth's surface; and*
>
> *To Robert P. Sharp, who taught me that basic physical principles could be used to understand these processes.*

Preface

One might well ask why one should write a book about so specialized a subject as glacier mechanics when there are already other good books on this subject written by eminent glaciologists. This text is an outgrowth of a course that I teach to students who, in many cases, do not have any background in continuum mechanics. Consequently, it was necessary to start at a level considerably less advanced than that at which other similar books begin, and to develop the theoretical principles one step at a time. Thus, unlike other books on the subject and the general scientific literature, in which space is at a premium, the steps leading from one equation to another are, in most cases, easily seen. In addition, qualitative interpretations of the equations are often provided to clarify the physics behind the mathematics. Capable students with a solid background in basic physics and in differential and integral calculus, and with some modest exposure to differential equations, will have little difficulty understanding the concepts and derivations presented.

My goal in writing this book was not to produce a comprehensive treatise on glacier mechanics, but rather to develop the basic foundation upon which the modern literature on this subject rests. Thus, many topics are not covered, or are treated in less detail than some readers might wish. However, students who have a full appreciation for the concepts in this book will have the background they need to understand most of the current literature.

Beginning students in glaciology will find that this book will save them many long hours of searching through the background literature to clarify basic concepts. Glacial geologists and geomorphologists will also find much of value, including applications of glacier physics to the origin of some glacial landforms. Structural geologists and others with interest in stress and deformation will likewise discover that glaciers are, in fact, monomineralic rock masses that are deforming at Earth's surface where they can be observed in detail.

The appendix contains a series of problems keyed to individual chapters. The endpapers provide a compilation of physical constants relevant to ice and a list of SI units and conversion factors.

The encouragement I have received in this undertaking from many present and former students, as well as from other glaciologists, has been a major stimulus in bringing this work to completion. I trust the final product is worthy of their confidence. The book has benefited from the critical comments of R.W. Baker, C.R. Bentley, G.K.C. Clarke, E.M. Grace, N.R. Iverson, M. Kuhn, M.F. Meier, C.F. Raymond, R.L. Shreve, J. Weertman, and especially B. Hanson, T. Jóhannesson, J.F. Nye, and I. Whillans.

Roger LeB. Hooke
Deer Isle, Maine

1

Why Study Glaciers?

Before delving into the mathematical intricacies with which much of this book is concerned, we should ask: Why are we pursuing the topic of glacier mechanics? For many who would like to understand how glaciers move, how they sculpt the landscape, how they respond to climatic change, mathematics does not come easily. I assure you that all of us have to think carefully about the meaning of the expressions that seem so simple to write out but so difficult to understand. Only then do they become part of our vocabulary, and only then can we make use of the added precision that mathematical analysis, properly formulated, is able to bring. Is it worth the effort? That depends upon your objectives; on why you chose to study glaciers.

There are many reasons, of course. Some are personal, some academic, and some socially significant. To me, the personal reasons are among the most important: Glaciers occur in spectacular areas, often remote, that have not been scarred by human activities. Through glaciology, I have had the opportunity to live in these areas; to drift silently in a kayak on an ice-dammed lake in front of our camp as sunset gradually merges with sunrise on an August evening; to marvel at the northern lights while out on a short ski tour before bedtime on a December night; and to reflect on the meaning of life and of our place in nature. Maybe some of you share these needs and will choose to study glaciers for this reason. I have found that many glaciologists do share them, and this leads to a comradeship which is rewarding in itself.

Academic reasons for studying glaciers are perhaps difficult to separate from socially significant ones. However, in three academic disciplines, the application of glaciology to immediate social problems is at least one step removed from the initial research. The first of these is glacial geology. Glaciers once covered 30% of the land area of Earth and left deposits of diverse shape and composition. How were these deposits formed, and what can they tell us about the glaciers that made them? The second discipline is structural geology; glacier ice is a metamorphic rock that can be observed in the process of deformation at temperatures close to the melting point. From the study of this deformation, both in the laboratory and in the field, much can be learned about the origin of metamorphic structures in other crystalline rocks that were deformed deep within Earth. The final discipline is paleoclimatology. Glaciers record climatic fluctuations in two ways: The deposits left during successive advances and retreats provide a coarse record of climatic change that, with careful study, a little luck, and a good deal of skill, can be placed in correct chronological order and dated. A more detailed record is contained in ice cores from polar glaciers such as the Antarctic and Greenland ice sheets. Isotopic and chemical variations in these cores reflect changes in the temperature and composition of the atmosphere.

Changes during the past several centuries to several millennia can be rather precisely dated using core stratigraphy. Changes further back in time are dated with less certainty using flow models.

Relatively recent changes in climate and in concentrations of certain anthropogenic substances in the atmosphere are attracting increasing attention as humans struggle with problems of feeding an expanding population and of maintaining a healthy living environment. Ice-core studies provide a baseline from which to measure these anthropogenic changes. Such studies can be of a basic or of a more applied nature.

Other applications of glaciology are not hard to find. An increasing number of people in northern and mountainous lands live so close to glaciers that their lives would be severely altered by ice advances comparable in magnitude to the retreats that have taken place during the past century in many parts of the world. Tales of glacier advances gobbling up farms and farm buildings and of ice falls smashing barns and houses are common from the seventeenth and eighteenth centuries, a period of ice advance as the world entered the Little Ice Age. Records tell of buildings being crushed into small pieces and mixed with "soil, grit, and great rocks" (Grove, 1988, p. 72). The Mer de Glace in France presented a particular problem during this time period, and several times during the seventeenth century exorcists were sent out to deal with the "spirits" responsible for its advance. They appeared to have been successful, as the glaciers there were then near their Little Age maxima and beginning to retreat.

Other people live in proximity to streams draining lakes damned by glaciers. Some of the biggest floods known from the geologic record resulted from failure of such ice dams, and smaller floods of the same origin have devastated communities in the Alps and Himalayas.

Somewhat further from human living environments, one finds glaciers astride economically valuable deposits, or discharging icebergs into the shipping lanes through which such deposits are moved. What complications would be encountered, for example, if mining engineers were to dig an open-pit mine through the edge of the Greenland Ice Sheet to tap an iron deposit? What is the possibility that Columbia Glacier in Alaska will retreat very rapidly, as have many neighboring glaciers, thus increasing perhaps tenfold, perhaps one hundred-fold, the flux of icebergs into the shipping lanes leading to the port of Valdez, at the southern end of the trans-Alaska pipeline? Were shipping to be halted there for an extended period so that the oil flow through the pipeline had to be stopped, oil would congeal in the pipe, making what one glaciologist referred to as the world's longest candle.

Glacier ice itself is an economically valuable deposit; glaciers contain 60% of the world's freshwater, and peoples in arid lands have seriously studied the possibility of towing icebergs from Antarctica to serve as a source of water. People in mountainous countries use the water not only for drinking but also as a source of hydroelectric power. By tunneling through the rock under a glacier and thence up to the ice-rock interface, they trap water at a higher elevation than would be possible otherwise, and thus increase the energy yield.

With the threat of global warming hanging over the world, the large volume of water locked up in glaciers and ice sheets represents a potential hazard for human activities in coastal areas. Collapse of the West Antarctic Ice Sheet could lead to a worldwide rise in sea level of 7 m in, perhaps, less than a century. Were this to be followed by melting of the East Antarctic Ice Sheet, sea level could rise an additional 50 m or so. Concern over these prospects has stimulated a great deal of research in the past decade.

Finally, we should mention a proposal to dispose of radioactive waste by letting it melt its way to the base of the Antarctic Ice Sheet. How long would such waste remain isolated from the biologic environment? How would the heat released affect the flow of the ice sheet? Might it cause a surge, with thousands of cubic kilometers of ice dumped into the oceans over a period of a few decades? This would raise sea level several tens of meters, with, again, interesting consequences! To accommodate these concerns, later versions of the proposal called for suspending the waste canisters on wires anchored at the glacier surface. The whole project was later abandoned, however, but not entirely on glaciological grounds. Rather, there seemed to be no risk-free way to transport the waste to the Antarctic.

A good quantitative understanding of the physics of glaciers is essential for rigorous treatment of a number of these problems of academic interest, as well as for accurate analysis of various engineering and environmental problems of concern to humans. The fundamental principles upon which this understanding is based are those of physics and, to a lesser extent, chemistry. Application of these principles to glacier dynamics is initially straightforward, but as with many problems, the more we seek to understand the behavior of glaciers the more involved and, in many respects, the more interesting the applications become.

Thus, we have answered our first question: We study glaciers for the same reasons that we study many other features of the natural landscape, but also for a special reason that I will try to impart to you, wordlessly, if you will stand with me looking over a glacier covered with a thick blanket of fresh powder snow to distant peaks, bathed in alpine glow, breathless from a quick climb up a steep slope after a day of work, but with skis ready for the telemark run back to camp. "Mäktig," my companion said—powerful.

2

Some Basic Concepts

This chapter introduces a few basic concepts that will be used frequently throughout this book. First, we review some commonly used classifications of glaciers by shape and thermal characteristics. We then consider the mathematical formulation of the concept of conservation of mass and, associated with it, the condition of incompressibility. This will appear again in Chapters 6 and 9. Finally, we discuss stress and strain rate, and we lay the foundation for understanding the most commonly used flow law for ice. Although a complete consideration of these latter concepts is deferred to Chapter 9, a modest understanding of them is essential for a fuller appreciation of some fundamental concepts presented in Chapters 4–8.

A NOTE ON UNITS

We will use SI (Système International) units in this book. The basic units of length, mass, and time are, respectively, the meter (m), kilogram (kg), and second (s) (MKS). Temperatures are measured in Kelvins (K) or in the derived unit, degrees Celsius (°C). Some other derived units and useful conversion factors are given in the Endpaper.

In comparison with the earlier glaciological literature, one of the most significant changes introduced by use of SI units is that from bars to pascals as the unit of stress. The bar (= 0.1 MPa ~ 1 atm) was a convenient unit because stresses in glaciers are typically ~1 bar.

GLACIER SIZE, SHAPE, AND TEMPERATURE

As humans, one way in which we try to organize knowledge and enhance communication is by classifying objects into neat compartments, each with its own label. The natural world persistently upsets these schemes by presenting us with particular items that fit neither in one such pigeonhole nor the next, but rather have characteristics of both, for continua are the rule rather than the exception. This is as true of glaciers as it is of other natural systems.

One way of classifying glaciers is by shape. We will be concerned here with only two basic shapes. Glaciers that are long and comparatively narrow, and that flow in basically one direction, down a valley, are called *valley glaciers*. When a valley glacier reaches the coast and interacts with the sea, it is called a *tidewater glacier*. (I suppose this name is appropriate even in circumstances in which the tides are negligible, although with luck no one will ever find a valley glacier encroaching on such a tideless marine environment.) Valley glaciers that are very short, occupying perhaps only a small basin in the mountains, are called *cirque glaciers*. In contrast to these forms are glaciers that spread out in all directions from a central dome. These are called either *ice caps*, or, if they are large enough, *ice sheets*.

There is, of course, a continuum between valley glaciers and ice caps or ice sheets. For example, Jostedalsbreen in Norway and some ice caps on islands in the Canadian arctic feed *outlet glaciers*, which are basically valley glaciers flowing outward from an ice cap. However, the end members, valley glaciers and ice sheets, typically differ in other significant ways (e.g., Fig. 3.1). Thus, a classification focusing on these two end members is useful.

Glaciers are also classified by their thermal characteristics, although once again a continuum exists between the end members. We normally think of water as freezing at 0°C, but may overlook the fact that once all the water in a space is frozen, the temperature of the resulting ice can be lowered below 0°C as long as heat can be removed from it. Thus, the temperature of ice in glaciers in especially cold climates can be well below 0°C. We call such glaciers *polar glaciers*. More specifically, polar glaciers are glaciers in which the temperature is below the melting temperature of ice everywhere, except possibly at the bed. In Chapter 6, we will investigate the temperature distribution in such glaciers in some detail.

Glaciers that are not polar are either *subpolar* or *temperate*. Subpolar glaciers, which are sometimes called *polythermal* glaciers, contain large volumes of ice that are cold, but also large volumes that are at the melting temperature. Most commonly, the cold ice is present as a surface layer, tens of meters in thickness, on the lower part of the glacier (the ablation area).

In simplest terms, a temperate glacier is one that is at the melting temperature throughout. However, the melting temperature, θ_m, is not easily defined. As the temperature of an ice mass is increased toward the melting point, veins of water form along the lines where three ice crystals meet (Fig. 8.1). At the wall of such a vein:

$$\theta_m = \theta_{TP} - \mathbb{C}P - \frac{\theta_m \gamma_{SL}}{L \rho_i r_p} - \Delta\theta \qquad (2.1)$$

(Raymond & Harrison, 1975). Here, θ_{TP} is the triple point temperature, 0.0098°C (Fig. 2.1); $\mathbb{C}$ is the depression of the melting point with increased pressure, P (Fig. 2.1); γ_{SL} is the liquid-solid surface energy, 0.034 J m^{-2}; L is the latent heat of fusion, 3.34×10^5 J kg^{-1}; ρ_i is the density of ice; r_p is the radius of curvature of liquid-solid interfaces; and $\Delta\theta$ is the depression of the melting point due to the presence of solutes. $\mathbb{C}$ is the Clausius-Clapeyron slope, given by

$$\mathbb{C} = \frac{d\theta}{dP} = \left(\frac{1}{\rho_i} - \frac{1}{\rho_w}\right)\frac{\theta_{TPK}}{L} = 0.0742 \ \text{K MPa}^{-1}. \qquad (2.2)$$

Here, ρ_w is the density of water and θ_{TPK} is the triple point temperature in Kelvin degrees. The third term on the right in equation (2.1) represents the change in melting temperature in the vicinity of veins.

Clearly, the melting temperature varies on many length scales in a glacier [equation (2.1)]. On the smallest scales, it varies within the veins that occur along crystal boundaries. On a slightly larger scale, it varies from the interiors of crystals to the boundaries because solutes become concentrated on the boundaries during crystal growth. On the largest scale, it varies with depth due to the change in hydrostatic pressure.

As a result of these variations, small amounts of liquid are apparently present on grain boundaries at temperatures as low as ~ − 10°C, and the amount of liquid increases as the temperature increases. This led Harrison (1972) to propose a more rigorous definition of a temperate glacier. He suggested that a glacier be considered temperate if its heat capacity is greater than twice the heat capacity of pure ice. In other words, this is when the temperature and liquid content of the ice are such that only half of any

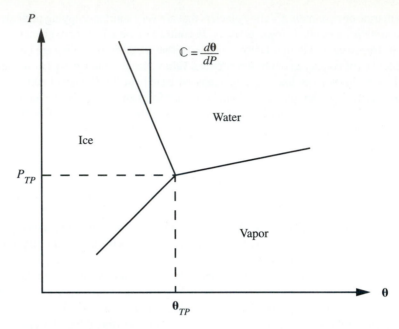

Figure 2.1   Schematic phase diagram for H_2O near the triple point, TP. At the triple point, liquid, solid, and vapor phases are in equilibrium. As long as all three phases are present, neither the pressure nor the temperature can depart from their triple-point values.

energy put into the ice is used to warm the ice (and existing liquid), while the other half is used to melt ice in places where the local melting temperature is depressed.

Harrison's definition, although offering the benefit of rigor, is not easily applied in the field. However, as we shall see in Chapter 4, relatively small variations in the liquid content of ice can have a major influence on its viscosity and crystal structure, among other things. Thus, this discussion serves to emphasize that the class of glaciers which we loosely refer to as temperate may include ice masses with a range of physical properties that are as wide as, or wider than, those of glaciers we refer to as polar.

Ice caps and ice sheets are commonly polar; valley glaciers are more often temperate. However, nothing in the respective classification schemes requires this. In fact, many valley glaciers in high Arctic areas and in Antarctica are at least subpolar, and some are undoubtedly polar. However, none of the major ice caps or ice sheets that exist today are temperate.

THE CONDITION OF INCOMPRESSIBILITY

Let us next examine the consequences of the requirement that mass be conserved in a glacier. In Figure 2.2 a control volume of size $dx \cdot dy \cdot dz$ is shown. The velocities into the volume in the x, y, and z directions are u, v, and w, respectively. The velocity out in the x-direction is

$$u + \frac{\partial u}{\partial x} dx$$

where $\partial u/\partial x$ is the velocity gradient through the volume, which, when multiplied by the length of the volume, dx, gives the change in velocity through the volume in the x-direction. The mass fluxes into and out of the volume in the x-direction are

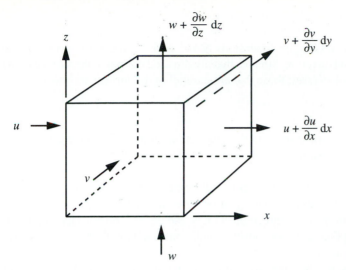

Figure 2.2 Derivation of the condition of incompressibility.

$$\rho \quad u \ dy \ dz \quad \text{and} \quad \left(\rho u + \frac{\partial \rho u}{\partial x} dx\right) dy \, dz$$

$$\frac{kg}{m^3} \frac{m}{a} m \ m = \frac{kg}{a}.$$

Here, ρ is the density of ice. The dimensions of the various parameters are shown beneath the left-hand term to clarify the physics. Similar relations may be written for the mass fluxes into and out of the volume in the y- and z-directions. Summing these fluxes, we find that the change in mass with time, $\partial m / \partial t$, in the control volume is

$$\frac{\partial m}{\partial t} = \rho u \, dy \, dz - \left(\rho u + \frac{\partial \rho u}{\partial x} dx\right) dy \, dz + \rho v \, dx \, dz - \left(\rho v + \frac{\partial \rho v}{\partial y} dy\right) dx \, dz$$

$$+ \rho w \, dx \, dy - \left(\rho w + \frac{\partial \rho w}{\partial z} dz\right) dx \, dy.$$

Note that each term on the right-hand side has the dimensions $M \cdot T^{-1}$, or in the units which we will use most commonly herein, kg a^{-1}. Simplifying by canceling like terms of opposite sign and dividing by $dx \cdot dy \cdot dz$ yields

$$-\frac{1}{dx \, dy \, dz} \frac{\partial m}{\partial t} = \frac{\partial \rho u}{\partial x} + \frac{\partial \rho v}{\partial y} + \frac{\partial \rho w}{\partial z}. \tag{2.3}$$

Ice is normally considered to be incompressible, which means that ρ is constant. This is not true near the surface of a glacier where snow and firn are undergoing compaction, but to a good approximation it is valid throughout the bulk of most ice masses. In this case, equation (2.3) becomes:

$$-\frac{1}{\rho \, dx \, dy \, dz} \frac{\partial m}{\partial t} = \frac{\partial u}{\partial x} + \frac{\partial v}{\partial y} + \frac{\partial w}{\partial z}. \tag{2.4}$$

The mass of ice in the control volume can change if the control volume is not full initially. When it is full of incompressible ice, however, $\partial m / \partial t = 0$, and equation (2.4) becomes:

$$\frac{\partial u}{\partial x} + \frac{\partial v}{\partial y} + \frac{\partial w}{\partial z} = 0. \tag{2.5}$$

This is the condition of incompressibility; it describes the condition that mass and density are not changing.

STRESSES AND STRAIN RATES

A stress is a force per unit area and has the dimensions N m^{-2}, or Pa. Stresses are vector quantities in that they have a magnitude and direction. Stresses that are directed normal to the surface on which they are acting are called *normal stresses*; those that are parallel to the surface are *shear stresses*.

Notation

Referring to Figure 2.3, σ_{xy} is the shear stress in the y-direction on the plane normal to the x-axis. Thus, the first subscript in a pair identifies the plane on which the stress acts, and the second gives the direction of the stress.

The sign convention used in such situations is as follows: Let **n** be the outwardly directed normal to a surface; **n** is positive if it is directed in the positive direction and conversely. σ_{yy} is a normal stress. If the normal stress is in the positive direction and **n** is also positive on this face, σ_{yy} is defined as positive. Thus, both σ_{yy}'s in Figure 2.3 are positive and, conversely, both σ_{xx}'s are negative. In other words, *tension is positive and compression is negative*.

Similarly, if a shear stress, σ_{yx}, is in the positive x-direction on a plane on which **n** is positive, that shear stress is considered to be positive. By this definition both shear stresses σ_{xy} and σ_{yx} in Figure 2.3 are positive.

As an example, consider the variation of u with depth in a glacier (Fig. 2.4). As depicted by the arrows above and below the box in Figure 2.4, the shear stress, σ_{yx}, is negative in the coordinate system shown. The velocity derivative, du/dy, is also negative (u decreases with increasing y). Thus, the negative shear stress results in a negative strain rate, as one would expect.

Tensors

The three-dimensional diagram in Figure 2.5 shows stress vectors on three faces of a cube. Similar stresses occur on the concealed faces, but they are in the opposite directions. The cube is considered to be infinitesimal, representing, say, a point in the glacier. Thus, stresses on any given face can be regarded as uniformly distributed and constant.

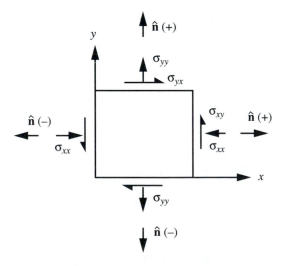

Figure 2.3   Sign convention for stresses in plane strain.

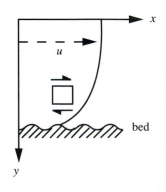

Figure 2.4   Vertical profile of horizontal velocity, u. Sense of shear stress, σ_{yx}, is shown by arrows above and below box.

To describe completely the state of stress at this point, we need nine stress components; thus,

$$\begin{matrix} \sigma_{xx} & \sigma_{xy} & \sigma_{xz} \\ \sigma_{yx} & \sigma_{yy} & \sigma_{yz} \\ \sigma_{zx} & \sigma_{zy} & \sigma_{zz} \end{matrix}$$

This assemblage of stress vectors is called a *second-rank tensor*. For comparison, to describe a first-rank tensor, a vector, we need its components along three coordinate axes.

For steady uniform motion, forces must be balanced. Thus, to ensure that there is no tendency for the cube in Figure 2.5 to rotate, it is necessary that $\sigma_{xy} = \sigma_{yx}$, $\sigma_{xz} = \sigma_{zx}$, and $\sigma_{yz} = \sigma_{zy}$. Such tensors are called *symmetric*.

When a tensor is symmetric, it is common to see, for example, xy used where, rigorously, yx might be more correct. In another common abbreviation often encountered, σ_x is written for σ_{xx}.

Strain Rates

In a deformable medium, stresses induce deformation or strain. The symbol $\dot{\varepsilon}$ is commonly used to denote the rate at which this deformation occurs. ε is the strain, and the dot superscript denotes the time derivative, making it a rate. As nine separate stress vectors are needed to describe fully the state of stress at a point, so also are nine strain rates needed to describe the state of straining at that point. This assemblage of strain rates is also a second-rank tensor, the strain-rate tensor. As was the case with the stress tensor, the strain-rate tensor is also symmetric, so $\dot{\varepsilon}_{xy} = \dot{\varepsilon}_{yx}$, and so forth.

In Chapter 9, we will show that

$$\dot{\varepsilon}_{xy} = \frac{1}{2}\left(\frac{\partial u}{\partial y} + \frac{\partial v}{\partial x}\right) \tag{2.6a}$$

and similarly for the other shear strain rates. When $x = y$, this becomes

$$\dot{\varepsilon}_{xx} = \frac{\partial u}{\partial x} \tag{2.6b}$$

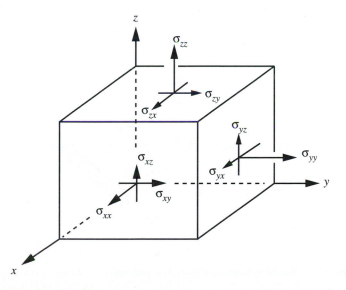

Figure 2.5   Stresses on a cube.

and so forth. Note that in terms of expressions like equation (2.6b), the incompressibility condition, equation (2.5), becomes

$$\dot{\varepsilon}_{xx} + \dot{\varepsilon}_{yy} + \dot{\varepsilon}_{zz} = 0 \tag{2.7}$$

Equations (2.6a and 2.6b) define strain rates in terms of differences in velocity between points that are an infinitesimal distance (e.g., dx) apart. However, when measuring strains or strain rates in the laboratory or field, it is technically impossible to resolve differences in velocity over "infinitesimal" distances. Thus, we make measurements over longer distances and use what is called *logarithmic strain*. The property measured is the change, $\Delta\ell$, in the distance between two points over a time interval, Δt. If the initial distance is ℓ_0 and the final distance is ℓ, then $\dot{\varepsilon}$ is defined as

$$\dot{\varepsilon} = \frac{1}{\Delta t} \ln \frac{\ell}{\ell_0}.$$

A more complete derivation of this relation is given in Chapter 9.

Yield Stress

In some materials no deformation occurs at stresses below a certain level, called the *yield stress*. The yield stress is a property of that particular material. In other materials, deformation rates are so low at low stresses that theoretical models *sometimes* assume the existence of a yield stress even though there may not actually be one. Ice is such a material.

Deviatoric Stresses

Ice does not deform in response to hydrostatic pressure alone. In other words, in a topographic depression containing ice (Fig. 2.6), the hydrostatic (or cryostatic) pressure would increase linearly with depth, z, at a rate $\rho g z$, where g is the acceleration due to gravity. As a rule of thumb, the pressure increases at a rate of 0.1 MPa for every 11 m of depth. Thus, it becomes quite high at large depths. However, if the surface of the ice in the depression is horizontal, as in a lake, the only deformation that would occur would be a relatively insignificant elastic compression.

On the other hand, if the ice surface slopes gently (Fig. 2.6, dashed line), and if points **A** and **B** are on a horizontal plane, then the pressure at **A** would be greater than the pressure at **B**. This pressure difference would result in a compressive strain between **A** and **B**. The strain rate would depend upon the small pressure difference and not, in any significant way, on the much larger hydrostatic pressure at depth z.

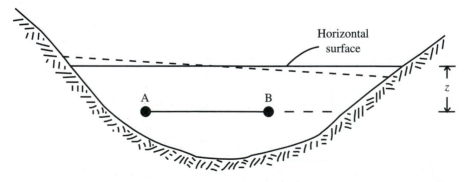

Figure 2.6   Sketch to illustrate nonhydrostatic pressure.

Because straining in glaciers is related to such stress differences, it is convenient to define a stress, called the *deviatoric stress*, which reflects this principle. The deviatoric normal stress in the *x*-direction is

$$\sigma'_{xx} = \sigma_{xx} - P \tag{2.8}$$

where P is the mean normal stress

$$P = \frac{1}{3}\left(\sigma_{xx} + \sigma_{yy} + \sigma_{zz}\right).$$

P is close to, but not necessarily equal to, the hydrostatic pressure. As P is a normal stress, it contributes only to the normal stresses, and not to the shear stresses in Figure 2.5. In other words, the deviatoric shear stresses are the same as their nondeviatoric or total counterparts, but the deviatoric normal stresses are very different from the total normal stresses, especially at depth.

Effective and Octahedral Shear Stresses

The most commonly used flow law for ice is Glen's flow law, named after John W. Glen upon whose experiments it is based (Glen, 1955). We will normally write Glen's flow law in the form:

$$\dot{\varepsilon}_e = \left(\frac{\sigma_e}{B}\right)^n \tag{2.9a}$$

where $\dot{\varepsilon}_e$ and σ_e are known as the *effective strain rate* and *effective shear stress*, respectively, B is a viscosity parameter that increases as the ice becomes stiffer, and n is an empirically determined constant, ≈ 3. An alternative form that is commonly used is

$$\dot{\varepsilon}_e = A\sigma_e^n. \tag{2.9b}$$

The effective shear stress and the effective strain rate are defined by

$$\sigma_e = \frac{1}{\sqrt{2}}(\sigma'^2_{xx} + \sigma'^2_{yy} + \sigma'^2_{zz} + \sigma^2_{xy} + \sigma^2_{yx} + \sigma^2_{xz} + \sigma^2_{zx} + \sigma^2_{yz} + \sigma^2_{zy})^{1/2} \tag{2.10}$$

and

$$\dot{\varepsilon}_e = \frac{1}{\sqrt{2}}(\dot{\varepsilon}^2_{xx} + \dot{\varepsilon}^2_{yy} + \dot{\varepsilon}^2_{zz} + \dot{\varepsilon}^2_{xy} + \dot{\varepsilon}^2_{yx} + \dot{\varepsilon}^2_{xz} + \dot{\varepsilon}^2_{zx} + \dot{\varepsilon}^2_{yz} + \dot{\varepsilon}^2_{zy})^{1/2}, \tag{2.11}$$

respectively. This reveals a fundamental tenet of Glen's flow law, namely that *the strain rate in a given direction is a function not only of the stress in that direction but also of all of the other stresses acting on the medium.* In the next several chapters we will be dealing with situations in which it is feasible to assume that one stress so dominates all of the others that the others can be neglected. However, the reader should be aware of the implications of this assumption.

In place of the effective stress and strain rate, some authors prefer to use the *octahedral shear stress* and *octahedral shear strain rate*, defined as

$$\sigma_o = \frac{1}{\sqrt{3}}(\sigma'^2_{xx} + \sigma'^2_{yy} + \sigma'^2_{zz} + \sigma^2_{xy} + \sigma^2_{yx} + \sigma^2_{xz} + \sigma^2_{zx} + \sigma^2_{yz} + \sigma^2_{zy})^{1/2} \tag{2.12}$$

and

$$\dot{\varepsilon}_o = \frac{1}{\sqrt{3}}(\dot{\varepsilon}^2_{xx} + \dot{\varepsilon}^2_{yy} + \dot{\varepsilon}^2_{zz} + \dot{\varepsilon}^2_{xy} + \dot{\varepsilon}^2_{yx} + \dot{\varepsilon}^2_{xz} + \dot{\varepsilon}^2_{zx} + \dot{\varepsilon}^2_{yz} + \dot{\varepsilon}^2_{zy})^{1/2} \tag{2.13}$$

respectively. In this case, B has to be adjusted appropriately. Note that if all the shear stresses are 0 in equation (2.12), we have

$$\sigma_o = \left(\frac{\sigma_{xx}^2 + \sigma_{yy}^2 + \sigma_{zz}^2}{3} \right)^{1/2}. \tag{2.14}$$

In Chapter 9 we will show that the situation in which all of the shear stresses vanish is a very important one, and we will give the name *principal stresses* to the remaining normal stresses. Thus, equation (2.14) shows that the octahedral shear stress is the root-mean-square of the principal stresses.
When the coordinate axes are aligned parallel to the principal stresses, the octahedral shear stress is
the resolved shear stress on the octahedral plane, a plane that intersects the three axes at points
equidistant from the origin (Fig. 2.7). Hence the name: octahedral shear stress.

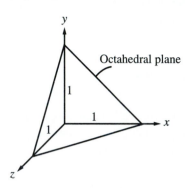

Figure 2.7   A plane that intersects the *x*, *y*,
and *z* axes at points equidistant from the
origin, in this case a unit distance, is called
the octahedral plane. If similar planes are
drawn involving the negative directions
along the axes, the solid figure formed is a
regular octahedron.

3

Mass Balance

Glaciers exist because there are areas, generally at high elevations or in polar latitudes, where snowfall during the winter exceeds melt (and other losses) during the summer. This results in net accumulation, and this part of the glacier is thus called the *accumulation area* (Fig. 3.1). As each snow layer is buried, the pressure of the overlying snow together with movement of molecules in the liquid and vapor phases results in compaction and snow metamorphism. Snow that is more than a year old, and has thus been altered by these processes, is called *firn*. The end result of the firnification process, normally after several years, is solid ice.

Where there are lower elevations to which this ice can move, gravitational forces drive it toward these areas. This eventually brings the ice into places where the annual melt exceeds snowfall. Here, all of the winter snow and some of the underlying ice melts during the summer. This is called the *ablation area*. The line separating the accumulation and the ablation areas at the end of a melt season is called the *equilibrium line*. Along the equilibrium line, melt during the just-completed summer exactly equaled net snow accumulation during the previous winter.

In this chapter we first discuss the transformation of snow to ice and show how the processes involved result in a physical and chemical stratigraphy that, under the right circumstances, can be used to date ice that is thousands of years old. We then explore the climatic factors that result in changes in the altitude of the equilibrium line, and hence in advance and retreat of glaciers.

THE TRANSFORMATION OF SNOW TO ICE

The first phase of the transformation of snow into ice involves diffusion of water molecules from the points of snowflakes toward the centers; the flakes thus tend to become rounded, or spherical (Fig. 3.2a), reducing their surface area and consequently their free energy. This is an example of a thermodynamic principle, namely that the free energy of a system tends toward a minimum. Such rounding occurs more rapidly at higher temperatures.

The closest possible packing of spherical particles would be one with a porosity of about 26%, the so-called rhombohedral packing. However, in natural aggregates of spheres of uniform diameter, the pore space is usually closer to 40%. In the case of firn, this corresponds to a density of ~550 kg m^{-3}.

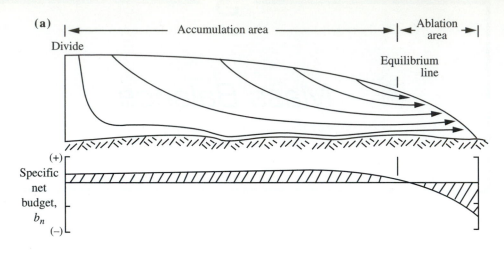

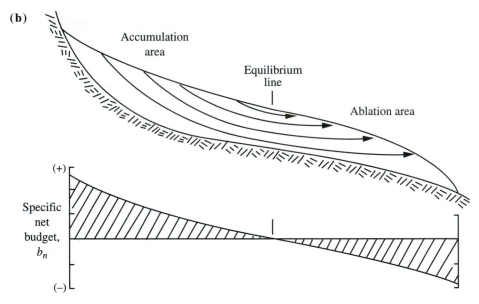

Figure 3.1 Cross sections of (**a**) a typical polar ice cap or ice sheet, and (**b**) a typical valley glacier, showing relation between equilibrium line and flowlines. Sketches are schematic, but relative proportions are realistic.

Further densification involves a process called *sintering* (Fig. 3.2b), which involves transfer of material by sublimation and by molecular diffusion within grains, nucleation and growth of new grains, and internal deformation of the grains (Fig. 3.2c). Sublimation is more important early in the transformation process when pore spaces are still large. Internal deformation increases in importance as the snow is buried deeper and pressures increase. In warm areas, the densification process is accelerated, both because grains may be drawn together by surface tension when water films form around them, and because percolating meltwater may fill air spaces and refreeze.

An important transition in the transformation process occurs at a density of ~830 kg m^{-3}. At about this density, pores become closed, preventing further air movement through the ice. Studies of the air thus trapped provide information on the composition of the atmosphere at the time of close

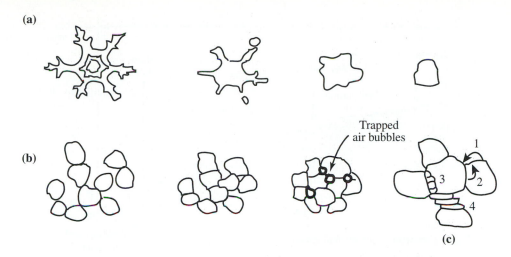

Figure 3.2 Transformation of snow to ice. (**a**) Modification of snowflakes to a subspherical form. (**b**) Sintering. (**c**) Processes during sintering: 1 = sublimation; 2 = molecular diffusion within grains; 3 = nucleation and growth of new grains; and 4 = internal deformation of grains. (Based on Sommerfeld & LaChanelle, 1970, Figs. 2, 16, and 17; and on Kinosita, 1962, as reported by Lliboutry, 1964, Fig. 1.14)

off (e.g., Raynaud et al., 1993). Measurements of the volume of such air per unit mass of ice yield estimates (albeit fairly crude given present technology) of the altitude of the pores at the time of close off (Martinerie et al., 1992). Knowing the depth in the glacier at which this occurs then permits an estimate of the elevation of the ice surface at that time. Pore close off can occur at depths of tens to over a hundred meters, depending on temperature (Paterson, 1994, Table 2.2).

SNOW STRATIGRAPHY

At high elevations on polar glaciers, such as the Antarctic or Greenland ice sheets, there are areas where no melting occurs during the summer. At slightly lower elevations, some melting does occur, and the meltwater thus formed percolates downward into the cold snow where it refreezes, forming lenses or glandlike structures. The higher of these two zones is called the *dry-snow zone* and the lower is the *percolation zone* (Fig. 3.3) (Benson, 1961; Müller, 1962). In keeping with stratigraphic terminology in geology, the parts of the annual snow pack on an ice sheet that have these respective properties are referred to as the *dry-snow facies* and the *percolation facies*, respectively. The boundary between these two zones or facies lies approximately at the elevation where the mean temperature of the warmest month is −6°C (Benson, 1962; cited by Loewe, 1970, p. 263).

At lower elevations, summer melting is sufficient to wet the entire snow pack. This is called the *wet-snow zone* (Fig. 3.3). When this snow refreezes, a firm porous layer is formed. In downglacier parts of this zone, the basal layers of the snow pack may become saturated with water. If the underlying ice is cold, this water-saturated snow can refreeze, forming what is called *superimposed ice*. As long as it is still undeformed, superimposed ice is readily recognized by its air bubbles, which are large and often highly irregular in shape.

At still lower elevations, only superimposed ice is present at the end of the melt season, and this is called the *superimposed ice zone*. The lower boundary of the superimposed ice zone at the end of the melt season is the equilibrium line.

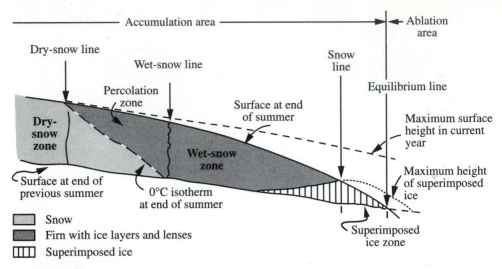

Figure 3.3 Variation in snow facies with altitude (After Benson, 1962). Horizontal distance from equilibrium line to dry-snow line is tens to hundreds of kilometers.

On typical alpine glaciers, the first water percolating into the cold snow at the beginning of the melt season may refreeze to form glands and lenses as on polar ice sheets. However, by the end of the melt season, the entire snow pack will have been warmed to the melting point. Thus, neither the dry snow nor the percolation facies are present on these glaciers. Furthermore, on a temperate glacier, heat conduction downward into the glacier beneath the snow pack is minimal, so little superimposed ice is formed.

Most of the warming of alpine snow packs is a result of the release of latent heat during refreezing of the first water to infiltrate. Freezing of 1 kg of water can warm 160 kg of snow 1°C. Conduction of heat from the surface is insignificant by comparison.

In addition to the zonation in snow stratigraphy with altitude (or temperature), particularly on polar glaciers, there is also a distinct vertical zonation at any given point on a glacier. Because the autumn snow in an annual layer is warmer than the overlying winter snow, the former has a higher vapor pressure. Thus, a vapor-pressure gradient exists, resulting in diffusion of molecules from the autumn to the winter snow. The autumn snow becomes coarser, and its density may decrease. These layers of coarse autumn snow are called *depth hoar*. Tabular crystals are the norm in depth hoar, but in extreme cases, large prism-shaped, pyramidal, or hollow hexagonal crystals develop.

Dating Ice Using Preserved Snow Stratigraphy

Depth hoar layers are important because they can be recognized at depth in snow pits and ice cores. Using such stratigraphic markers, glaciologists have been able to determine accumulation rates, averaged over several years or decades, in many areas of Antarctica and Greenland, and in some cases over millennia in deep cores from these ice sheets.

In one of the most remarkable examples of the use of such physical stratigraphy, Alley and others (1993) found that the accumulation rate high on the Greenland Ice Sheet increased by approximately a factor of two at the end of the Pleistocene, and that the change took place in a time span of only 3 or 4 years! This increase in the accumulation rate was attributed to a warming of the climate, and it was this warming that caused retreat of the ice sheets.

Dating of ice can also be accomplished by detailed laboratory studies of cores or of samples from pit walls. The most commonly used technique for this purpose involves measuring $\delta^{18}O$ variations. Because the air is colder during the winter, $\delta^{18}O$ values in winter snow are more negative (the snow contains less of the heavier isotope of oxygen, ^{18}O) than in summer snow. Thus, a series of samples taken from a single annual layer will show a roughly sinusoidal variation in $\delta^{18}O$ (Fig. 3.4a). A prodigious number of samples must be analyzed when this technique is used to date very old ice. However, the potential is there; annual layers, much compressed but still recognizable by their isotopic variations, have been identified in ice more than 8000 years old (Fig. 3.4b).

The electrical conductivity of ice and the concentration of microparticles in ice also vary seasonally, presumably because the aerosol content of the snow varies, and because winds entrain dust during the summer when blowing over snow-free outwash plains and similar surfaces, respectively (e.g., Thompson et al., 1986). These variations are also used for dating.

When such techniques are employed to date relatively old ice, errors accumulate because some annual layers either lack a variation of the parameter being used, or on occasion have two cycles of

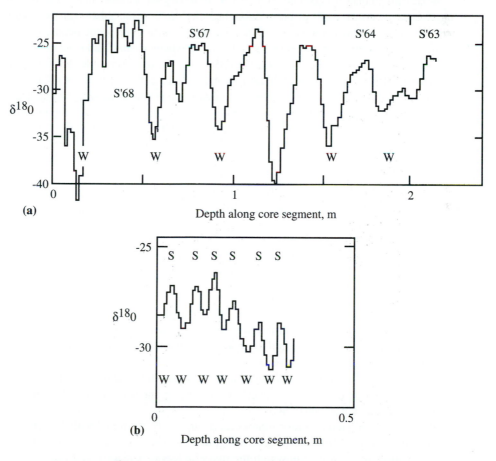

Figure 3.4 $\delta^{18}O$ variations in the Camp Century, Greenland, ice core. (a) Ice from 1963 to 1968. (b) Ice that is approximately 8300 years old, and in which seasonal variations can still be detected. S = summer, W = winter. (Adapted from Johnsen et al., 1972. Reproduced with permission of the authors and *Nature*.)

variation. However, layers of volcanic ash are frequently found in cores, and when these can be associated with an eruption of known age, an absolute date can be assigned to the ice containing the ash. In this way, the age of ice that is thousands of years old has been established quite accurately.

MASS-BALANCE PRINCIPLES

A number of terms are used to describe different aspects of the mass balance on a glacier. The *winter balance* is the amount of snow that accumulates during the winter months. Conversely, the *summer balance*, a negative quantity, is the amount of snow and ice lost by melt. Over the course of a balance year, which is commonly taken to extend from the end of one melt season to the end of the next, the sum of the winter and summer balances is the *net balance*. Normally, these balances are expressed in terms of the thickness of a layer of water, or in water equivalents. When referred to a specific place on the glacier, they are expressed in m a^{-1}, or kg a^{-1}m^{-2}, and are called *specific* balances. Sometimes the word *budget* is used instead of balance, particularly when referring to the net balance.

Significant amounts of accumulation can occur during the summer in the accumulation areas of polar glaciers; conversely, melt can occur throughout the winter in the ablation areas of some temperate glaciers. The terms summer and winter balance are applied with some poetic license in these instances. The most extreme example of this is on tropical glaciers where accumulation and melt can alternate on a time scale of hours to days. Despite these complications, the basic principles are still applicable.

There are a number of ways of measuring mass balance, but we will not go into them all here. Perhaps the most common method, and the one that is easiest to visualize, is to measure the height of the snow or ice surface on stakes placed in the glacier in holes drilled for the purpose. The measurements are made first at the end of one melt season, then at the end of the following winter to obtain the winter balance, and finally at the end of the next melt season to obtain both the summer and the net balances. Snow-density measurements must also be made to convert the winter accumulation and summer snowmelt to water equivalents.

We define $b_s(x,y,z)$ as the specific summer balance, $b_w(x,y,z)$ as the specific winter balance, and $b_n(x,y,z)$ as the specific net balance. Clearly,

$$b_n = b_s + b_w \qquad (3.1)$$

and the overall state of health of the glacier can be evaluated from

$$B_n = \int_A (b_s + b_w)\, dA \qquad (3.2)$$

where A is the area of the glacier and B_n is the net balance. B_n is often normalized to the area of the glacier, thus: $\bar{b}_n = B_n/A$. When B_n or $\bar{b}_n$ is positive, the glacier is said to have a positive mass balance; if this condition persists for some years, the glacier will advance, and conversely. Thus, B_n is an important parameter to measure and to understand, and to this end we now consider meteorological factors influencing its components, b_s and b_w.

It is convenient to restrict our discussion to variations in b_s and b_w with elevation, z. This is not normally valid in practice because of the effects of drifting and shading, which result in lateral variations in both accumulation and melt.

Often, $b_n(z)$ is plotted as a function of elevation; this is illustrated with data from a valley glacier in the Austrian Alps, Hintereisferner, in Figure 3.5a. The curve labeled "o" in this figure represents the situation during a year in which the mass budget is balanced, or $B_n = 0$. [Despite the low values of b_n at higher elevations, equation (3.2) is satisfied in this instance because, as is true of most valley glaciers, the width of Hintereisferner increases with elevation.] Curves labeled "+" and "−" represent years of exceptionally positive or negative mass balance, respectively. Note that melting normally increases systematically with decreasing elevation, so the lower parts of the curves in Figure 3.5a are relatively straight. However, at higher elevations in this particular case, snowfall decreases with elevation, resulting in curvature in the upper parts of the plot.

The differences between the "o" curve and the "–" and "+" curves are shown in Figure 3.5b, and c, respectively. These differences are referred to as the budget imbalance, b_i (Meier, 1962). In years of exceptionally negative b_n (Fig. 3.5b), b_i increases with decreasing elevation; this means that such years are normally a consequence of unusually high summer melt. Conversely, unusually positive budget years commonly result from exceptionally high winter accumulations (Fig. 3.5c). Budget years that are only moderately positive or negative can result from deviations of either accumulation or melt from their values in years when the budget is balanced.

CAUSES OF MASS-BALANCE FLUCTUATIONS

Let us assume that b_w is composed of precipitation and drifting alone, thus ignoring mass additions by condensation and avalanching. Likewise, we take b_s to be a function only of surface melt, ignoring mass losses by evaporation, calving, bottom melting, and so forth. Note that although mass additions and losses by condensation and evaporation, respectively, are thus assumed to be negligible, the energy involved in these phase-change processes is taken into consideration below.

Surface melting is controlled by the available energy:

$$\Sigma Q = R + H + V \tag{3.3}$$

where Q is the energy in kJ m^{-2} d^{-1}, R is the net radiation, H is the sensible heat input, and V is the latent heat input due to condensation, or loss due to evaporation (Kuhn, 1981). Then:

$$-b_s = \frac{\mathbf{T}}{L} \Sigma Q \tag{3.4}$$

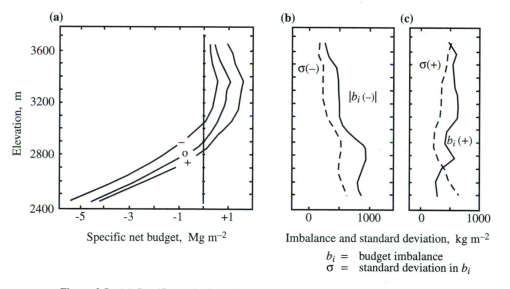

Figure 3.5 (a) Specific net budget, b_n, plotted against elevation for Hintereisferner. Curve "o" is for a year of balanced mass budget, whereas curves "–" and "+" are for years of exceptionally negative or positive budget, respectively. (b) and (c) Difference between curve "o" and curves "–" and "+", respectively. (After Kuhn, 1981, Fig. 1. Reproduced with permission of the author and the International Association of Hydrological Sciences.)

where $\mathbf{T}$ is the length of the melt season and L is the latent heat of fusion, 334 kJ kg^{-1}. (In the remainder of this discussion it will be convenient to use kg m^{-2} a^{-1} for the units of mass balance.)

Assume further that the transfer of sensible and latent heat to the glacier surface is proportional to the temperature difference between the air and the glacier surface; thus,

$$H + V = \gamma(T_a - T_s) \tag{3.5}$$

where T_a and T_s are the temperatures of the air and the glacier surface, respectively, and γ is a constant of proportionality. Kuhn (1989) suggests that γ lies between 0.5 and 2.7 MJ m^{-2}d^{-1}°C^{-1}; a value frequently found for firn is 1.0 MJ m^{-2} d^{-1} °C^{-1}, while a reasonable mean value for glacier ice is ~1.7 MJ m^{-2}d^{-1}°C^{-1}. The range of values reflects the fact that the actual heat transfer is strongly influenced by such factors as wind speed and the roughness of the glacier surface.

Combining equations (3.1), (3.3), (3.4), and (3.5), rearranging terms, and writing all of the parameters as functions of elevation, z, yields

$$b_w(z) = \frac{\mathbf{T}}{L}[R(z) + \gamma(T_a(z) - T_s)] + b_n(z). \tag{3.6}$$

As we are dealing with melting conditions, $T_s = 0$, and does not vary with z.

Our objective now is to study quantitatively how changes in winter precipitation, summer temperature, and radiation balance affect a glacier's mass balance. Curves "−", "o", and "+" in Figure 3.5a are nearly parallel to one another, suggesting that one may be "derived" from another simply by a lateral translation. Such a translation, however, results in a change in the equilibrium line altitude (ELA), represented in Figure 3.5a by the point where the curves cross the 0-specific net balance line. This suggests that changes in equilibrium line altitude may be a fairly good measure of the impact of climate variations. The effects of changes in the principal measures of climate, namely precipitation and temperature, on the ELA are best studied with the use of perturbation theory, a technique used by Kuhn (1981), whose approach we adopt herein.

At the equilibrium line, $b_n(z) = 0$ by definition. Thus, if h is the elevation of the equilibrium line and h_o is its elevation in a year of balanced mass budget, equation (3.6) can be rewritten as

$$b_w(h_o) = \frac{\mathbf{T}}{L}[R(h_o) + \gamma(T_a(h_o) - T_s)]. \tag{3.7}$$

The standard approach in perturbation theory is to rewrite equation (3.7) for a situation in which the equilibrium line is at an elevation, h, which is slightly higher or lower than in the "o" state, and then to subtract equation (3.7) from this new relation, which we now proceed to do. Let primed values represent the perturbed state; thus,

$$b'_w(h) = \frac{\mathbf{T}}{L}[R'(h) + \gamma(T'_a(h) - T_s)]. \tag{3.8}$$

Subtracting:

$$b'_w(h) - b_w(h_o) = \frac{\mathbf{T}}{L}[R'(h) - R(h_o) + \gamma(T'_a(h) - T_a(h_o))]. \tag{3.9}$$

Any of the primed parameters in equation (3.9) that vary with z can be represented by perturbation equations of the form, using T_a as an example:

$$T'_a(h) = T_a(h_o) + \frac{\partial T_a}{\partial z}\Delta h + \delta T_a. \tag{3.10}$$

Here, Δh may be an observed change in altitude of the equilibrium line, so $\dfrac{\partial T_a}{\partial z}\Delta h$ represents the change in temperature that would be expected simply because the ELA changed. However, a change in mean summer air temperature may have been partially responsible for the change in ELA, and this part of the change in T_a is represented by δT_a. Figure 3.6 is a graphical representation of equation (3.10). Writing equations similar to (3.10) for b_w and R, rearranging them, and substituting into equation (3.9) yields

$$\frac{\partial b_w}{\partial z}\Delta h + \delta b_w = \frac{\mathbf{T}}{L}\left[\frac{\partial R}{\partial z}\Delta h + \delta R + \gamma\left(\frac{\partial T_a}{\partial z}\Delta h + \delta T_a\right)\right]. \tag{3.11}$$

The significance of this relation can be elucidated with the use of a numerical example. Suppose $\partial b_w/\partial z = 1$ kg m^{-2} m^{-1}, $\mathbf{T} = 100$ d, $\gamma = 1.7$ MJ m^{-2} d^{-1}°C^{-1}, and the lapse rate, $\partial T_a/\partial z$, is -0.006°C m^{-1}. Suppose further that $\partial R/\partial z = 0$, as the radiation input does not vary significantly with elevation. Now consider an increase in the ELA of 100 m $(= \Delta h)$ in a particular year. Calculate the changes in b_w, R, and T_a that would be sufficient, if they occurred alone, to cause this change in ELA. The reader is encouraged to carry out this calculation to gain familiarity with the relation. Answers are given in Table 3.1.

To place the results of this calculation in perspective, at 3050 m on Hintereisferner, an elevation that is slightly above the normal position of the equilibrium line, the mean winter snowfall is 1620 kg m^{-2} and its standard deviation is 540 kg m^{-2}. Likewise, the mean summer temperature is +0.4°C, and its standard deviation is 0.8°C. Comparing these standard deviations with the values of δb_w and δT_a in Table 3.1, it is clear that a 100-m change in the ELA could result, with nearly equal likelihood, either from a change in b_w or from a change in T_a. Similarly, the total radiation input is ~46 MJ m^{-2} d^{-1}, whereas the loss is ~40 MJ m^{-2} d^{-1}, leaving a mean radiation balance, R, of ~6 MJ m^{-2} d^{-1}. Changes of 1.35 MJ m^{-2} d^{-1}—due to changes in cloud cover, for example—are small compared with the total radiation budget, and thus are not unreasonable.

For comparison, the mean winter balance on Barnes Ice Cap on Baffin Island is ~400 kg m^{-2} (Hooke et al., 1987). Here, a δb_w of -400 kg m^{-2} is highly improbable, as this would mean virtually no accumulation. Thus, in this case, a 100-m change in the ELA would most likely be a result of a change in T_a.

This comparison illustrates a fundamental difference between glaciers in relatively dry but cold areas, areas that we refer to as having a *continental* climate, and glaciers in warmer and wetter *maritime* climates. Glaciers in continental settings owe their existence to low temperatures, and fluctuations in their mass budgets are strongly (inversely) correlated with mean summer temperature.

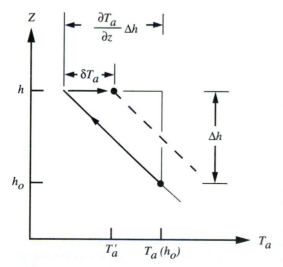

Figure 3.6 Sketch illustrating parameters in equation (3.10). Consider a year during which the equilibrium line is at elevation h, Δh m above its elevation, h_o, in years when the mass budget is balanced. The lapse rate during a year of balanced mass budget and during the year in question are represented by the slopes of the slanting solid and dashed lines, respectively. The mean air temperature at elevation h during this warm summer, T_a', is less than $T_a(h_o)$ by an amount $\left(\frac{\partial T_a}{\partial z}\right)\Delta h$ but higher than the temperature at elevation h during a year of balanced mass budget by an amount δT_a.

In the case shown, because T_a' is lower than $T_a(h_o)$, some factor(s) in addition to the increase in temperature, δT_a, must have contributed to the increase in ELA.

TABLE 3.1 Possible causes of a 100-m increase
in ELA

$\delta b_w = -400 \text{ kg m}^{-2}$	if $\delta T_a = \delta R = 0$
$\delta R = +1.35 \text{ MJ m}^{-2}\text{d}^{-1}$	if $\delta b_w = \delta T_a = 0$
$\delta T_a = +0.8°C$	if $\delta b_w = \delta R = 0$

Conversely, glaciers in maritime settings form in response to high winter snowfall; on such glaciers, the mass balance is less well correlated with T_a alone, and correlations can be improved significantly by adding winter precipitation to the regression. In fact, on some maritime glaciers the correlation of net balance with b_w alone is quite good (Walters & Meier, 1989, p. 371).

In the above analysis, T_a and R have been treated as independent variables. This is not strictly correct because an increase in T_a of 1°C increases R by about 0.3 MJ m^{-2} d^{-1} (Kuhn, 1981). This is a result of the increase in "black body" radiation, which varies as T^4. Incorporating this effect into the above calculation (Table 3.1) reduces δT_a to +0.7°C.

THE BUDGET GRADIENT

Recall that curve "o" in Figure 3.5a represents the distribution of b_n in a year in which the mass budget is balanced. The slope of this curve, $\partial b_{no}/\partial z$, is known as the *budget gradient*. High budget gradients represent situations in which there is a lot of accumulation above the equilibrium line and a lot of ablation below the equilibrium line, and conversely (Fig. 3.7). High budget gradients are thus indicative of high flow rates, as a lot of ice must be transferred from the accumulation area to the ablation area to maintain a steady-state profile. For this reason, Shumskii (1964, p. 442) referred to $\partial b_{no}/\partial z$ as the energy of glaciation, and Meier (1961) called it the activity index.

The budget gradient tends to be high on glaciers in maritime settings and low in continental settings. Typical values might be 10 kg m^{-2}m^{-1}a^{-1} in the former and 3 kg m^{-2}m^{-1}a^{-1} in the latter (Haefeli, 1962).

To explore factors controlling $\partial b_{no}/\partial z$, rearrange equation (3.6) and take its derivative, noting again that $T_s = 0$ on a melting-glacier surface:

$$\frac{\partial b_{no}}{\partial z} = \frac{\partial b_w}{\partial z} - \frac{\text{T}}{L}\left[\frac{\partial R}{\partial z} + \gamma\frac{\partial T_a}{\partial z}\right]. \tag{3.12}$$

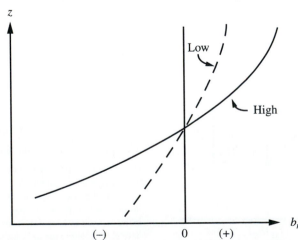

Figure 3.7 Sketch illustrating difference between low and high budget gradients.

Thus, $\frac{\partial b_{no}}{\partial z}$ depends upon $\frac{\partial b_w}{\partial z}$, $\frac{\mathbf{T}}{L}\left(\frac{\partial R}{\partial z}\right)$, and $\frac{T\gamma}{L}\left(\frac{\partial T_a}{\partial z}\right)$. The precipitation gradient, $\frac{\partial b_w}{\partial z}$, is commonly almost negligible. However, it may become significant if snow drift is important at higher elevations. It is also larger in areas where a significant amount of the summer precipitation occurs as snow at high elevations and rain at low elevations. In the Alps, where this is commonly the case, Kuhn (1981) suggests that a value of 0.5 kg m^{-2}m^{-1}a^{-1} is reasonable.

The net radiation gradient, $\partial R/\partial z$, is small as long as snow covers the ablation area. However, once ice is exposed, particularly if it has a thin dirt cover, the albedo drops and there is a significant change in R across the firn edge, or boundary between firn and ice. The first ice to become exposed is normally near the snout of the glacier, and the firn edge rises as the melt season progresses. Taking this into consideration, $\frac{\mathbf{T}}{L}\left(\frac{\partial R}{\partial z}\right)$ may be as high as ~7 kg m^{-2}m^{-1} over a 120-d melt season (Kuhn, 1981).

The lapse rate, $\partial T_a/\partial z$, is limited by the dry adiabatic rate, ~0.010°C m^{-1}, but a more realistic free-air lapse rate along a glacier surface is ~0.007°C m^{-1}. Thus, for a 120-d melt season, $\frac{T\gamma}{L}\left(\frac{\partial T_a}{\partial z}\right)$ is ~4.3 kg m^{-2}m^{-1}.

Clearly, the dominant controls on $\partial b_{no}/\partial z$ are the terms involving the lapse rate and, below the equilibrium line, the radiation balance. As both $\partial T_a/\partial z$ and $\partial R/\partial z$ are likely to be comparable in maritime and continental settings, we have to appeal to differences in the length of the melt season, $\mathbf{T}$, to explain the difference in budget gradient. Melt seasons in high arctic continental settings may be a half to a third as long as those in, say, the Alps. Glaciers in continental settings also tend to be cleaner, thus reducing the albedo contrast across the firn edge. Finally, summer rain is likely to be less of a factor in continental climates.

As noted, the imbalance in the net budget, $b_{ni}(z)$, is the difference between the "o" curve in Figure 3.5a and a similar curve for a year in which the mass budget is not balanced, so the glacier has either gained or lost mass. To a good approximation, the average value of $b_{ni}(z)$, which we will denote by $\bar{b}_{ni}(z)$, is equal to the value of b_{ni} at the equilibrium line, $b_{ni}(h)$. This, in turn, is related to changes in the ELA, Δh, (Fig. 3.8) by

$$b_{ni}(h) = -\left(\frac{\partial b_{no}}{\partial z}\right)\Delta h. \qquad (3.13)$$

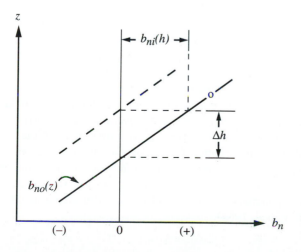

Figure 3.8 Relation between budget imbalance and changes in the ELA. The sloping solid line represents a small segment of the "o" curve in Figure 3.5a. During a year of negative mass balance, this curve is shifted upward by an amount Δh as a consequence of a negative budget imbalance, b_{ni}. As we are interested in changes in the ELA, the relevant value of b_{ni} is that at the equilibrium line, $b_{ni}(h)$, which is approximately equal to the average value, $\bar{b}_{ni}(z)$.

Thus, with the use of the budget gradient, we can estimate the amount by which the net budget must differ from b_{no} (that is, estimate b_{ni}) to produce a given change in the equilibrium-line altitude. This is relevant to estimates of the changes in climate necessary to produce observed changes in the extent of glaciers in the past.

EFFECT OF WEATHER PATTERNS ON MASS-BALANCE FLUCTUATIONS

There are two spatial scales of variation in coherence of glacier mass-balance patterns. On the one hand, there are worldwide climatic changes such as those that resulted in the major ice advances of the Pleistocene and the minor advances of the Little Ice Age. These are both well-known and poorly understood, except that variations in Earth's orbit that affect the timing and amount of solar radiation received at higher latitudes appear to modulate the longer cycles (Hays et al., 1976).

On a smaller scale are regional variations in weather that may cause glaciers only a few hundred kilometers apart to behave differently. We consider here some examples of such regional-scale variations.

During most years, the net balances of glaciers in Alaska are out of phase with those of glaciers in southwestern Canada and adjacent areas in the United States. When glaciers in one area have a relatively good year, those in the other area normally have a bad year. Walters and Meier (1989) studied this and found that when the atmospheric low pressure region that lies off the Aleutian Islands, the Aleutian Low (Fig. 3.9), is normal in the fall and winter, storms are deflected into Alaska, resulting in high winter balances there. However, when this low is not as deep as it usually is, storm tracks remain farther south and accumulation rates are high in Washington and British Columbia.

There is a fascinating relation between the depth of the Aleutian Low and the presence or absence of El Niño/Southern Oscillation, or ENSO, events. During ENSOs, the normal upwelling of cold water off Peru is damped, and the ocean there becomes warmer. The air over the water is thus warmed, decreasing the pressure gradient between this area and the western Pacific. Westerly air flow is thus decreased, and the region of heavy rainfall in the western Pacific shifts eastward. This, in turn, shifts the position of the jet stream, and hence the magnitude of the Aleutian Low, causing storms to enter North America hundreds of kilometers south of their normal entry points (Rasmussen, 1984).

Summer balances in western North America are likewise affected by the summer low along the west coast. When this low is relatively deep and there is a corresponding high over British Columbia, conditions tend to be hot and dry, leading to large negative summer balances.

Asynchroneity of mass balances can also result from the scale of pressure patterns, as reflected in the height of the 500-mb surface, or the surface on which the atmospheric pressure is 500 mb (about half the pressure at Earth's surface). Sentinel Glacier in British Columbia responds to small-scale disturbances that are related to migratory perturbations imbedded in larger-scale air flows. Low-pressure disturbances of this scale in the winter result in cyclonic storms, characterized by counterclockwise winds around a low-pressure area. Such storms increase the winter balance. Conversely, a high frequency of anticyclonic patterns, or those resulting from high pressure and thus accompanied by clockwise winds, inhibits accumulation in winter and increases melt in summer. In contrast, Peyto Glacier, which lies about 500 km east of Sentinel Glacier, is affected only by larger disturbances related to long-wave patterns over the North Pacific. Storms from smaller disturbances do not penetrate that far inland (Yarnal, 1984).

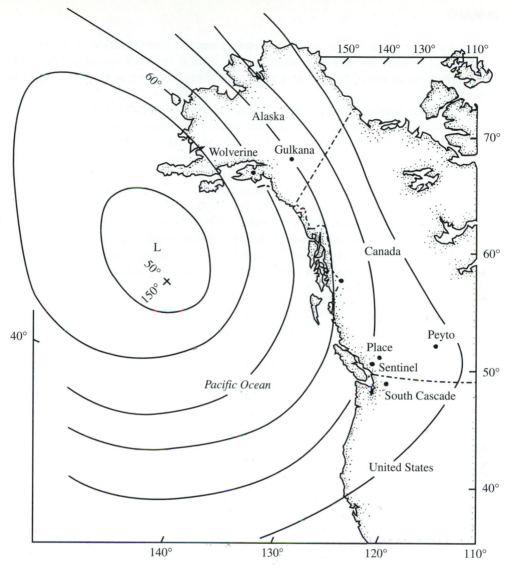

Figure 3.9   Map of the west coast of North America showing locations of some glaciers for which there are good mass-balance records, and the Aleutian Low. (Based on Walters & Meier, 1989, Figs. 1 and 9.)

Clearly, to acquire a detailed understanding of regional variations in mass balance one must have extensive knowledge of atmospheric circulation patterns on a variety of spatial scales. The database necessary for such studies has only been available for a few years, and much will be learned as glaciologists and meteorologists begin to exploit it. Particularly intriguing are the teleconnections between oceanic and atmospheric circulation that are beginning to appear. Beyond this, however, is the question of what controls variations in atmospheric and oceanic circulation on time scales of decades to centuries.

SUMMARY

In this chapter we discussed snow accumulation and the transformation of this snow to ice. We found that in polar environments where there is little if any melting, the physical and chemical stratigraphy in an annual layer of snow persists for many thousands of years and can be used to date the ice.

We then defined some terms used to discuss mass balance, particularly summer, winter, and net balance, and we used a perturbation approach to study the influence of winter balance, temperature, and radiation on net balance. By comparing observed variations in these parameters with calculated values, it became clear that the net balance of glaciers in continental environments was sensitive, primarily, to summer temperature, whereas that of glaciers in maritime areas was sensitive to both winter balance and summer temperature. Radiation balance, principally due to differences in cloud cover, could play a role in either environment. The lower budget gradient, and consequently the more sluggish behavior of continental glaciers compared with their maritime counterparts, turned out to be largely related to the shorter melt season in continental environments.

Finally, we learned that variations in intermediate and large-scale weather patterns that we are just beginning to understand can result in asynchronous mass-balance patterns on glaciers only a few hundred kilometers apart.

4

Flow of a Crystalline Material

Before proceeding to a more theoretical discussion of glacier mechanics, it is worthwhile presenting a brief introduction to the voluminous literature on deformation or creep of ice. We will begin by looking at deformation processes on an atomic scale and then introduce empirical and semi-empirical relations that provide a macroscopic description of the deformation.

CRYSTAL STRUCTURE OF ICE

There are nine known crystalline forms of ice, but seven of them are stable only at pressures in excess of about 200 MPa, and the eighth, a cubic form, ice Ic, is stable only at temperatures below about –100°C (Fig. 4.1). As the highest pressures and lowest temperatures in glaciers on Earth are about 40 MPa and –60°C, respectively, these eight forms need not concern us. We thus restrict our attention to the common form of terrestrial ice, ice Ih.

The structure of ice Ih is shown, in stereoscopic view, in Figure 4.2a. It is a hexagonal mineral (hence the "h") with a rather open structure in which every oxygen atom, represented by the large circles in Figure 4.2a, is bonded to four additional oxygen atoms at the corners of a tetrahedron. The tetrahedra are joined together in such a way that the oxygens form hexagonal rings with the O=O bonds zigzagging lightly up and down as one progresses around the ring (Fig. 4.2b); three of the oxygens thus lie 0.09 nm above the other three. The plane of these rings is called the *basal plane* of the crystal structure.

The fourth oxygen atom in the tetrahedron is ~0.28 nm above or below that in the center of the tetrahedron. A line parallel to this bond, and hence normal to the basal plane, is called the *c-axis*.

It is evident from Figure 4.2b that there are many more O=O bonds within a basal plane than there are between basal planes. Thus, bonding between basal planes is much weaker than that within the basal plane.

Around each oxygen atom there are, of course, two hydrogen atoms. The hydrogen atoms, represented by the small circles in Figure 4.2a, lie on the bonds between the oxygen atoms. They are situated close to the oxygen to which they are bonded. As each oxygen atom is bonded to four others, only two of these hydrogen sites, selected at random, are occupied.

There are two hydrogen sites along each O=O bond. Normally only one of these is occupied. Situations in which neither site is occupied are called *Bjerrum L defects*, and situations in which both sites are occupied are called *Bjerrum D defects*, or just L and D defects, respectively.

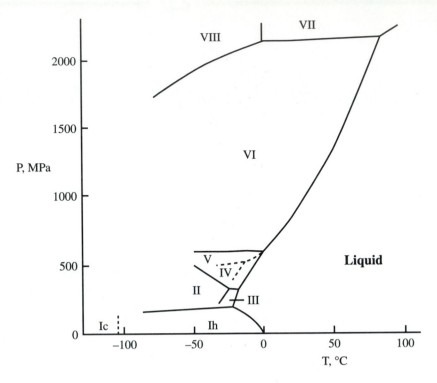

Figure 4.1   Part of the phase diagram of water (After Kamb, 1965, Fig. 1). The various polymorphs of ice are designated by roman numerals. Ice IV is a metastable phase, unstable everywhere with respect to ice V. Ice Ic is also metastable with respect to ice Ih.

DISLOCATIONS

Another type of defect in a crystal is the *dislocation*. Dislocations are places where the crystal structure is discontinuous or offset in some way. The two basic types of dislocation, the *edge dislocation* and the *screw dislocation*, are illustrated in Figure 4.3. Virtually all crystalline materials contain dislocations.

Dislocations play a vital role in the deformation or creep of crystalline materials. If we tried to deform the perfect crystal in Figure 4.3a by shearing the top three layers of atoms over the bottom two, the stress required would be enormous, as every one of the bonds indicated by an "x" would have to be broken *simultaneously*. In contrast, the crystal in Figure 4.3b would deform much more easily because the bonds could be broken one at a time. Thus, the bond between E and F could be broken and a new bond formed between D and F. Calculations show that, in the absence of dislocations, crystalline materials could not possibly deform under the stresses at which they are observed to deform. In fact, it was through such theoretical studies that the existence of dislocations was first inferred.

Upon application of a stress, the number of dislocations in a crystal increases. Some of these new dislocations are generated at *Frank-Reed sources*. A Frank-Reed source consists of a dislocation lying between two points at which the dislocation is fixed, called *pinning points* (Fig. 4.4). Impurities or immobile tangles of dislocations may serve as pinning points. When a stress is applied, this dislocation is bowed out until it meets itself at "a." At this point, the dislocations coming from opposite

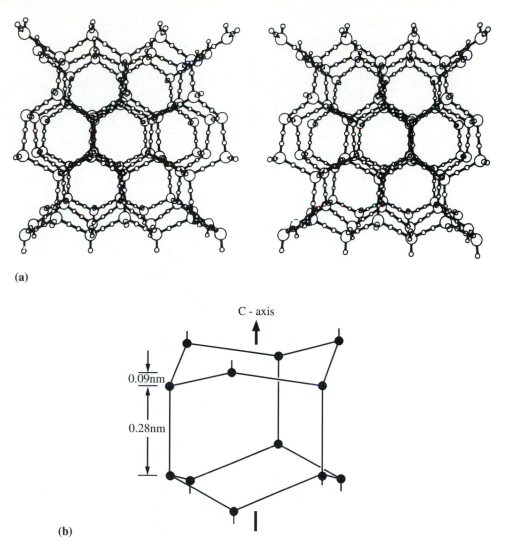

(a)

(b)

Figure 4.2 (a) Stereographic view of the structure of ice Ih viewed down the *c*-axis. Only half of the possible hydrogen sites, indicated by small circles, are occupied. (After Hamilton & Ibers, 1968.) (b) Structure of ice 1h viewed normal to the *c*-axis. Two of the hexagonal rings are shown. Short lines leading upward and downward from these rings are bonds to rings above and below. (Modified from Hobbs, 1974, Fig. 1.7.)

directions are of opposite sign, and the dislocation is locally annihilated, leaving a ring and a new dislocation between the pinning points. This new dislocation can then repeat the process, so there is a continuous source of dislocations. The dislocation is of the edge type ahead of and behind the source, of the screw type at the sides, and of mixed type at intermediate positions. Dislocations can also multiply by a process called *multiple cross glide*, in which dislocations from a Frank-Reed source never complete a full cycle but rather spread to neighboring planes (Hull, 1969, pp. 165–167).

Dislocations also form at points of stress concentration on grain boundaries. For example, shear along a discrete atomic plane in one crystal will result in an offset of the crystal boundary. To

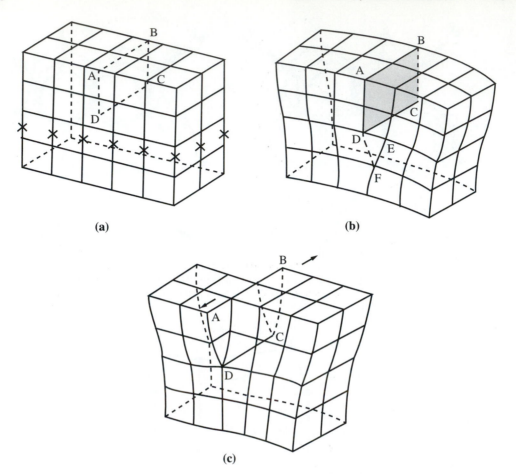

(a)   **(b)**

(c)

Figure 4.3   (**a**) A perfect crystal. (**b**) An edge dislocation. (**c**) A screw dislocation. (Modified from Hull, 1969, p. 17.)

accommodate this offset, the neighboring crystal must also yield, so dislocations are formed at the boundary and move into this crystal.

Slip along grain boundaries is believed to occur at temperatures above −10°C. This, too, will result in stress concentrations in neighboring crystals, and hence in generation of dislocations in these crystals (Fig. 4.5).

Dislocations move by formation of kink pairs (Fig. 4.6), followed by lateral movement of the kinks. As long as the bond formed by movement of a kink is a normal bond, with only one hydrogen between two oxygens, the kink can move readily. However, if the movement would result in a Bjerrum defect the energy required for movement would be much higher. It is presumed that in such situations, movement of the kink is delayed until diffusion or rearrangement of the hydrogen atoms (proton rearrangement) results in a geometry such that the kink can migrate without formation of a Bjerrum defect. As shown in Figure 4.6, there may be a number of kinks along a dislocation line. Two kinks moving toward one another will annihilate each other when they meet, resulting in advance of the dislocation line.

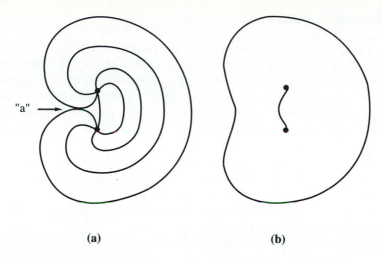

(a)   (b)

Figure 4.4   Generation of dislocations at a Frank-Reed source. Each line in (**a**) represents
a successive position of a dislocation as it is bowed out between two pinning points. (**b**) This
shows the final stage, with the new dislocation expanding outward and another dislocation
between the pinning points.

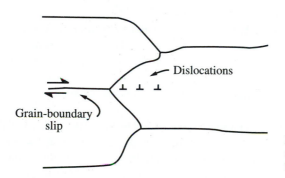

Figure 4.5   Generation of dislocations at
a three-grain intersection due to grain-
boundary slip.

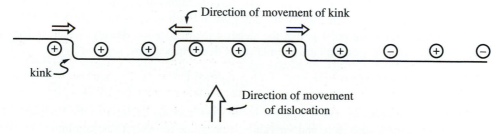

Figure 4.6   Lateral movement of kinks causing forward movement of a dislocation. The
symbol ⊕ represents situations in which a normal bond would form by passage of a kink.
The symbol ⊖ represents situations in which a Bjerrum defect would be formed.

RATE-LIMITING PROCESSES

The rate of deformation of a crystal or of a polycrystalline aggregate depends on how rapidly dislocations can move. This, in turn, may depend upon a number of factors: effectiveness of the mechanisms resisting motion; the ability of a dislocation to move from one atomic plane to another; the orientation of the atomic plane in which the dislocation is moving. Usually, one process is significantly more important than the others, principally because it is more effective than the others in retarding dislocation motion. This process is called the *rate-controlling* or *rate-limiting* process.

In any mechanistic approach to the study of creep, it is obvious that emphasis must be placed on identifying the rate-limiting process. This, however, is difficult because the rate-limiting process is different in different materials, and within any one material it changes with temperature and stress and possibly also impurity content. Furthermore, the rate-limiting process may be different in single crystals and in polycrystalline aggregates.

We describe below some possible rate-limiting processes in ice, and the evidence for them.

Drag as the Rate-Limiting Process

At the moderate stress levels normally found in glaciers, dislocations moving in a crystallographic plane are restrained in their motion by a number of drag mechanisms. The velocity of such a dislocation is given by

$$v = c\sigma e^{-\frac{Q}{R\theta_K}} \tag{4.1}$$

where c is a parameter that is dependent upon factors such as the atomic spacing in the crystal (the Burgers vector) and Boltzmann's constant; σ is the stress; R is the gas constant; θ_K is the temperature in Kelvins; and Q is the activation energy for the rate-limiting deformation mechanism (Weertman, 1983). Unfortunately, derivation of this equation, as well as of a few of those to follow, is beyond the scope of this book. Noteworthy, however, is the dependence on very fundamental physical parameters.

A brief comment on activation energy is required. The activation energy is the magnitude of an energy barrier that must be overcome for a kinematic process to occur. Each kinematic process has its own activation energy, so there is an activation energy for self-diffusion of hydrogen and oxygen in ice ($\approx$60 kJ/mol), an activation energy for creep of ice ($\approx$79 kJ/mol), and so forth.

The creep rate, $\dot{\varepsilon}$, resulting from movement of dislocations in a crystallographic plane, is given by

$$\dot{\varepsilon} = \alpha\rho_d v \tag{4.2}$$

(Weertman, 1983) where α is a constant with the dimensions of length and ρ_d is the dislocation density; α depends on the orientation of the slip planes, but it is approximately 4.5×10^{-10} m. For the moment, the interesting point is that the steady-state dislocation density, which reflects a balance between the applied stress and the internal stress from dislocations, is given by

$$\rho_d = \frac{1}{\alpha}\left(\frac{\sigma}{\mu}\right)^2 \tag{4.3}$$

(Alley, 1992; Weertman, 1983) where μ is the shear modulus. (Refer to the Endpapers for definitions of parameters such as μ, and for values appropriate for ice.) Thus, combining equations (4.1), (4.2), and (4.3) leads to:

$$\dot{\varepsilon} \propto \sigma^3 \tag{4.4}$$

at constant temperature. In other words, the velocity of dislocations varies with σ, and their density varies with σ^2, leading to a cubic dependence of $\dot{\varepsilon}$ on σ.

A large volume of experimental data on ice deformed in the laboratory and on natural ice in glaciers and ice shelves yields such a cubic dependence, particularly for stresses above 0.1 MPa. This agreement supports the conclusion that the theoretical model presented above is fundamentally sound, and that drag mechanisms which determine the velocities of dislocations are the rate-limiting factors.

When a single crystal of ice is stressed in such a way that there is an appreciable component of the stress on the basal (0001) plane, the deformation rate increases with time (Fig. 4.7). (Note that the deformation rate is proportional to the slope of the ε-t curve in Fig. 4.7.) This can be explained by the fact that the dislocation density is normally low in unstressed crystals, and increases gradually to its steady-state value, given by equation (4.3), after the stress is applied. This further supports the suggestion that drag is the rate-limiting process.

Climb as the Rate-Limiting Process

An additional dislocation process must also be considered. As dislocations multiply, they may begin to interfere with one another, resulting in gridlock. These dislocation tangles inhibit deformation. Under such conditions, application of a stress to a previously unstrained sample results in a deformation rate that initially decreases with time (Fig. 4.8). At sufficiently low temperatures this decreasing creep rate may continue indefinitely, but at higher temperatures *recovery processes* involving diffusion of atoms away from (or vacancies toward) dislocations can result in movement of the dislocation from one crystallographic plane to another. For example, if the atom at D in Figure 4.3b diffused away, the dislocation would move upward. These processes, called *dislocation climb* if the dislocations are of the edge type, and *cross slip* if they are of the screw type, relieve the tangles and, after an initial transient period, allow deformation to continue at a more-or-less steady rate (central part of curve in Fig. 4.8). As climb is the recovery process requiring more energy, it would be rate-controlling.

It is noteworthy that samples of polycrystalline ice deformed in compression invariably go through a transient phase of decelerating creep and then a period of nearly constant creep rate. [If the test is continued long enough the creep rate may begin to increase again (Fig. 4.8); this is attributed to recrystallization, which will be discussed further below.] This supports the suggestion that climb is the rate-limiting process. Furthermore, it may be significant that in such experiments the activation energy for creep (79 kJ/mol) is close to that for self-diffusion (60 kJ/mol), again suggesting that climb (which is a result of diffusion) is the rate-limiting process. However, the difference between these two activation energies seems to be larger than can be explained by experimental error, so other processes may also be involved.

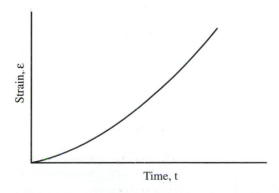

Figure 4.7   Variation of total strain with time during deformation of a single crystal oriented so that glide occurs on the basal plane (easy glide).

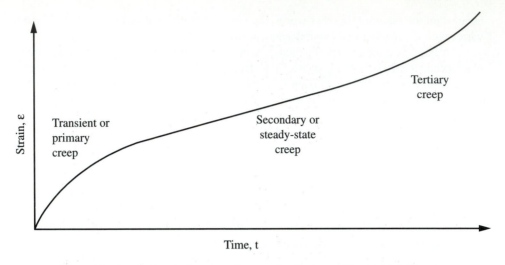

Figure 4.8 A typical strain-time curve for a sample of polycrystalline ice loaded in uni-axial compression.

Slip on the Pyramidal Plane

Another candidate for the rate-limiting role in the creep of polycrystalline ice is slip on the pyramidal plane (Fig. 4.9). The argument is as follows: In a polycrystalline aggregate if one crystal deforms, the neighboring crystals must also deform to preserve continuity of the medium. Thus, all crystals must take part in the deformation. Most crystals will be oriented so that the applied stress on that crystal will have a component parallel to the basal plane. However, a few crystals

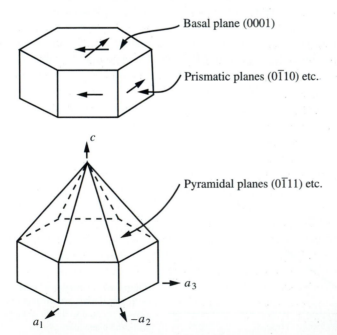

Basal plane (0001)

Prismatic planes (0$\bar{1}$10) etc.

Pyramidal planes (0$\bar{1}$11) etc.

Figure 4.9 Crystallographic planes in a hexagonal crystal.

will not be so oriented, and these must therefore deform in some direction that is not parallel to the basal plane.

Slip on the prismatic plane (Fig. 4.9) is one possibility. However, the most likely slip direction in the prismatic planes is parallel to the basal plane, as shown by the arrows in Figure 4.9. These slip directions would not accommodate stresses normal to the basal plane.

Slip on pyramidal planes (Fig. 4.9), if the slip direction is not parallel to the basal plane, could accommodate stresses normal to the basal plane. If such a slip system exists, it is probably harder than slip on the prismatic planes and certainly harder than slip on basal planes. It is therefore conceivable that slip on the pyramidal plane is the rate-limiting process in the deformation of polycrystalline ice.

Duval and others (1983) suggest an alternative to this. They observe that the above considerations are based on the assumption of homogeneous strain. However, in polycrystalline ice, the stress on any single crystal will not equal the stress on the bulk sample, as crystals that slip easily in one direction but not in others will transmit the stress nonuniformly. If it is assumed, instead, that the strain is inhomogeneous, plastic strain is possible without slip on either prismatic or pyramidal planes. This is because climb of dislocations normal to the basal plane can accommodate stresses normal to this plane.

INTERNAL STRESSES

The inhomogeneity of internal stresses in a polycrystalline sample was mentioned briefly above. This problem merits further discussion.

When an ice sample is loaded, it first deforms elastically. Then, grains that are favorably oriented begin to deform by slip on the basal plane. The load is thus transferred to grains less favorably oriented, and high stresses can develop in such grains. As deformation continues the inhomogeneity of the stress distribution becomes more pronounced, with peak stresses that may be many times the mean stress in the sample (Duval et al., 1983). The wavelength of these stress variations is comparable to the grain size.

It has been found experimentally that the resistance to shear on nonbasal planes in monocrystals (single crystals) is many times the resistance to shear on the basal planes (Fig. 4.10). This is due to the paucity of bonds between basal planes mentioned earlier (Fig. 4.2b). For example, to produce a strain rate of 10^{-8} s^{-1} at $-10°C$ requires a stress of about 0.01 MPa in a monocrystal oriented for easy glide, and a stress of about 1 MPa in a monocrystal oriented for hard glide. Thus, within a sample of polycrystalline ice the stress may vary over a range of as much as a factor of 100. Not surprisingly, the bulk strain rate of the polycrystalline sample lies between the values for the monocrystal in the easy-glide and hard-glide orientations (Fig. 4.10).

A consequence of this inhomogeneous internal stress distribution is that grains which are not favorably oriented for basal glide may accommodate some of the plastic deformation in adjacent crystals by deforming elastically. These crystals, therefore, have a fair amount of stored elastic strain energy. Upon release of the load, this elastic energy exerts a "back stress" on the crystals that previously deformed plastically, and a certain amount of reverse creep results. Such reverse creep has been measured by Duval (1978) (Fig. 4.11). This reverse creep is not instantaneous, as would be the case in a purely elastic medium, because the grains favorably oriented for basal glide under the original loading must now creep "backwards" to relieve the stored elastic energy. As might have been anticipated, the initial rate of reverse creep upon unloading was almost the same as the creep rate when the sample was first loaded.

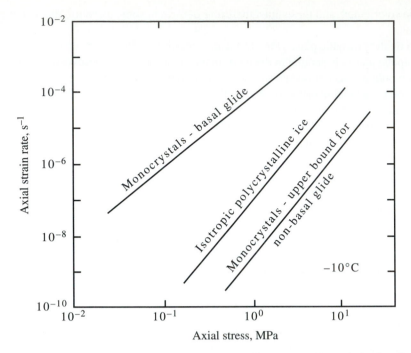

Figure 4.10   Stress-strain rate data for ice at −10°C (Adapted with permission from Duval et al., 1983, Fig. 2. Copyright 1983 American Chemical Society.)

RECRYSTALLIZATION

Crystals of glacier ice vary in size and also in the degree to which they are interlocked. If there were no bonding across grain boundaries, for example, some polycrystalline ice samples would fall apart into a pile of roughly equant grains, up to a few millimeters in maximum dimension, while other samples would hang together like a three-dimensional jigsaw puzzle. We will use the term *texture* to refer to these characteristics of crystal size and shape. In addition, under prolonged strain the *c*-axes of crystals in glacier ice develop a variety of preferred orientations, or *fabrics*. Both texture and fabric affect the rheology of ice.

Through a series of processes which we will refer to, collectively, as *dynamic recrystallization*, texture and fabric are altered during deformation. Dynamic recrystallization, or simply recrystallization, is a consequence of both the high local internal stresses mentioned above and the resulting widely differing internal energies in adjacent grains.

One or more of three processes may be involved in recrystallization. In order of increasing energy difference between grains, these are *grain growth, polygonization*, and *nucleation of new grains* (Duval & Castelnau, 1995). Grain growth results from relatively slow migration of grain boundaries. The migration is driven by curvature of the boundaries (Alley, 1992). Higher pressures occur on the concave sides of such boundaries, which is commonly the side of the smaller grain, and molecules tend to move from the high pressure to the low pressure side of the boundary. Thus, smaller crystals disappear. Polygonization involves the alignment of dislocations to form a new grain boundary within a bent crystal. The crystal is thus divided into two grains with nearly the same orientation. Nucleation of new grains entails the appearance of small grains that are oriented for easy glide. At sufficiently high internal energies, such nucleation is followed by comparatively rapid

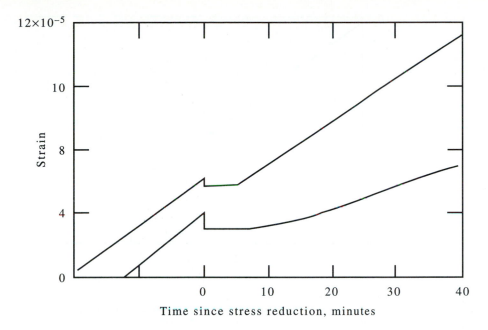

Figure 4.11 Reverse creep due to the release of stored elastic strain energy upon unloading of a sample. The original loading of the sample was 1 MPa. The stress was reduced 0.06 and 0.13 MPa, respectively, at time 0. Horizontal parts of curves reflect a balance between further creep under the reduced stress and delayed release of elastic strain energy. (Adapted with permission from Duval et al., 1983, Fig. 5. Copyright 1983 American Chemical Society.)

migration of the boundaries of these new grains into adjacent, larger, highly stressed grains. Such grain growth occurs predominantly at temperatures above ~−10°C.

During recrystallization, crystals that are well-oriented for easy glide, the orientation with the lower free energy, tend to grow at the expense of those that are not so oriented. This results in development of fabrics with preferred orientations of *c*-axes, and in a concomitant increase in the creep rate (Fig. 4.8). The fabrics that develop reflect the recrystallization mechanisms acting in the ice. The strongest fabrics result from nucleation processes followed by rapid grain-boundary migration. These processes require an energy density on the order of 0.01 to 0.1 MJ m^{-3}. This energy density is three to four orders of magnitude greater than that which would be expected in polycrystalline ice deforming with a uniform internal stress distribution, but it is comparable to the local energy densities that can develop due to the stress concentrations just mentioned (Duval et al., 1983; Pimienta & Duval, 1987).

The increase in creep rate resulting from dynamic recrystallization can be observed in laboratory tests carried to sufficiently high strains. It generally becomes apparent once the ice has undergone a strain of about 1% (Fig. 4.12), suggesting that a strain of this amount is necessary to establish the required energy difference. Of interest is the fact that this critical total strain seems to be independent of temperature.

After a total strain of about 10% is reached, the fabric and texture are approximately in equilibrium with the applied stress (or resulting strain rate), and no further acceleration occurs (Budd & Jacka, 1989). This is known as *steady-state tertiary* creep. Budd and Jacka suggest that this steady state involves a balance between recrystallization of grains in orientations that favor deformation and rotation of grains out of such favorable orientations.

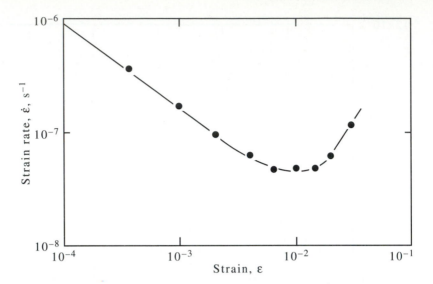

Figure 4.12 Laboratory test on a sample of polycrystalline ice at 1.02 MPa and −10°C showing increase in strain rate after about 1% strain. (Adapted with permission from Duval et al., 1983, Fig. 6. Copyright 1983 American Chemical Society.)

Several varieties of preferred-orientation fabric are found in glaciers. Ice with a single-maximum fabric, in which many crystals are oriented with their c-axes approximately normal to the plane of simple shear (Fig. 4.13a), is common. However, other fabrics are also found, particularly in ice that is deformed at temperatures above −10°C. Two-maximum and four-maximum fabrics (Fig. 4.13b,c) occur in ice subjected to simple shear, and conical distributions (Fig. 4.13d) are found in ice in uniaxial compression (Hooke & Hudleston, 1980).

DEFORMATION MECHANISM MAPS

Our discussion so far has focussed on the type of creep most commonly observed in glaciers, called *power-law creep* because the creep rate is proportional to the stress raised to some power >1 [see equation (4.4)]. For completeness, some other types of creep should be mentioned.

In recent years, scientists working on ice deformation mechanisms have found it useful to plot "maps" showing the deformation mechanisms operating at different temperatures and stresses (Fig. 4.14). The temperature is usually normalized by dividing by the melting temperature in Kelvins, θ_{Km}. This is called the *homologous temperature*. Similarly, the stress is normalized by dividing by Young's modulus. In Figure 4.14 the stress used is $\sqrt{3}\,\sigma_e$.

The heavy lines in Figure 4.14 divide the diagram into fields in which a single deformation mechanism is dominant. Power-law creep occupies much of the right side of the diagram. Below and to the left of the power-law creep field is the field of diffusional flow. In this type of flow, atoms move from crystal boundaries that are under compression to ones that are under tension. At low temperatures, atoms are believed to move along grain boundaries (grain boundary or Coble creep), whereas at higher temperatures they probably move through the crystal lattice (lattice diffusion or Nabarro-Herring creep). These two diffusional creep fields are separated by a vertical dashed line at about 0.8 θ_{Km} (Fig. 4.14). The shading along the heavy lines separating these two diffusional creep fields from each other and from the power-law field represents the zone of overlap of the fields. On

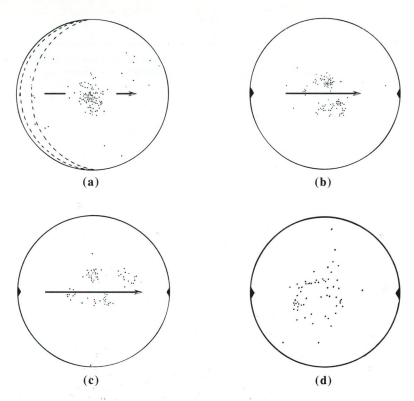

Figure 4.13 Examples of crystallographic fabrics observed in ice. Plots are projections of lower hemisphere of an equal-area net. Dashed lines show attitude of foliation. Triangles on edges show direction of bubble elongation, which is presumed to be parallel to the direction of shear shown by arrows. Fabrics in (**a**), (**b**), and (**c**) are a result of shear in plane of diagram. Fabric in (**d**) is a result of uniaxial compression normal to the plane of the diagram. [(a), (b), and (c) from Hooke & Hudleston, 1980; (d) from Hooke & Hudleston, 1981.]

the right edge of the shaded zone, power-law creep contributes 90% of the deformation, and on the left edge, diffusional creep contributes 90%.

At stresses between 2 and 10 MPa, cleavage fracture also contributes to the deformation. The onset of fracturing occurs at lower stresses in tension than in compression, so this field has an appreciable width. Under sufficiently large confining pressures, fracture is suppressed entirely, and the upper part of the diagram in Figure 4.14 is then accessible.

Another field shown on the diagram is that labeled "grain-boundary mobility." At temperatures above about $-10°C$, the increase in creep rate with temperature is substantially more rapid than predicted by an activation energy of 79 kJ/mol. In addition, there is a rapid increase in the electrical conductivity of ice (Mellor & Testa, 1969), and a less striking but still significant increase in the heat capacity (Harrison, 1972). Finally, multiple maximum c-axis fabrics are common in glacier ice deformed at temperatures above $-10°C$, but rare or absent in ice below $-10°C$ (Hooke & Hudleston, 1980).

The first three of these phenomena can be explained if grain-boundary melting begins at about $-10°C$. Grain-boundary melting involves the formation of a widened zone with a liquidlike structure at grain boundaries and particularly at multiple grain junctions (Duval et al., 1983; de La Chapelle et al., 1995), and is probably due to a higher concentration of impurities, which lower the melting

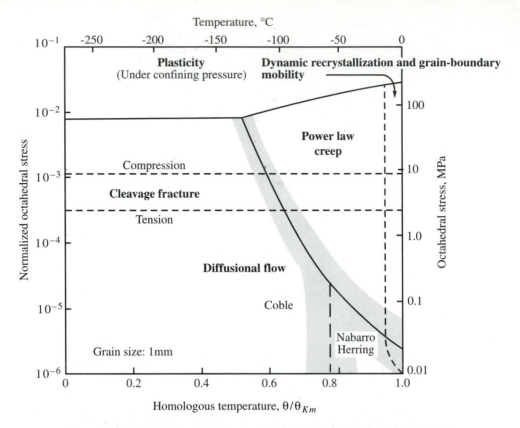

Figure 4.14   Deformation mechanism map for ice with a grain size of 1 mm. (Adapted with permission from Duval et al., 1983, Fig. 1. Copyright 1983 American Chemical Society.)

point, at grain boundaries. This can explain the increase in creep rate (as the liquid phase reduces grain interactions and thus attenuates the internal stress field), the increase in electrical conductivity (as impure water has a much higher conductivity than pure ice), and the increase in heat capacity (as some heat is absorbed by melting). Whether it can explain the development of multiple maximum fabrics is uncertain, as no adequate theory explaining these fabrics is available.

In glaciers, stresses rarely exceed 0.2 MPa, and temperatures are rarely below –50°C, so for our purposes only the lower right-hand corner of the deformation mechanism map is of interest. It appears from this part of the diagram (Fig. 4.14) that both diffusional and power-law creep should occur, as long as the grain size of the ice is about 1 mm. However, this is actually a lower limit for the grain size in glacier ice. Grain sizes of 10 to 30 mm are common in polar and subpolar glaciers, and even larger crystals can be found in temperate glaciers. As grain size increases, the power-law creep field increases at the expense of the diffusional creep field. This is intuitively reasonable because the diffusion paths become longer as the grain size increases.

In diffusional creep, the strain rate is linearly proportional to stress, in contrast to the situation with power-law creep. At present, no unequivocal field evidence suggests that diffusional creep is important in glaciers. It may be important in the coldest parts of the Antarctic and Greenland ice sheets, but then principally near the surface where stresses are low. However, in a study of floating ice shelves bordering the Antarctic Ice Sheet, Thomas (1973b) and Jezek and others (1985) found that power-law creep with $n \approx 3$ seemed to prevail at temperatures of $\sim -15°C$ and stresses of 0.04 to 0.06 MPa.

A FLOW LAW FOR GLACIER ICE

Thus far in this chapter we have looked at some details of the mechanics of the deformation process, and we have found some uncertainty, particularly in attempts to identify the rate-limiting process. In the remainder of this book, we will frequently need a simple yet reasonably accurate expression relating stress and strain rate in ice. In general, we will use the expression

$$\dot{\varepsilon}_e = \left(\frac{\sigma_e}{B}\right)^n \tag{4.5}$$

which, as mentioned briefly in Chapter 2, is often referred to as Glen's flow law, as it was first suggested by John Glen (1955) on the basis of his early uniaxial compression experiments on ice.

As noted, the exponent, n, depends on the creep mechanism operating. A substantial body of experimental data on laboratory ice and on natural glacier ice supports adopting a value of 3 for this exponent (Hooke, 1981), and this seems to be consistent with theoretical expectations (equation 4.4). Older experimental data suggesting a lower value at low stresses are often questionable because tests were not continued long enough to be sure that the transient phase of creep was complete. However, recent experiments by Pimienta and Duval (1987) again raise the possibility of values between 1 and 2 for deformation at low stresses, temperatures, and cumulative strains. They argue that grain-boundary migration is particularly efficient under these conditions so internal strain energy is small, and the density of dislocations therefore does not increase as rapidly with stress as suggested by equation (4.3). Healing of dislocations by diffusional processes may also limit the density (Alley, 1992). Some data from borehole deformation experiments support Pimienta and Duval's interpretation. Other borehole deformation data, on the other hand, suggest values of $n > 3$. However, analysis in these cases was done assuming that the viscosity parameter, B, was constant, independent of depth. This may not be the case; if B actually decreases with depth due to the development of a nonrandom c-axis fabric, to an increase in interstitial water content, or for other reasons, artificially high values of n will be obtained when B is assumed to be constant.

As we have just noted, B is a measure of the viscosity of the ice. Viscosity depends upon a number of factors; we have already mentioned the importance of temperature and fabric, and have hinted at the role of water content, particularly along grain boundaries. Other factors of possible significance are pressure, texture, and other molecular or macro-scale structural features such as dislocation density or grain-boundary structure. Thus, equation (4.5) with a constant value of B is useful only in situations in which an average value of B can be chosen that is reasonably representative of the viscosity of the ice mass as a whole. In the remainder of this section we will explore modifications of equation (4.5) that incorporate some of these variables.

The easiest variable to incorporate is temperature; thus,

$$\dot{\varepsilon}_e = \left(\frac{\sigma_e}{B_o}\right)^n e^{-\frac{Q}{R\theta_K}} \tag{4.6}$$

where, as before, Q is the activation energy for creep. B_o is now a reference parameter; literally, it is the viscosity at $\theta_K = \infty$, but this is physically meaningless. B_o is still a function of the other parameters listed above, such as fabric, texture, microstructure, and so forth.

The activation energy for a process can be determined by running a series of tests at constant stress but varying temperature. In the case of the activation energy for power-law creep, the parameter measured is the strain rate, $\dot{\varepsilon}$, and Q is determined from

$$\frac{\dot{\varepsilon}_1}{\dot{\varepsilon}_2} = e^{\frac{Q}{R}\left(\frac{1}{\theta_{K2}} - \frac{1}{\theta_{K1}}\right)} \tag{4.7}$$

which is readily derived from equation (4.6). Figure 4.15 shows a plot of ln $\dot{\varepsilon}$ vs. $1/T$ in such a series of tests. As expected, many of the data points, in particular those at lower temperatures, fall on a straight line with slope Q/R. However, at temperatures above $-10°C$, the points deviate from a straight line in a direction implying a higher activation energy. In other words, the increase in creep rate with temperature is greater than expected. This is because the grain-boundary processes discussed above begin to influence the creep rate. In particular, the decrease in grain-boundary interactions with increasing water content probably allows the creep rate to increase faster than is the case at lower temperatures.

For some applications, the temperature effect can be represented by

$$\dot{\varepsilon} = \dot{\varepsilon}_o e^{k\theta_n} \tag{4.8}$$

where k is a "constant" that is typically between 0.1 and 0.25°C^{-1}, θ_n is the temperature in Celsius degrees (hence a negative number), and $\dot{\varepsilon}_o$ is a reference strain rate. This relation has the benefit of mathematical simplicity, and it is a reasonable approximation in situations in which the temperature varies within a restricted range. However, k varies slowly with temperature and with Q, so the approximation becomes increasingly imperfect as the temperature range increases.

The next parameter that we will incorporate is hydrostatic pressure, P; thus,

$$\dot{\varepsilon}_e = \left(\frac{\sigma_e}{B_o}\right)^n e^{-\frac{Q+PV}{R\theta_K}}, \tag{4.9}$$

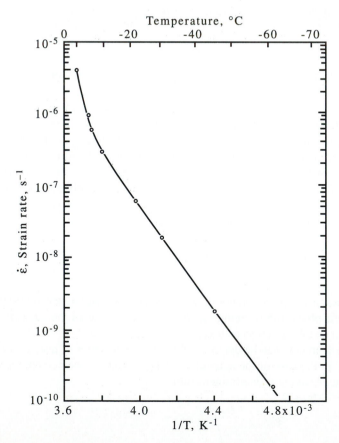

Figure 4.15 Results of a laboratory experiment on the variation of strain rate with temperature. All experiments were run at the same stress. The slope of the straight part of the curve is proportional to the activation energy. (After Mellor & Testa, 1969, Fig. 3. Reproduced with permission of the International Glaciological Society.)

where V is the activation volume for self-diffusion, and the quantity $(Q + PV)$ is the activation enthalpy. In ice, it turns out that V is very low, and it is not clear whether it is positive or negative. Rigsby (1958) conducted experiments at constant temperature, which suggested a slight increase in creep rate with P, implying a negative V. However, when he varied the temperature in such a way that the difference between the experimental temperature and the melting point temperature remained constant, the creep rate was essentially independent of pressure. In other words, V was 0. Later experiments suggested that it might be slightly positive. In any case, it is small enough to be neglected in most if not all applications.

Some experiments have been interpreted as indicating that $\dot{\varepsilon}$ increases with increasing ice crystal size, or grain size (Baker, 1978). As grain size is believed to influence the creep rate in metals, this was not unexpected. However, subsequent experiments failed to find a dependence on grain size (e.g., Jones & Chew, 1983; Jacka, 1984). It was thus suggested that Baker's (1978) samples were too small, and that samples with larger crystals may have been able to creep faster because too much of the strain was occurring in single crystals, favorably oriented for basal glide, that were not buttressed by surrounding crystals with unfavorable orientations. In more recent experiments, however, Baker (1981, 1982) used larger samples, deformed in simple shear, and still found a clear dependence of $\dot{\varepsilon}$ on crystal size.

It is true that crystal size varies directly with temperature and inversely with strain rate. Crystals in rapidly deforming cold ice may be in the millimeter to submillimeter size range, whereas those in slowly deforming temperate ice may exceed a decimeter. The key question is whether these variations in size are merely a consequence of the temperature and strain rate, or alternatively whether they actually influence the strain rate.

The effect of crystal orientation on the strain rate is normally included in the flow law by multiplying the right-hand side by a factor, E, called the *enhancement factor*; thus,

$$\dot{\varepsilon}_e = E \left(\frac{\sigma_e}{B_o} \right)^n e^{-\frac{Q}{R\theta_K}} . \tag{4.10}$$

Rigorously, however, Glen's flow law is based on the assumption that the material is isotropic (see Chapter 9). Thus, adding an enhancement factor in this way to accommodate anisotropy is tacit admission of the failure of this assumption.

We do not yet have enough understanding of the recrystallization process to write an empirical relationship between E and the factors such as temperature, strain rate, and cumulative strain on which it depends (Hooke & Hudleston, 1980). Therefore, selection of the appropriate values of E to use in any given situation is largely subjective.

Laboratory experiments provide some basis for estimating E. Russell-Head and Budd (1979) and Baker (1981, 1982) studied natural ice with a single-maximum fabric. When deformed in simple shear in the laboratory, with the sample oriented so that the c-axes were approximately perpendicular to the shear plane, this ice deformed ~4 times faster than comparable samples with random c-axis orientations. Russell-Head and Budd also found that a section of a borehole in Law Dome, Antarctica, which passed through ice with a single-maximum fabric, deformed ~4 times faster than it would have in ice without such a fabric. However, other experiments in glaciers have been less successful in demonstrating a clear relation between fabric and deformation rate (Hooke, 1981). More recently, Budd and Jacka (1989) have suggested that enhancement factors of ~3 are reasonable for ice in uniaxial compression, and that factors of 8 to 10 may be appropriate for ice in simple shear.

Finally, we return to the effect of the water content of ice on the creep rate. This was studied by Duval (1977) in a pioneering set of very sophisticated experiments. His results, expressed in

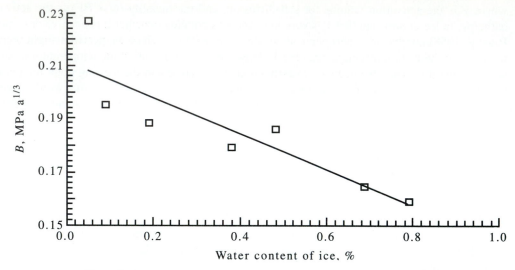

Figure 4.16 Variation in B with water content of the ice. (Data obtained by Duval, 1977, and reported by Lliboutry, 1983.)

terms of the decrease in B with increasing water content, are shown in Figure 4.16. With an increase in water content from 0% to 0.8%, B decreases from ~0.21 to ~0.16 MPa $a^{1/3}$. Lliboutry (1983) reported that the water content of basal ice of temperate glaciers typically varies between 0.6% and 0.95%, corresponding to a variation in B from 0.167 to 0.154, and hence in $\dot{\epsilon}$ of a factor of 1.3. Lower water contents, and hence higher values of B, are likely in temperate ice of subpolar glaciers.

SUMMARY

In this chapter we first reviewed the crystal structure of ice, and we noted that there are imperfections in this structure, called dislocations, that allow ice (and other crystalline materials) to deform under stresses that are low compared with the strength of individual molecular bonds. The rate of deformation is limited by the speed with which dislocations move. Processes that might limit this speed are those which inhibit motion in a single crystallographic plane (drag), those which prevent dislocations from climbing from one crystallographic plane to another to get around tangles, and those which impede motion on certain crystallographic planes.

Experimental data do not, at present, provide a basis for choosing among these possible rate-limiting processes. However, the drag mechanism does provide a theoretical basis for the commonly observed value of the exponent, n, in the flow law: Dislocation speed is proportional to stress, and dislocation density is proportional to stress squared, so deformation rate is proportional to stress cubed.

Because some crystals in a polycrystalline aggregate are not oriented for easy glide, stress concentrations develop. These result in recrystallization by three distinct processes. Recrystallization leads to preferred orientations of c-axes, and hence to more rapid deformation.

To place the creep processes in ice in a more general framework, we introduced a deformation mechanism map in which we displayed the range of temperatures and stresses under which different deformation processes occur. Within the temperature and stress ranges normally found in glaciers, we found that power-law creep was likely to be the dominant process, although diffusional creep may occur in some low-stress situations.

Finally, we introduced Glen's flow law, and we related the exponent, n, in the flow law to the creep mechanisms discussed earlier. Then we considered how temperature, pressure, texture, fabric, and water content affect the viscosity parameter, B. Temperature and pressure effects may be incorporated into the flow law by rigorous, physically based modifications, whereas *ad hoc* procedures based on empirical evidence are used to incorporate the other effects.

5

The Velocity Field in a Glacier

Many problems in glaciology require an understanding of the flow field in a glacier. For example, the way in which flow redistributes mass determines the shape of a glacier, and also the rapidity with which glaciers respond to climatic change. Flow also redistributes energy and thus affects the temperature distribution. This, in turn, has important implications for the nature of the coupling with the glacier bed. Spatial variations in speed, or strain rates, are of concern to structural geologists using glaciers as analogs for deformation of rocks. From a geomorphic perspective, the entrainment of debris and the character of moraines constructed from this debris are dependent upon the flow field. In short, understanding the flow field is fundamental to the analysis of many problems in glacier mechanics.

For a full description of the flow field in a glacier, we need the horizontal and vertical components of the velocity at every point. By making several assumptions, we can obtain approximate solutions to this problem that will give insights into certain characteristics of glaciers and the landforms they produce. Initially, we will limit the analysis to two dimensions and also assume a steady state.

We will begin by studying the distribution of horizontal velocity. Given the pattern of mass accumulation and loss over a glacier, we can use conservation of mass to determine the mean (depth-averaged) horizontal velocity. Then, using conservation of momentum and a simplified version of the flow law [equation (4.5)], we study the variation in horizontal velocity with depth in an ice sheet and in a valley glacier. Differences between these solutions and measured velocity distributions reveal inadequacies of the theory, and they draw attention to the need for a better understanding of the basal boundary condition. Finally, by integrating the velocity over depth, we calculate the mass flux, and also obtain an expression for the mean velocity in terms of the glacier thickness.

The vertical velocity field is treated next. Again we will use the steady-state assumption and the pattern of mass balance (conservation of mass) to determine the vertical velocity at the surface. By assuming, as a first approximation, that the longitudinal strain rate, or rate of stretching in the longitudinal direction, is independent of depth, we then find that we can calculate the variation in vertical velocity with depth, and hence obtain an approximation to the full velocity field.

We will also discuss the role of drifting on the flow field, and hence the topography of a glacier surface. This will lead to a better understanding of the processes of formation of certain types of moraine. Finally, we will explore the reasons for inhomogeneous flow in ice sheets, as manifested by ice streams.

BALANCE VELOCITY

In a general sense, flow paths and speeds in a glacier are determined by the net budget. Consider an idealized glacier that, over a period of years, is in a steady state so its thickness (or surface profile) does not change. Then, at some distance, x, from the divide, the mean horizontal velocity averaged over depth is

$$\bar{u} = \frac{1}{h(x)} \int_0^x b_n(x)\,dx, \tag{5.1}$$

where, $h(x)$ is the glacier thickness, and for convenience, the units of b_n are taken to be meters of ice per year (Fig. 5.1). This equation is an expression of the principle of *conservation of mass* in an incompressible medium; as much mass must be moved out of the control volume, V, by flow, $\bar{u} \cdot h(x)$, as enters it by accumulation, $\int b_n(x)\,dx$, on the surface.

SHEAR-STRESS DISTRIBUTION

To determine the velocity distribution at depth in a glacier, we will find, below, that we need an expression for σ_{zx} as a function of z. Thus, we digress briefly from the principal objectives of this chapter to derive two similar expressions for σ_{zx} that are commonly used in the literature. The derivations differ only in the orientation of the axes and of the plane on which σ_{zx} operates.

Consider, first, the situation in Figure 5.2a. The x-axis is taken parallel to the surface, which slopes at an angle α. The z-axis points downward, normal to the surface. We want the shear stress on a plane at depth h. The column of ice is taken to be 1 m on a side. Therefore, the weight of the column is $\rho g h$, where ρ is the density of ice and g is the acceleration due to gravity. The component of that weight parallel to the plane of interest is, thus, $\rho g h \sin\alpha$. For static equilibrium, this must be balanced by a shear stress on the plane; thus,

$$\sigma_{zx} = -\rho g h \sin\alpha. \tag{5.2a}$$

The shear stress is negative because it acts in the negative x-direction on a plane with an outwardly directed normal in the positive z-direction (see Chapter 2).

Next, consider the situation in Figure 5.2b. The x-axis is now horizontal and the z-axis vertical. The column is again of unit cross-sectional area. The plane of interest is at depth h below the surface

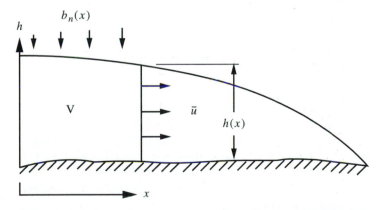

Figure 5.1 Schematic diagram illustrating dependence of horizontal velocity on accumulation rate on an ice sheet.

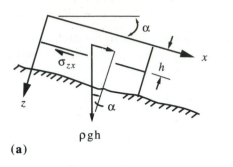

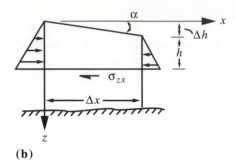

(a)   (b)

Figure 5.2   Derivations of two alternative expressions for the shear stress, σ_{zx}, on a plane at depth h.

on the right side of the column. To a good approximation, the hydrostatic (or lithostatic) pressure at this depth is $\rho g h$ and the pressure varies linearly with depth. Thus, the mean pressure is $\frac{1}{2}\rho g h$, and the force on this side is $\frac{1}{2}\rho g h\, h$, where the second h represents the area of the face. Similarly, on the left side, the force is $\frac{1}{2}\rho g (h + \Delta h)^2$. The force on the plane of interest is $-\sigma_{zx}\Delta x$. Summing forces on the column yields

$$\frac{1}{2}\rho g(h + \Delta h)^2 - \frac{1}{2}\rho g h^2 - (-\sigma_{zx}\Delta x) = 0.$$

Expanding the first term, neglecting the term in Δh^2, noting that $\tan \alpha = \Delta h/\Delta x$, and rearranging terms leads to

$$\sigma_{zx} = -\rho g h \tan \alpha. \tag{5.2b}$$

This equation is appropriate for a situation in which both the x-axis and the plane of interest are horizontal, but the glacier surface is sloping.

Because glacier-surface slopes are normally small, $\sin \alpha \approx \tan \alpha \approx \alpha$, so equations (5.2a) and (5.2b) are nearly interchangeable. Thus, we commonly write

$$\sigma_{zx} = -\rho g h \alpha \tag{5.2c}$$

for small slopes. Note that in a coordinate system in which the z-axis is directed upward, σ_{zx} would be positive.

HORIZONTAL VELOCITY AT DEPTH IN AN ICE SHEET

Demorest (1941, 1942) argued that the horizontal velocity in a glacier should increase with depth. He thought that the pressure of the overlying ice would soften the deeper ice, making it flow faster. Nye (1952a), however, pointed out that this concept was physically unsound because the faster-moving deeper ice would exert a shear stress on the overlying ice, and there would be no corresponding resisting forces to oppose this shear stress. Therefore, the overlying ice must move at least as fast as that below. We now know from numerous borehole deformation experiments that Nye's analysis was basically correct.

To pursue Nye's reasoning quantitatively, we start with the flow law, equation (4.5), and assume that strain rates other than $\dot{\varepsilon}_{xz}$ and $\dot{\varepsilon}_{zx}$ and stresses other than σ_{xz} and σ_{zx} are negligible. Then, using equations (2.10) and (2.11), and making use of the symmetry of the tensor so that $\dot{\varepsilon}_{xz} = \dot{\varepsilon}_{zx}$ and $\sigma_{xz} = \sigma_{zx}$, we obtain

$$\dot{\varepsilon}_{zx} = \left(\frac{\sigma_{zx}}{B}\right)^n. \tag{5.3}$$

Using an equation analogous to (2.6a) and assuming that all of the shear takes place in the plane normal to z so $\partial w / \partial x = 0$, a mode of deformation known as *simple shear*, this becomes

$$\frac{du}{dz} = 2\left(\frac{\sigma_{zx}}{B}\right)^n. \tag{5.4}$$

To integrate this, σ_{zx} must be expressed as a function of z. We will use the coordinate system of Figure 5.2a and equation (5.2a), and will integrate from the surface down to depth h (Fig. 5.3); thus,

$$\int_{u_s}^{u_h} du = -2\left(\frac{\rho g \sin \alpha}{B}\right)^n \int_o^h z^n dz. \tag{5.5}$$

Carrying out the integration and rearranging terms yields:

$$u(h) = u_s - \frac{2}{n+1}\left(\frac{\rho g \sin \alpha}{B}\right)^n h^{n+1}. \tag{5.6}$$

This is the desired solution for the velocity profile. It was first derived by Nye (1952b). Knowing u_s, B, and α, we can calculate the velocity as a function of depth, $u(h)$. If the total thickness, H, is known, we can solve equation (5.6) for the velocity at the bed, u_b; thus,

$$u_b = u_s - \frac{2}{n+1}\left(\frac{\rho g \sin \alpha}{B}\right)^n H^{n+1}. \tag{5.7}$$

Because $n \simeq 3$, the velocity at the bed is quite sensitive to the values of α, B, and H.

It is important to emphasize that this derivation is rigorously correct only for a glacier that is in the form of a slab of infinite extent on a uniform slope. If the glacier is bounded laterally, drag on the sides must be considered in calculating σ_{zx}. We will take this up in the next section. If the thickness is not uniform in the longitudinal direction, there are likely to be gradients in the longitudinal stresses that either augment or diminish σ_{zx} relative to the values calculated from any of equations (5.2). Normally, these gradients are sufficiently small that this source of error is not of major concern in comparison with some others. This is discussed further below and in Chapter 10.

A velocity profile for a glacier 300 m thick with a surface slope of 2.2°, $n = 3$, and $B = 0.2$ MPa a$^{1/n}$, calculated from equation (5.6), is shown in Figure 5.4. The profile has a distinctive form; the velocity is nearly independent of depth in the upper part of the glacier, and then decreases rapidly near the bed. For comparison, the dashed line shows the profile for a linearly viscous ($n = 1$)

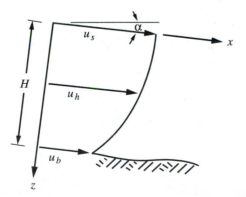

Figure 5.3 Parameters involved in integrating equation (5.4).

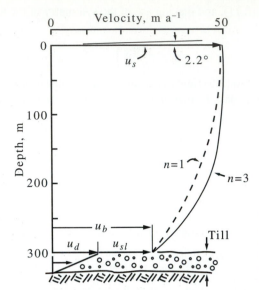

Figure 5.4 Velocity profile for an ice sheet with a surface slope of 2.2° and $B = 0.2$ MPa a$^{1/n}$. A profile for a linearly viscous material is shown for comparison. The thickness of the till layer at the base is greatly exaggerated.

material, with the value of B adjusted to give the same velocity at the bed. The distinctive form of the $n = 3$ profile is a consequence of the "high" value of n.

Note also in Figure 5.4 that the velocity at the bed, u_b, is composed of two components. If the glacier is at the pressure melting point at the bed, it can slide over its substrate (with speed u_{sl}), whether that substrate be hard bedrock or unconsolidated material. If the substrate is unconsolidated material such as glacial till, it may also deform. This adds a speed u_d to the total. These contributions to the speed of a glacier are discussed in detail in Chapter 7.

HORIZONTAL VELOCITY IN A VALLEY GLACIER

In a valley glacier, some of the resistance to flow, or drag, is provided by valley sides. To see how this alters the situation, consider first a glacier in a semicircular valley of radius R (Fig. 5.5) and slope α. Balancing forces on a cylindrical surface of radius r and of unit length parallel to the flow gives

$$\sigma_{rx}(\pi r) = -\rho g\left(\frac{\pi r^2}{2}\right)\sin\alpha. \tag{5.8}$$

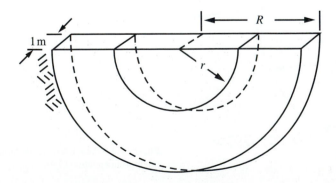

Figure 5.5 Cross section of a glacier in a semicircular valley.

Here, πr is the area of the surface and $\rho \pi r^2/2$ is the mass of ice inside the surface. The latter, multiplied by $g \sin\alpha$, is the total force parallel to the surface that must be resisted by a shear stress, σ_{rx}, on the surface. Thus, now:

$$\sigma_{rx} = -\frac{1}{2}\rho g r \sin\alpha. \tag{5.9}$$

Inserting this in equation (5.4) with r in place of z as the depth dimension, and integrating as before, yields

$$u(r) = u_{sc} - \frac{2}{n+1}\left(\frac{\rho g \sin\alpha}{2B}\right)^n r^{n+1}, \tag{5.10}$$

where u_{sc} is the velocity at the surface on the centerline. But for the change to a cylindrical coordinate system, this result differs from that of equation (5.6) only in the factor of 2 in the denominator of the term in brackets. However, as $n \simeq 3$, the difference in velocity between the surface and a given depth is a factor of 8 less in the valley glacier. This represents the effect of drag on the valley sides.

Semicircular cross sections are not common in nature, so it is worthwhile considering a more realistic shape (Fig. 5.6). By analogy with equation (5.8) we write

$$\overline{\tau_b}P = -\rho g A \sin\alpha. \tag{5.11}$$

Here, $\overline{\tau_b}$ is the drag exerted on the glacier by the bed, averaged over the length of the ice-bed interface, P, and A is the cross-sectional area of the glacier. Although τ_b is a force per unit area and is often called the basal shear stress, it is confined to a plane and is thus a vector, not a tensor quantity. Therefore, we will use the term *drag* and the symbol τ for it. Dividing by P and multiplying the top and bottom of the right-hand side by the thickness of the glacier at the centerline, H, yields

$$\overline{\tau_b} = -\rho g \frac{A}{PH}H\sin\alpha = -S_f\rho g H \sin\alpha. \tag{5.12}$$

Here, we have defined $A/PH = S_f$. S_f is known as the shape factor. The reader will readily see that S_f is 1 for an infinitely wide glacier and $\frac{1}{2}$ for a semicircular glacier.

We now make the assumptions:

$$\tau_{b,\mathfrak{C}} = \overline{\tau_b} \tag{5.13}$$

and

$$\sigma_{zx} = \frac{z}{H}\tau_{b,\mathfrak{C}} = \frac{z}{H}\overline{\tau_b}. \tag{5.14}$$

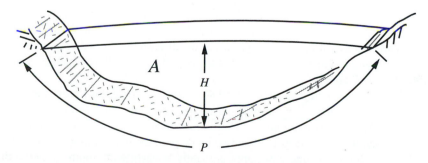

Figure 5.6 Cross section of a valley glacier.

Assumption (5.13) says that the basal drag at the centerline is equal to the average over the cross section, and assumption (5.14) says that the shear stress at the centerline varies linearly with depth and approaches τ_b at the bed without any discontinuity. With these assumptions, equation (5.12) can be rewritten as

$$\sigma_{zx} = -S_f \rho g z \sin \alpha. \tag{5.15}$$

Inserting this in equation (5.4) and integrating as before yields

$$u(h) = u_s - \frac{2}{n+1} \left(\frac{S_f \rho g \sin \alpha}{B} \right)^n h^{n+1}. \tag{5.16}$$

If one knows u_s and can make a reasonable estimate of H, equations (5.15) and (5.16) can be used to calculate the basal drag and speed, respectively. With such calculations, Nye (1952b, p. 86) argued that a large fraction of the movement of temperate valley glaciers was due to sliding (or till deformation) at the bed, and that despite a large variation in thickness and surface slope, basal drags fell within a relatively narrow range: $0.5 < \tau_b < 1.5$. In practice, however, values of u_b thus calculated are not very reliable because small errors in τ_b are amplified when it is raised to the n power [equation (5.16)].

The narrow range in τ_b is a consequence of the nonlinearity of the flow law. Small increases in H result in comparatively large increases in u_s, and hence in the rate at which mass is transferred from the accumulation area to the ablation area. Thus, positive net balances may lead to significant advances of glaciers with only modest increases in thickness, and conversely. Following retreat of a glacier, evidence for this effect is commonly seen in vegetation boundaries, called *trimlines*, that reflect the former position of the ice surface. Near the terminus, such trimlines are typically high above the present glacier surface, and they merge with the valley bottom well down valley from the terminus. However, when traced up-valley, they become quite close to the level of the present surface.

Comparison with Measurements

It is revealing to compare the velocity distribution measured in a temperate glacier with one calculated on the basis of arguments similar to those leading to equation (5.16). The measurements in Figure 5.7a were made by Raymond (1971) on Athabasca Glacier in the Canadian Rocky Mountains and are compared with calculations (Fig. 5.7b) made by Nye (1965a) for a glacier in a cylindrical parabolic channel with a similar aspect ratio. Nye assumed that the basal sliding speed was uniform over the cross section.

There are some interesting discrepancies between the observed and calculated distributions. First, the observed basal velocity is 80% to 90% of the surface velocity over the central section of the glacier, and then decreases rapidly toward the valley sides. These large lateral gradients in u_b conflict with Nye's assumption. The gradients are attributed to lateral variations in water pressure at the ice-rock contact. The role of water pressure in sliding and till deformation will be discussed further in Chapters 7 and 8.

Second, in the field measurements, $\partial u/\partial y > \partial u/\partial z$ (where the y-axis is transverse), whereas in the theoretical model the reverse is true. Unless the ice is quite anisotropic, the contrast in strain rates indicates a similar contrast in stress. The higher shear-strain rates near the margin of Athabasca Glacier indicate that the glacier is supported by drag on the margin more than by drag on the bed.

Third, although σ_{zx} increases approximately linearly with depth, as in the theory, $\tau_{b,\ell} < \tau_b$, contrary to our assumption [equation (5.13)]. In other words, if S_f is calculated from its definition,

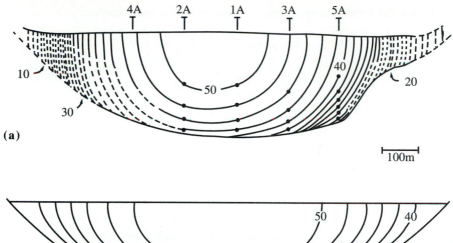

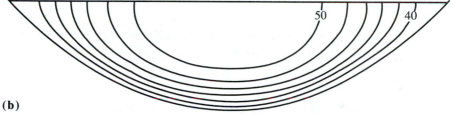

Figure 5.7 (a) Longitudinal velocity distribution, in $m\,a^{-1}$, on Athabasca Glacier. Dots indicate points of measurement. Dashed contours are extrapolated. (b) Theoretical distribution of longitudinal velocity in a parabolic channel, scaled to cover approximately the observed range of velocities in (a) (After Raymond, 1971, Fig. 10. Reproduced with permission of the author and the International Glaciological Society.)

A/PH, and if this is then used to calculate σ_{zx} at the centerline, the value of σ_{zx} will be too high. Put differently, S_f is actually less than A/PH. The fact that the velocity contours are nearly semicircular in shape, which is quite different from the shape of the margin, suggests that $S_f = \frac{1}{2}$ would give a better estimate of τ_b.

MEAN VELOCITY AND ICE FLUX

The ice flux per unit width, q, is readily obtained by integrating the velocity-profile [equations (5.6), (5.10), or (5.16)] over depth. We will illustrate this with equation (5.16); thus,

$$q_{\mathfrak{k}} = \int_o^H u(z)\,dz = \int_o^H \left[u_s - \frac{2}{n+1}\left(\frac{S_f \rho g \sin \alpha}{B} \right)^n z^{n+1} \right] dz$$

$$= u_s H - \frac{2}{(n+1)(n+2)}\left(\frac{S_f \rho g \sin \alpha}{B} \right)^n H^{n+2}. \tag{5.17}$$

Possibly of greater use is the mean velocity over depth, $\bar{u} = q_{\mathfrak{k}}/H$:

$$\bar{u} = u_s - \frac{2}{(n+1)(n+2)}\left(\frac{S_f \rho g \sin \alpha}{B} \right)^n H^{n+1}. \tag{5.18}$$

Combining this with equation (5.16) and simplifying the result leads to

$$\bar{u} = u_b + \frac{2}{(n+2)}\left(\frac{S_f \rho g \sin \alpha}{B}\right)^n H^{n+1}$$

(5.19)

$$= u_b + \frac{n+1}{n+2}(u_s - u_b)$$

$$= \frac{4}{5}u_s + \frac{1}{5}u_b,$$

where the numerical values are calculated assuming $n = 3$. This relation will be of use in Chapter 12.

VERTICAL VELOCITY

Let us now consider the variation in vertical velocity with depth. Because we are dealing with a two-dimensional situation, the incompressibility condition [equation (2.5)], becomes

$$\frac{\partial u}{\partial x} + \frac{\partial w}{\partial z} = 0.$$

(5.20)

Let us ignore the compressibility of firn near the surface, and also assume that the longitudinal strain rate, $\partial u/\partial x$, is independent of depth and thus equal to a constant, say $-c$. Equation (5.20) then reduces to $\partial w/\partial z = c$. Finally, we assume that $w = 0$ on the bed, $z = 0$, thus ignoring any contribution from melting or refreezing, the rates of which are normally small. It thus becomes convenient to adopt a coordinate system in which the origin is on the bed and the z-axis is positive upward, normal to the bed. Then:

$$\int_0^w dw = c\int_0^z dz$$

or

$$w = cz.$$

At the surface, $z = H$, we have $w = w_s$ so $c = w_s/H$. Therefore,

$$w = \frac{z}{H}w_s.$$

(5.21)

In other words, the vertical velocity decreases linearly with depth.

The key assumption in this derivation is that $\partial u/\partial x$ is independent of depth. Clearly, if the ice is frozen to the bed, $\partial u/\partial x = 0$ at the bed. Thus, if it is nonzero higher in the glacier, it must decrease (in absolute value) with depth. A necessary (but not sufficient) condition for $\partial u/\partial x = c$ is that $\partial u_b/\partial x = \partial u_s/\partial x$. Then, any change in u_s in the longitudinal direction would have to be matched by an identical change in u_b. However, $u_s = u_b + (u_s - u_b)$ and $(u_s - u_b) \propto \tau_b^3$ so any downglacier change in τ_b, which, while likely to change u_b (see Chapter 7), would change u_s more than u_b. Thus, $\partial u_b/\partial x \neq \partial u_s/\partial x$.

We might guess that the variation with depth of $\partial u/\partial x$ was roughly proportional to that in u in many situations. Then, because u is nearly independent of depth in the upper part of a glacier (Fig. 5.4), the assumption that $\partial u/\partial x$ is independent of depth over the full thickness becomes a reasonable first approximation. This can be appreciated by visualizing deformation profiles in two locations some distance apart in the longitudinal direction. This argument is stronger in polar glaciers because the ice near the surface is colder, and hence more viscous (higher B). The shear-strain rate, $\dot{\varepsilon}_{zx}$, is thus lower, so u is nearly constant over a greater fraction of the ice thickness.

In view of these rationalizations and the simplicity of equation (5.21), this approximation is widely used in calculations, as we shall see in Chapter 6 and elsewhere.

SUBMERGENCE AND EMERGENCE VELOCITIES

Earlier [equation (5.1)], we gained insight into the magnitude of the horizontal velocity by considering a glacier in a steady state, such that its surface profile remained unchanged. Let us now use this idealization to study vertical velocities. In such a steady state, the surface in the accumulation area must everywhere be sinking at a rate that balances accumulation, and conversely in the ablation area. Thus, the vertical velocity at the surface, w_s, is clearly related to the net balance rate, b_n. Remembering that a point on the surface is also moving with a horizontal velocity, u_s, and that the surface has a slope, α, we find that the appropriate relation is (Fig. 5.8):

$$b_n = -w_s + u_s \tan \alpha. \qquad (5.22)$$

In writing equation (5.22), we have taken the x-axis as horizontal and positive in the downglacier direction, and the z-axis as positive upward. Thus, in the accumulation area, both w_s and α are negative and, owing to the relative magnitude of the two terms on the right-hand side of the equation (see Fig. 5.8a), the minus sign in the equation makes the right-hand side positive. In the accumulation area, the right-hand side is called the *submergence velocity*.

Equation (5.22) also applies in the ablation area (Fig. 5.8b), except that here w_s is positive so both terms on the right-hand side take on negative values. Thus, b_n is negative, reflecting ablation. Here, the right-hand side is called the *emergence velocity*.

Clearly, the submergence and emergence velocities are defined for any point on a glacier surface. However, they equal b_n only in the idealized steady-state situation that we have specified. This is because b_n varies from year to year, and because, even averaged over several years, glaciers are rarely in a steady state. Put differently, if the accumulation rate consistently exceeds the submergence velocity and the ablation rate consistently falls short of the emergence velocity, the glacier is becoming thicker and will advance, and conversely.

Other possibilities can also be visualized. For example, if the equality in equation (5.22) holds everywhere except in the lower part of the ablation area where the ablation rate exceeds the emergence velocity, the glacier may be in the final stages of adjustment to a climatic warming. The implication of such a situation would be that the accumulation area has essentially adjusted to the warming, but the glacier is still retreating slightly.

We have shown that on a glacier which is in a steady state and which has a balanced mass budget, the velocity field at the surface is related to b_n. It is instructive to consider in greater detail the physical mechanisms behind this relation. In this case, b_n is the independent variable and the velocity field is the dependent variable. (In a larger system involving glacier-climate interactions, b_n would be dependent upon the climate.) The physical mechanism by which b_n and the velocity field are related is viscous flow, in which the flow rate increases with the driving stress. If the velocities are, say, too low (in absolute value), the submergence velocity will be less than the accumulation rate so the glacier will become *thicker* in the accumulation area (Fig. 5.8a). Similarly, the emergence velocity will

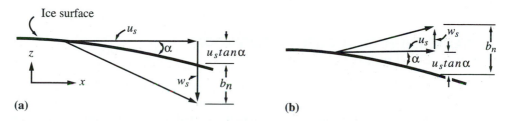

Figure 5.8   Diagrams illustrating (**a**) submergence and (**b**) emergence velocities.

be less than the ablation rate, so the glacier will become *thinner* in the ablation area (Fig. 5.8b). The slope of the glacier surface thus increases. The increase in slope, coupled with the increase in thickness in the accumulation area, increases the driving stress and hence u_s. Because $u = 0$ at the head of the glacier and at the terminus, an increase in u_s in the middle makes $\partial u/\partial x$ more extending in the accumulation area and more compressive in the ablation area. Thus, by the arguments leading to equation (5.22), $|w_s|$ increases. The increases in both u_s and $|w_s|$ tend to restore the steady state.

FLOW FIELD

We now have the tools needed to make a first-order estimate of the flow field in a glacier, given $b_n(x)$. In a steady-state situation, equation (5.1) gives the depth-averaged horizontal velocity, $\bar{u}(x)$, which is probably sufficient for most applications. However, various levels of sophistication could be added; equations (5.16) (with $h = H$) and (5.19) could be solved simultaneously for u_s and u_b, and equation (5.16) could then be used to estimate the variation in u with depth. This would give $u(x,z)$. Then, equations (5.21) and (5.22) provide a reasonable first estimate of $w(x,z)$. Hence, one could plot vectors u and w at a large number of points in a glacier cross section and sketch flowlines based on these vectors. The result would be flowlines much like those in Figure 3.1.

It may be worthwhile studying Figure 3.1 in connection with the above equations to develop an intuitive sense of why the flowlines appear as they do. From equation (5.22) and Figure 5.8, it is clear that the vertical velocity must be downward in the accumulation area and upward in the ablation area. Owing to the slope of the glacier surface, the location where $w_s = 0$ is not at, but rather slightly downglacier from, the equilibrium line [equation (5.22)]. From equation (5.1) it is obvious that u increases outward from the divide, reaching a maximum again not at, but rather slightly downglacier from, the equilibrium line. Because the variation in w_s is small compared with that in u_s, the resultant velocity vectors plunge most steeply near the divide and ascend most steeply near the margin, as u_s is low in these locations. In fact, at the divide on a polar ice sheet, $u_s = 0$ so the vector points directly downward. Near the bed, w_s is low so the vectors approach parallelism with the bed.

In summary, flowlines tend to be downward in the accumulation area and upward in the ablation area. At the equilibrium line of our idealized steady-state glacier, they are parallel to the surface, and everywhere along the bed they are parallel to the bed. The flow line starting at the divide on an ice sheet will go straight downward until it reaches the bed, and then will follow the bed, remaining strictly parallel to the bed in the absence of melting or refreezing, until it emerges at the margin.

In the steady state, the volume of ice moving between two adjacent flowlines remains constant along the full length of these two flowlines. This is true by definition; material cannot cross a flowline, so all material that starts between two flowlines must remain between them. A consequence of this is that the velocity will be highest where the flowlines are closest together, as the ice is assumed to be incompressible. Thus the highest velocities will be near the equilibrium line, as is evident from equation (5.1).

TRANSVERSE PROFILES OF SURFACE ELEVATION
ON A VALLEY GLACIER

In the ablation area of a valley glacier, transverse profiles of surface elevation are commonly convex upward (Fig. 5.9a), whereas in the accumulation area they are concave upward (Figure 5.9b). This can be understood by considering the emergence and submergence velocities. In a steady-state situation, w_s cannot be 0 along the margins of a glacier in either the accumulation area or the abla-

(a)

(b)

(c)

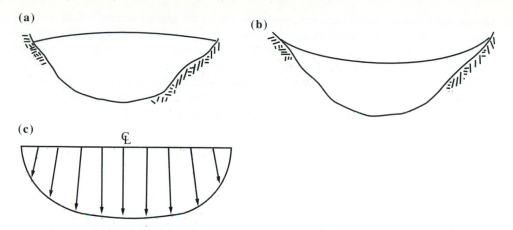

Figure 5.9 Schematic cross sections of a valley glacier in (**a**) the ablation area, and (**b**) the accumulation area; (**c**) a plan view of the glacier showing the transverse variation in u_s in the ablation area.

tion area because there is accumulation or ablation, respectively, in these locations. However, the ice thickness goes to 0 at the margin. Thus, to provide a downward w_s near the margin in the accumulation area, ice must be drawn away from the valley sides, and conversely in the ablation area. The transverse surface slopes, toward the center of the glacier in the accumulation area and away from the center in the ablation area, provide the forcing for this flow.

Consideration of transverse variations in the emergence and submergence velocities provides insight into lateral variations in w_s and into the mechanism of adjustment of transverse profiles. Consider the ablation area first. Horizontal velocities are normally highest near the center of a glacier and decrease toward the margins due to drag on the valley sides (Fig. 5.9c), in much the same way that velocities decrease with depth due to drag on the bed. The ablation rate, however, is normally approximately constant across the glacier. It may, in fact, be somewhat higher near the margins due to heat radiated or advected from dark rocks of the valley walls. The longitudinal surface slope, α, will also be approximately constant across the glacier. Thus, from equation (5.22) or Figure 5.8b, it is clear that w_s must be higher near the margins than along the centerline, as illustrated in Figure 5.10.

The process by which the transverse profile in the ablation area is adjusted is easy to understand. Consider what would happen were the profile flat and w_s constant along this profile. Suppose the ablation rate equals the emergence velocity at the centerline. Along the margins where u_s is

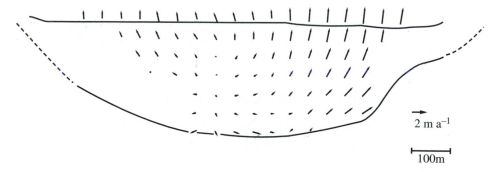

2 m a^{-1}

100m

Figure 5.10 Transverse velocity field measured in a cross section of Athabasca Glacier by Raymond (1971, Fig. 12). (Reproduced with permission of the author and the International Glaciological Society.)

lower, melt would exceed the emergence velocity, so the glacier surface would decrease in elevation, leading to the convexity that is commonly observed. The resulting transverse surface slope would force a transverse component to the flow. Of potential interest in trying to understand landforms produced by glacial erosion is the fact that this transverse component of the flow is apparently greatest near the bed, according to measurements made by Raymond (1971) on Athabasca Glacier (Fig. 5.10). Because the valley walls inhibit such flow, a transverse compression develops, thus increasing the rate of vertical extension near the sides of the glacier (assuming no compensating change in the longitudinal strain rate). The magnitude of the transverse slope increases, thus increasing w_s, until the emergence velocity equals the ablation rate.

In the accumulation area, the situation is reversed. As in the ablation area, u_s is lower near the margin. Thus, if the longitudinal component of α is approximately constant across the glacier, the flow field would have to develop in such a way that w_s was actually somewhat less negative (downward) near the margins than near the centerline [equation (5.22) or Fig. 5.8a]. However, the ice is thinner near the margin so longitudinal stretching is normally not sufficient to produce the required thinning rate (or w_s). Furthermore, accumulation rates are likely to be higher along the margin owing to drifting and avalanching. The concave cross-valley profiles (Fig. 5.9b) typical of accumulation areas compensate for these two effects. The transverse surface slope results in a transverse component of flow toward the center of the glacier, and because glaciers normally increase in thickness rapidly away from the margins, stretching rates are high along flowlines which diverge from the margin. The resulting transverse stretching provides the more negative w_s required near the margins.

To emphasize the physical process of adjustment of the surface topography to the local mass budget in the accumulation area, consider again a hypothetical case in which the transverse profile is initially flat, rather than concave upward. In this case, there would be no transverse component to the surface slope, and hence little or no transverse component to the flow. Consequently, flowlines would be parallel to the glacier margin, and longitudinal strain rates along such flowlines would be too low to provide the negative w_s needed to balance the accumulation. In other words, the left-hand side of equation (5.22) would be larger than the right-hand side, and the glacier would become thicker in this area. This thickening along the margins would continue as long as the imbalance persisted, thus establishing the characteristic concave-upward transverse profile.

If transverse profiles are generally convex upward in the ablation area and concave upward in the accumulation area, it is interesting to consider exactly where the transition between the two types of profile should occur. Let us return to our idealized steady-state glacier with a balanced mass budget. At the equilibrium line on this glacier, $b_n = 0$, so from equation (5.22) $w_s = u_s \tan \alpha$. As $\alpha < 0$ and $u_s > 0$, w_s must still be somewhat negative (downward), as mentioned previously. $w_s = 0$ somewhat downglacier from the equilibrium line where $b_n = u_s \tan \alpha$, and it is approximately at this point that the transition should occur. It is not realistic to carry this analysis further, however, because u_s varies across the glacier, and slight variations in longitudinal strain rate can have a significant effect on the small values of w_s involved.

EFFECT OF DRIFTING SNOW ON THE VELOCITY FIELD

Glaciers flow over irregular beds, and thus have undulating surface profiles. Furthermore, wind patterns and solar radiation may be influenced by nunataks or irregular valley walls. Patterns of both accumulation and ablation thus can be uneven owing to drifting and to shading from the sun during the melt season.

To understand how this influences the flow field and surface profile, consider the hypothetical situation shown in Figure 5.11 in which a glacier flows over a convexity in the bed, resulting in a similar

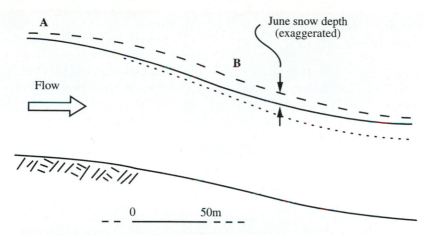

Figure 5.11   Effect of drifting snow on the surface profile of a glacier. Owing to the additional accumulation in the lee of the surface convexity at **A**, w_s does not need to be as high at **B** as otherwise would be the case.

convexity in the surface. Owing to drifting in the lee of the surface convexity, the normal June snow depth at **B** is, say, 2 m, while that at **A** is only 1 m. During a normal melt year, all of the snow and 0.5 m of the underlying ice melts at A, whereas at **B**, melting removes only the snow cover. The emergence velocity at **A** must thus be 0.5 m a^{-1}, whereas at **B** it is 0. At **A**, therefore, the longitudinal strain rate must be compressive, whereas at **B** it is 0 or even slightly extending (so $w_s = u_s \tan \alpha$). In the absence of this extra accumulation at **B**, the glacier would probably be thinner there, as shown schematically by the dotted line in Figure 5.11. The greater surface slope between **A** and **B** would then provide the increased longitudinal compression needed to develop a positive emergence velocity at **B**.

Thule-Baffin moraines (Fig. 5.12) provide a rather dramatic example of the importance of drifting snow. These moraines, first studied in detail by Goldthwait (1951), consist of a superficial layer of till, usually no more than a meter or so in thickness, overlying dirty ice with quite variable debris concentrations. The dirt is typically segregated into laminae or folia, millimeters in thickness, that dip steeply upglacier (Fig. 5.13b). The wedge of ice downglacier from the moraine is clean. It is too thin to flow at an appreciable rate, and is frozen to its bed, preventing sliding. The low flow rate, in conjunction with the observed dip of the foliation, led to the mistaken impression that the foliation planes were actually shear planes, and that the dirty ice was actively shearing over the wedge of clean ice in such a way that debris, entrained at the bed, was carried to the surface on these planes.

The geometry of this type of glacier margin can be understood in terms of the concepts we have been discussing. At "a" in Figure 5.13b, the mean June snow cover is about 1 m thick (Fig. 5.13d). During an average summer, this snow and ~0.55 m of the underlying ice melt. Longitudinal compression here (Fig. 5.13c) results in a positive (upward) w_s, and the emergence velocity is ~0.50 m a^{-1} (Fig. 5.13e). This is slightly less than the net balance, but high enough that a slight change in regime could bring about a steady state with equation (5.22) satisfied.

At "b" in Figure 5.13b, the snow cover is 0, but so is the ablation rate as the till layer insulates the ice. On a ridge like this, it is difficult to know what is meant by "emergence velocity" as the ridge has both up- and downglacier slopes. However, because there is still some longitudinal compression, the ridge is gradually increasing in height. This is a non-steady-state process that does not continue indefinitely. As the height of the ridge increases, it becomes steeper until, eventually, till on the downglacier face slumps and exposes the underlying ice. This ice then melts rapidly due to the lack of snow cover and to the thin covering of dirt that remains on it, decreasing its albedo.

Figure 5.12 Thule-Baffin moraines (on skyline) in one of the type areas, Thule, Greenland. Note folding (**arrow**) in foliation (sedimentary stratification) in superimposed ice in center of photograph (From Hooke, 1970, Fig. 8. Reproduced with permission of the International Glaciological Society.)

At "c" the June snow cover is nearly 2.5 m thick. During an average summer, this snow melts, but essentially no ice is lost; $b_n \approx 0$. Thus, this sloping margin can exist despite the fact that the emergence velocity in it is negligible. During a series of cool summers this sloping margin becomes a local accumulation area at the edge of the glacier. If cool climatic conditions persist long enough, the glacier may advance, overriding and deforming this accumulation of superimposed ice as shown in Figure 5.14. Recognition of this process provided an alternative to the shearing mechanism proposed by Goldthwait.

Three lines of evidence support the origin of the Thule-Baffin moraines shown in Figure 5.14. First, the less deformed superimposed ice is fine grained (1–2 mm) and lacks any development of a deformation fabric such as would be present in highly deformed basal ice. Second, oxygen isotope ratios show that this ice accumulated under conditions broadly similar to those prevailing today, yet it underlies ice with isotopic ratios characteristic of Pleistocene ice. Third, folding of the downglacier dipping layers of superimposed ice occasionally can be observed in ice cliffs with the proper orientation (Fig. 5.15). Further discussion of this process and the evidence for it is presented by Hooke (1970, 1973a, 1976) and Hooke and Clausen (1982).

Moraines are formed by the process illustrated in Figure 5.14 only under relatively cold climatic conditions. For example, in the Søndreström area of Greenland, a few hundred kilometers south of Thule, summer temperatures are warm enough to melt all of the snow at the margin, even though drifting can result in a rather thick June snow cover there. Thus, there is no marginal zone of superimposed ice. However, the process of moraine formation shown in Figure 5.14 was probably important in northern Minnesota, North Dakota, Alberta, and Saskatchewan during the maximum ice advances of the Late Wisconsin. Evidence for this is found in the multiplicity of features indicating that temperatures

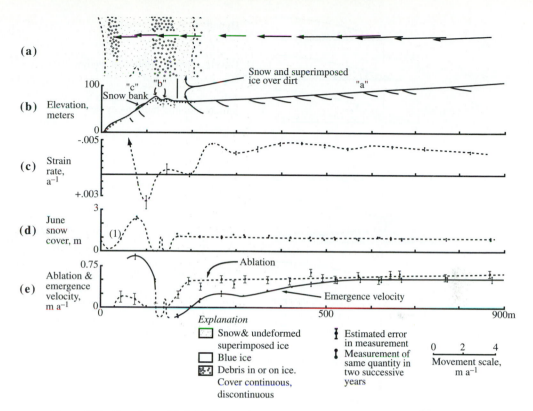

Figure 5.13   Data from a Thule-Baffin moraine on Barnes Ice Cap, Baffin Island. (**a**) Map showing moraine and line of stakes used for velocity and mass-balance measurements. Velocities are shown by arrows. (**b**) Surface profile along stake line, showing dip of foliation at surface and inferred dip beneath surface. (**c**) Strain rates. (**d**) June snow depth. (**e**) Ablation and emergence velocity along stake line. (Modified from Hooke, 1973, Fig. 3D. Reproduced with permission of the Geological Society of America.)

were cold, and that, despite the lack of nunataks projecting above the ice surface, large volumes of till somehow accumulated on the surface (Moran et al., 1980).

INHOMOGENEITIES IN THE FLOW FIELD IN LARGE ICE SHEETS: ICE STREAMS

In the mid-1980s, glaciologists became aware that the flow field in large ice sheets was not as homogeneous as previously believed. In particular, several linear zones of accelerated flow were found in an area of West Antarctica, known as the Siple Coast, that drains to the Ross Ice Shelf. These features, called ice streams, are tens of kilometers wide and hundreds of kilometers long. While some ice streams occupy distinct bedrock valleys, the channels beneath those on the Siple Coast are shallow and poorly defined, and their sides do not always coincide with the edges of the ice streams (Shabtaie & Bentley, 1988).

The best studied ice stream, Ice Stream B on the Siple Coast, has a maximum speed of about 825 m a^{-1} while ice on either side of it typically moves only 10 to 20 m a^{-1} (Whillans & van der Veen,

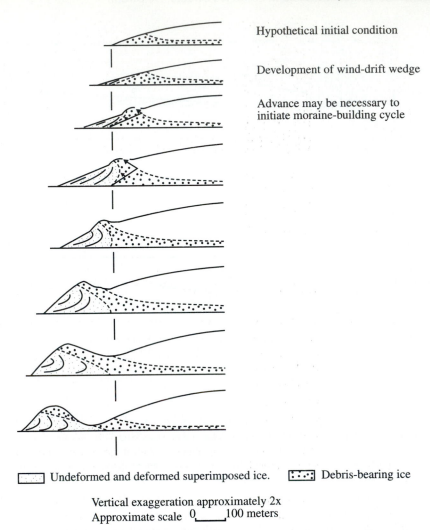

Hypothetical initial condition

Development of wind-drift wedge

Advance may be necessary to
initiate moraine-building cycle

☐ Undeformed and deformed superimposed ice.   ☐ Debris-bearing ice

Vertical exaggeration approximately 2x
Approximate scale 0_____100 meters

Figure 5.14 Sequential cross sections showing schematically the processes by which superimposed ice is overridden during an advance of a glacier that is frozen to its bed at the margin. Light lines in superimposed ice show deformation of sedimentary layering. The last two cross sections show how the moraine becomes separated from the glacier during subsequent retreat. (Modified from Hooke, 1973, Fig. 1. Reproduced with permission of the Geological Society of America.)

1993). The high-velocity gradients across the lateral boundaries result in intensely crevassed shear zones, up to 5 km wide, that are visible from the air and on satellite images. It was the existence of these shear zones that, in part, led to the discovery of the ice streams.

Despite thicknesses approaching 1000 m, surface slopes of ice streams on the Siple Coast are so low that driving stresses are only about 20 kPa. These driving stresses do not differ appreciably from those in intervening areas. Thus, differences in driving stress cannot account for the great difference in speed. Rather, conditions at the bed are inferred to be responsible. Ice Stream B is underlain by till with a high clay content, and water pressures at the ice-till interface are very close to overburden pressure (Engelhardt et al., 1990). The high water pressures either

Figure 5.15   Ice cliff near Thule, Greenland. Sedimentary bedding on left is wrinkled and overturned by ice advancing from right. The boundary between the active and the less rapidly deforming superimposed ice is marked by a dirt band that can be traced into a Thule-Baffin moraine. (From Hooke, 1970, Fig. 9. Reproduced with permission of the International Glaciological Society.)

decouple the ice from the bed, allowing it to slide rapidly, or they weaken the till allowing it to deform, or both.

Ice Stream C on the Siple Coast is also bounded by shear zones, but in this case the crevasses, recognized only on radar images, are buried beneath several meters of snow. The shear zones indicate that Ice Stream C must once have been active, but velocity measurements show that it is now virtually stagnant. Based on the thickness of the snow cover over the crevasses and the accumulation rate, it is estimated that it has been inactive for about 250 years (Shabtaie & Bentley, 1987). The reason for this non-steady-state behavior constitutes one of the great mysteries of ice-stream behavior.

Understanding ice streams has taken on new urgency because rising sea levels induced by global warming could destabilize the Ross Ice Shelf, causing it to break up. Should this occur, ice streams that are currently buttressed by the ice shelf might accelerate, causing a rapid draw-down of the West Antarctic Ice Sheet and a further rise in sea level, perhaps amounting to several meters over the next century.

Quaternary geologists have become fascinated by the possibility that ice streams may have been present in the Laurentide Ice Sheet that covered the northern part of North America during the Wisconsin glaciation. Widely discussed is the possibility that the Laurentide was drawn-down rapidly and repeatedly by an ice stream discharging through Hudson Strait. Several layers of ice-rafted detritus identified in cores from the North Atlantic Ocean are inferred to have been deposited by icebergs that originated in such relatively catastrophic discharges. Broecker (1994) summarizes the evidence for these events, and gives a number of references. It has also been suggested, on the basis of the distribution of distinctive rock types in till in central and northern Canada, that ice streams may have been present in the southern part of the Laurentide Ice Sheet as well. Such ice streams would be unusual in that the bottoms of most known ice streams are well below sea level.

An exciting development that is likely to lead to a much better understanding of flow fields in large ice sheets is the use of satellite images taken at, say, intervals of a year or so, to determine the velocity field. The images are coregistered based on the assumption that certain features are stationary or moving only slowly compared with others. Then, cross-correlation techniques are used to compare crevasse patterns or other recognizable moving features on the two images and to determine how far they have moved between the dates of the respective images (Bindschadler & Scambos, 1991; Whillans & Tseng, 1995). High-speed digital computers are needed to make these comparisons. In this way, a detailed quantitative map of the flow field can be produced. This is revolutionizing the measurement of velocities that, a scant 30 years ago, could only be obtained by tedious precision surveying from fixed points located off the glacier.

SUMMARY

In this chapter we developed a broad picture of the flow field in a glacier or ice sheet using a conservation-of-mass approach and considering the distribution of income from accumulation and loss from ablation over the surface. We found that this leads to a flow field in which vertical velocities are downward in the accumulation area and upward in the ablation area, and horizontal velocities increase with distance from the head of the glacier, reaching a maximum just below the equilibrium line, and then decrease again. In developing this model, we assumed a steady state. Steady-state conditions are rarely if ever strictly achieved in nature, but deviations from a steady state are usually sufficiently small that gross patterns of flow are approximated well by this model.

We also used conservation of momentum principles to make more precise calculations of the distribution of both horizontal and vertical velocity with depth. The former calculation depends upon being able to choose a viscosity parameter, B, that is reasonably representative for the entire thickness of the glacier (or being able to specify a variation with depth), and also assumes that all deformation takes place as simple shear. Complications resulting from the presence of valley sides and a nonuniform basal boundary condition were discussed. Our calculation of the variation in w with depth depended upon the assumption that $\partial u/\partial x$ was independent of depth.

We then explored the effects of drifting snow, and found that this could lead to significant differences in the flow field near the terminus in polar and temperate environments. In polar environments, such snow accumulation can lead to the development of ice-cored moraines some distance upglacier from the margin.

Finally, we noted that inhomogeneous bed conditions can lead to streaming flow within ice sheets. This is a topic of considerable current interest.

6

Temperature Distribution in Polar Ice Sheets

In this chapter we will derive the energy-balance equation for a polar ice sheet. Solutions of this equation yield the temperature distribution in the ice sheet and the rate of melting or refreezing at its base. We will study some analytical solutions of the equation for certain relatively simple situations. A solution of the full equation is possible, however, only with numerical models. This is because: (1) ice sheets have irregular top and bottom surfaces; (2) the boundary conditions—that is, the temperature or temperature gradient at every place along the boundaries—vary in space and time; and (3) there may be extension or compression transverse to the flowline, which makes the problem three-dimensional. Furthermore, because the temperature distribution is governed, in part, by ice flow (or advection), and conversely, because the flow rate is strongly temperature dependent, a full solution requires coupling of the energy and flow equations.

The thermal conditions in and at the base of an ice sheet are of interest not only to the glacier modeler, concerned with flow rates and the possibility of sliding, but also to the glacial geologist with interest in the erosive potential of the ice and the processes of subglacial deposition.

ENERGY BALANCE IN AN ICE SHEET

Advection

Consider a control volume of length dx, width dy, and height dz, as shown in Figure 6.1. This volume represents an element of space within an ice sheet. Ice flows into the volume from the left with a velocity u, and out on the right with the same velocity. The temperature of the ice flowing into the volume is θ, and that of the ice flowing out is $\theta + (\partial\theta/\partial x)\, dx$. The rate of energy transfer into the control volume, measured in joules per year, is

$$(u\,dy\,dz)\ \rho\quad C\quad \theta$$

$$\frac{\mathrm{m}^3}{\mathrm{a}}\quad \frac{\mathrm{kg}}{\mathrm{m}^3}\ \frac{\mathrm{J}}{\mathrm{kgK}}\ \mathrm{K} = \frac{\mathrm{J}}{\mathrm{a}}$$

where ρ is the density of ice and C is the heat capacity or specific heat. Here, as in some of the equations in earlier chapters and in some to follow, the dimensions of the terms are written beneath the equation to clarify the physics. A similar expression can be written for the rate of energy transfer out

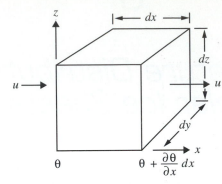

Figure 6.1   Parameters used in derivation of the advection term in the energy-balance equation.

of the volume at temperature $\theta + (\partial\theta/\partial x)\, dx$. The change in energy within the volume per unit time, $\partial q/\partial t$, is the difference between these two expressions, or

$$\frac{\partial q}{\partial t} = u\,dy\,dz\,\rho C\left[\theta - \left(\theta + \frac{\partial\theta}{\partial x}\,dx\right)\right]$$

$$= -u\,dx\,dy\,dz\,\rho C\frac{\partial\theta}{\partial x}.$$

To obtain the change of temperature in the volume per unit time, it is clear from the dimensions of the terms that it is necessary to divide by $\rho\, C\, dx\, dy\, dz$; thus

$$\frac{\partial\theta}{\partial t} = \frac{1}{\rho C\,dx\,dy\,dz}\frac{\partial q}{\partial t} = -u\frac{\partial\theta}{\partial x}$$

$$\frac{1}{\dfrac{kg}{m^3}\dfrac{J}{kg\,K}}\frac{J}{m^3}\frac{}{a} = \frac{K}{a}. \tag{6.1}$$

Here, $\partial\theta/\partial t$ is the rate of change of temperature in the volume as a result of the fact that ice is being *advected*, or moved, into the volume at a temperature that is different from that of ice leaving it. Similar equations may be written for the y- and z-directions, and the results summed to obtain the total change in temperature per unit time in the control volume.

Note that we have been careful to emphasize changes in a particular element of space, the control volume, as distinct from those in an element of ice moving through space. This is because we are using an Eulerian coordinate system, with the coordinate axes fixed in space. Sometimes it is more convenient to use a Lagrangian coordinate system in which an element of ice is followed as it moves through space.

Conduction

The energy content of the control volume may also change as a result of conduction of heat. Consider the situation depicted in Figure 6.2 in which the temperature gradient across the left-hand face, *dy dz*, is $\partial\theta/\partial x$, and that across the corresponding right-hand face is

$$\frac{\partial\theta}{\partial x} + \frac{\partial}{\partial x}\left(\frac{\partial\theta}{\partial x}\right)dx.$$

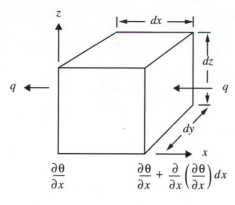

$$\frac{\partial \theta}{\partial x} \qquad \frac{\partial \theta}{\partial x} + \frac{\partial}{\partial x}\left(\frac{\partial \theta}{\partial x}\right) dx$$

Figure 6.2 Parameters used in derivation of the conduction term in the energy-balance equation.

The heat flux is proportional to the temperature gradient. The constant of proportionality is K, the thermal conductivity of ice. Thus, on the left-hand face there is a heat flux:

$$q = \quad K \frac{\partial \theta}{\partial x} dy\, dz$$

$$\frac{J}{maK} \frac{K}{m} m\, m = \frac{J}{a}. \tag{6.2}$$

Heat flows from warm areas to cold areas, which means that for positive $\partial \theta / \partial x$, the heat flux is to the left, or out of the left-hand side of the control volume in Figure 6.2.

As before, we write a similar expression for the heat flux into the control volume, and subtract the flux out from the flux in; thus:

$$\frac{\partial q}{\partial t} = \left[K \frac{\partial \theta}{\partial x} + \frac{\partial}{\partial x}\left(K \frac{\partial \theta}{\partial x}\right) dx - K \frac{\partial \theta}{\partial x}\right] dy\, dz$$

$$= \frac{\partial}{\partial x}\left(K \frac{\partial \theta}{\partial x}\right) dx\, dy\, dz$$

$$= \left(K \frac{\partial^2 \theta}{\partial x^2} + \frac{\partial K}{\partial x}\frac{\partial \theta}{\partial x}\right) dx\, dy\, dz.$$

The change in temperature in the control volume is then

$$\frac{\partial \theta}{\partial t} = \frac{K}{\rho C}\frac{\partial^2 \theta}{\partial x^2} + \frac{1}{\rho C}\frac{\partial K}{\partial x}\frac{\partial \theta}{\partial x} \tag{6.3}$$

$K/\rho C$ is called the thermal diffusivity, κ, so equation (6.3) becomes

$$\frac{\partial \theta}{\partial t} = \kappa \frac{\partial^2 \theta}{\partial x^2} + \frac{\partial \kappa}{\partial x}\frac{\partial \theta}{\partial x}. \tag{6.4}$$

Thus, the change in temperature with time in the control volume due to conduction is related to the changes, as one moves from one side of the volume to the other, in the temperature gradient, $\partial \theta / \partial x$, and in κ. Again, similar equations can be written in the y- and z-directions.

Strain Heating

Finally, a certain amount of heat is generated within the control volume owing to straining of the ice. During deformation, the energy expenditure is the work done divided by the time required to do the work, and work is force times distance; thus:

$$\frac{Work}{time} = \frac{Force \times distance}{time}. \tag{6.5}$$

In simple shear (Fig. 6.3), the average distance moved in a unit time is one-half the displacement of the top of the control volume with respect to the bottom, or $\frac{1}{2}(\partial u/\partial z)dz$, and the force exerted is $\sigma_{zx}\, dx\, dy$. In Chapter 2 [equation (2.6)] we noted that strain rates may be defined in terms of velocity derivatives, thus:

$$\dot{\varepsilon}_{zx} = \frac{1}{2}\left(\frac{\partial u}{\partial z} + \frac{\partial w}{\partial x}\right).$$

As $\partial w/\partial x = 0$ in simple shear, equation (6.5) becomes

$$\frac{Work}{time} = \sigma_{zx}dx\, dy\, \dot{\varepsilon}_{zx}dz$$

$$\frac{N}{m^2}\, m\, m\, \frac{1}{a}\, m = \frac{N-m}{a} = \frac{J}{a}.$$

Dimensionally, this is seen to be a rate of energy expenditure, so again we divide by $\rho\, C\, dx\, dy\, dz$ to obtain the rate of change of temperature per unit volume:

$$\frac{\partial \theta}{\partial t} = \frac{\sigma_{zx}\dot{\varepsilon}_{zx}}{\rho C}. \tag{6.6}$$

Equation (6.6) was derived for a situation in which deformation was restricted to simple shear in the x-z plane. In the general case, other components of the stress tensor will be different from 0, so other deformations will be occurring. With a little more background, it is relatively easy to show that the general form of equation (6.6) is

$$\frac{\partial \theta}{\partial t} = \frac{\sigma_e \dot{\varepsilon}_e}{\rho C}, \tag{6.7}$$

but we will not do this here. For convenience, we may often use Q to represent the heat production instead of $\sigma_e \dot{\varepsilon}_e$ or $\sigma_{zx}\dot{\varepsilon}_{zx}$, thus:

$$\frac{\partial \theta}{\partial t} = \frac{Q}{\rho C}. \tag{6.8}$$

If deformation is restricted to simple shear, we can approximate σ_{zx} by $\rho g d\alpha$ and $\dot{\varepsilon}_{zx}$ by $(\sigma_{zx}/B_o)^n e^{k\theta}$ [see equations (4.6) through (4.8)]. Then:

$$\frac{Q}{\rho C} = \frac{(\rho g d\alpha)^{n+1} e^{k\theta}}{\rho C B_0^n}. \tag{6.9}$$

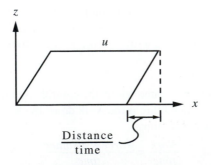

z

u

x

$\dfrac{\text{Distance}}{\text{time}}$

Figure 6.3 Work done in simple shear.

The Generalized Heat-Flow Equation

The rate of change of temperature in the control volume is the sum of the changes represented by equations (6.1), (6.4), and (6.8), plus changes resulting from heat advection and conduction in the y- and z-directions; thus:

$$\frac{\partial\theta}{\partial t} = \kappa\left[\frac{\partial^2\theta}{\partial x^2} + \frac{\partial^2\theta}{\partial y^2} + \frac{\partial^2\theta}{\partial z^2}\right] + \frac{\partial\kappa}{\partial x}\frac{\partial\theta}{\partial x} + \frac{\partial\kappa}{\partial y}\frac{\partial\theta}{\partial y} + \frac{\partial\kappa}{\partial z}\frac{\partial\theta}{\partial z} - u\frac{\partial\theta}{\partial x} - v\frac{\partial\theta}{\partial y} - w\frac{\partial\theta}{\partial z} + \frac{Q}{\rho C}. \tag{6.10}$$

As equation (6.10) is rather cumbersome, it is often convenient to simplify it by using the del operator, defined by:

$$\nabla = \frac{\partial}{\partial x}\hat{\mathbf{i}} + \frac{\partial}{\partial y}\hat{\mathbf{j}} + \frac{\partial}{\partial z}\hat{\mathbf{k}}, \tag{6.11}$$

where $\hat{\mathbf{i}}$, $\hat{\mathbf{j}}$, and $\hat{\mathbf{k}}$ are unit vectors in the x-, y-, and z-directions, respectively. When applied to scalar quantities such as either κ or θ, the del operator gives a gradient, which is a vector quantity. Accordingly, the fourth through sixth terms and the seventh through ninth terms, respectively, on the right in equation (6.10) become scalar or dot products of two vectors; thus:

$$\frac{\partial\theta}{\partial t} = \kappa\nabla^2\theta + \nabla\kappa\cdot\nabla\theta - \vec{u}\cdot\nabla\theta + \frac{Q}{\rho C}. \tag{6.12a}$$

Here, $\vec{u}$ is the vector velocity. The first term on the right in equation (6.12a) also represents a scalar product: $\nabla\cdot\nabla\theta$.

It is sometimes convenient to define

$$\frac{D\theta}{Dt} = \frac{\partial\theta}{\partial t} + \vec{u}\cdot\nabla\theta$$

in which case, equation (6.12a) becomes

$$\frac{D\theta}{Dt} = \kappa\nabla^2\theta + \nabla\kappa\cdot\nabla\theta + \frac{Q}{\rho C}. \tag{6.12b}$$

Equation (6.12a) is the Eulerian form of the equation, in which the coordinates are fixed in space, whereas equation (6.12b) is the Lagrangian form in which the coordinate system is moving with the ice. $D\theta/Dt$ is known as the substantial or Lagrangian derivative.

DEPENDENCE OF κ ON TEMPERATURE

The thermal diffusivity of ice is ~37 m^2a^{-1} at 0°C and ~58 m^2a^{-1} at −60°C. Thus, to the extent that the temperature varies in any of the coordinate directions, κ also varies. This effect is normally neglected, except in relatively sophisticated numerical models, and we will follow this custom. Equation (6.10) thus becomes

$$\frac{\partial\theta}{\partial t} = \kappa\left[\frac{\partial^2\theta}{\partial x^2} + \frac{\partial^2\theta}{\partial y^2} + \frac{\partial^2\theta}{\partial z^2}\right] - u\frac{\partial\theta}{\partial x} - v\frac{\partial\theta}{\partial y} - w\frac{\partial\theta}{\partial z} + \frac{Q}{\rho C}. \tag{6.13}$$

Neglecting the temperature dependence of κ is reasonable because the effect is relatively small, which, in combination with small temperature gradients, makes these terms negligible in comparison with the others in equation (6.10).

THE STEADY-STATE TEMPERATURE PROFILE
AT THE CENTER OF AN ICE SHEET

Our next task is to solve equation (6.13) for some relatively simple situations. The first is that at an ice divide, at the center of an ice sheet, a problem first investigated by Robin (1955). The following development follows his closely. The coordinate system we will use is shown in Figure 6.4: x is horizontal and directed downglacier, and z is vertical and positive upward; $z = 0$ is at the bed.

Simplifying Assumptions

At an ice divide, there is no flow in the horizontal directions, and the temperature field is assumed to be symmetrical about the divide. Thus, $u = v = \partial/\partial x = \partial/\partial y = 0$. We further assume that strain rates are small, so strain heating can be neglected. Finally, we seek a steady-state solution so $\partial\theta/\partial t = 0$. Equation (6.13) now becomes

$$0 = \kappa\frac{\partial^2\theta}{\partial z^2} - w\frac{\partial\theta}{\partial z}. \tag{6.14}$$

To integrate this, w must be expressed as a function of z. To do this we assume, as in Chapter 5, that ice is incompressible, that the longitudinal strain rate is independent of depth, and that $w = 0$ on the bed, thus obtaining [equation (5.21)]:

$$w = \frac{z}{H}w_s. \tag{6.15}$$

We have not yet specified either the sign or the magnitude of w_s. At an ice divide, the vertical velocity is downward (Fig. 3.1a), so the sign of w_s is negative in the coordinate system of Figure 6.4, and in the steady state $|w_s| = b_n$, the accumulation rate. Thus, replacing w_s with $-b_n$ in equation (6.15), combining it with (6.14), and rearranging, we obtain

$$0 = \frac{\partial^2\theta}{\partial z^2} + \frac{b_n z}{\kappa H}\frac{\partial\theta}{\partial z}. \tag{6.16}$$

To calculate the temperature distribution, this equation must be integrated twice.

The First Integration

For the first integration, let $2\zeta^2 = b_n/\kappa H$ and $\beta = \partial\theta/\partial z$. Equation (6.16) then becomes

$$0 = \frac{\partial\beta}{\partial z} + 2\zeta^2 z\beta. \tag{6.17}$$

Separating variables, we obtain

$$\int\frac{\partial\beta}{\beta} = -2\zeta^2\int z\,dz,$$

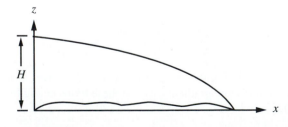

Figure 6.4   Coordinate system used in calculating the steady-state temperature profile at the center of an ice sheet.

which may be integrated to yield

$$\ln \beta = -\zeta^2 z^2 + c$$

or:

$$\beta = e^c e^{-\zeta^2 z^2}. \tag{6.18}$$

The next task is to evaluate the constant of integration, e^c.

The Basal Boundary Condition

The constant of integration may be evaluated by using the boundary condition $\beta = \beta_o$ on $z = 0$. In other words, we presume that the temperature gradient at the bed, β_o, is known or can be estimated. Making these substitutions in equation (6.18) yields $e^c = \beta_o$. Thus, replacing e^c with β_o and β with $\partial\theta/\partial z$ in equation (6.18) yields

$$\frac{\partial\theta}{\partial z} = \beta_o e^{-\zeta^2 z^2}. \tag{6.19}$$

This is a solution for the temperature gradient as a function of elevation above the bed.

The requirement that the temperature gradient in the basal ice be known is fundamentally unavoidable. However, this is not as serious a problem as one might, at first, expect. In the steady state, β_o is adjusted so that all of the heat coming from within the earth, the geothermal flux, can be conducted upward into the ice. Thus, if the geothermal flux can be estimated, β_o can be calculated because the constant of proportionality between the two, the thermal conductivity of ice, K, is known.

To clarify the physical processes by which β_o is adjusted, consider a non-steady-state situation in which β_o is too low. Some of the geothermal heat would then remain at the ice-rock interface where it would warm the ice. Because the temperature decreases upward in the glacier, the ice being colder than Earth's interior, such warming would increase β_o until all of the heat could be conducted upward into the ice, thus tending to re-establish the steady state. (For the moment, we will neglect basal melting.)

Geothermal heat is produced by radioactive decay in the crustal rocks as well as by residual cooling of the mantle and core. Numerous measurements of the geothermal flux have been made, so we have a fair idea of its magnitude in different geological terranes. Geophysicists use the *heat flow unit*, or *HFU*, to describe this flux: 1 HFU is 1 μcal cm^{-2} s^{-1}. The worldwide average geothermal flux is 1.2 HFU. Conveniently, this can be converted to a temperature gradient in ice by dividing by 53 (as the thermal conductivity of ice is 5.3×10^{-3} cal cm^{-1} s^{-1} K^{-1}). Thus, 1.2 HFU corresponds to a basal ice temperature gradient of 0.0226 K m^{-1}. In general, geothermal fluxes are highest in volcanic terranes, high in geologically young terranes, and lowest in geologically ancient terranes. A few examples of geothermal fluxes in glaciated areas are given in Table 6.1.

TABLE 6.1 Geothermal fluxes in some geological terranes in which glaciers exist now or existed in the past

Locality	Heat flux (HFU)	Reference
Canadian Shield	0.8	
World average	1.2	
East Antarctica	1.2[a]	Budd et al., 1971
Baffin Bay	1.35	
West Antarctica	1.4[a]	Budd et al., 1971

[a]Estimated.

In the discussion above, it was stated that knowledge of β_o was "fundamentally unavoidable." It is true, of course, that a boundary value problem such as this could be solved with some other basal boundary condition, such as the basal temperature. (This will be left as an exercise for the reader.) However, as the basal temperature is one of the quantities that we are particularly eager to determine, and as basal temperatures are much harder to estimate from existing data than are basal temperature gradients, choosing β_o as the basal boundary condition is the only logical choice in most situations.

The Second Integration

To obtain the actual temperature distribution, it is necessary to integrate equation (6.19). Separating variables as before yields

$$\int_{\theta(h)}^{\theta_H} d\theta = \beta_o \int_h^H e^{-\zeta^2 z^2} dz. \tag{6.20}$$

Here, the integration is from some level, $z = h$, in the glacier, where the temperature is $\theta(h)$, to the surface at $z = H$ where the temperature is θ_H. [Note that in this case, rather than solve equation (6.19) as an indefinite integral and then evaluate a constant of integration by applying a boundary condition, it is more convenient to express the integrals as definite integrals. Thus the boundary condition, $\theta = \theta_H$ on $z = H$, is incorporated into the limits of integration. Further discussion of this boundary condition is deferred for the moment.]

The integral on the right-hand side of equation (6.20) does not have a solution in closed form. However, it occurs frequently, and thus has been tabulated. In addition, many computer statistical packages have solutions. The challenge is to express it in the terms used in these tables.

We first express the integral on the right-hand side as the difference between integrals over the range $0 \rightarrow H$ and $0 \rightarrow h$; thus,

$$\theta_H - \theta(h) = \beta_o \left[\int_0^H e^{-\zeta^2 z^2} dz - \int_0^h e^{-\zeta^2 z^2} dz \right], \tag{6.21}$$

and then make the substitution: $\zeta z = t$, whence $dz = dt/\zeta$, and $t = \zeta h$ on $z = h$. We also multiply and divide by $\sqrt{\pi}/2$; thus,

$$\theta_H - \theta(h) = \frac{\sqrt{\pi}}{2} \frac{\beta_o}{\zeta} \left[\frac{2}{\sqrt{\pi}} \int_0^{\zeta H} e^{-t^2} dt - \frac{2}{\sqrt{\pi}} \int_0^{\zeta h} e^{-t^2} dt \right]. \tag{6.22}$$

By definition:

$$erf(x) = \frac{2}{\sqrt{\pi}} \int_0^x e^{-t^2} dt, \tag{6.23a}$$

where $erf(x)$ is called the error function of x. Thus, our final solution for the temperature at depth h, $\theta(h)$, is

$$\theta(h) = \theta_H - \frac{\sqrt{\pi} \beta_o}{2 \zeta} [erf(\zeta H) - erf(\zeta h)]. \tag{6.24}$$

As noted, $erf(x)$ has been tabulated, so values of it can be looked up, much as can values of a sine or cosine. Caution is required, however, as some tables define $erf(x)$ slightly differently than we have in equation (6.23a), and thus require a different set of substitutions in equation (6.21). A common alternative definition is

$$erf(x) = \frac{1}{\sqrt{2\pi}} \int_0^x e^{-\frac{t^2}{2}} dt, \tag{6.23b}$$

which requires the substitution: $\zeta z = t/\sqrt{2}$. This leads to other changes in equations (6.22) and (6.24). Budd (1969) uses a less common definition, namely

$$erf(x) = \int_0^x e^{-t^2} dt \qquad (6.23c)$$

and has plotted this (Figure 6.5) along with another function that arises in calculations of temperature distributions in ice sheets. With this definition, equation (6.22) can be written

$$\theta(h) = \theta_H - \frac{\beta_o}{\zeta}[erf(\zeta H) - erf(\zeta h)]. \qquad (6.25)$$

The Boundary Condition at the Surface

As noted, the boundary condition at the surface, $z = H$, is the ice temperature, θ_H, and this must be known in order to calculate a temperature profile from equations (6.24) or (6.25). It has been shown that the temperature at a depth of about 10 m in a glacier is very close to the mean annual atmospheric temperature, θ_a, so it is normally assumed that $\theta_H = \theta_a$. However, it may be well to note some situations in which this approximation may not be very good.

Temperatures in the ablation zones of some glaciers can be somewhat warmer than the mean annual temperature. This is because snow insulates the ice during the winter, preventing cooling. In addition, percolating meltwater reaches the snow/ice interface soon after melting starts in the spring, thus warming the ice faster than would be the case with conduction alone. On Barnes Ice Cap, these two effects result in near-surface ice temperatures that are about 2°C above the mean annual temperature (Hooke et al., 1983).

Somewhat higher on a glacier, near and above the equilibrium line, percolating meltwater can penetrate into the firn of prior years. When this water refreezes the heat of fusion is released at a significant depth in the glacier, not just at the snow/ice interface. The warming effect is thus much enhanced, and ice temperatures in this zone may be several degrees warmer than the mean annual temperature.

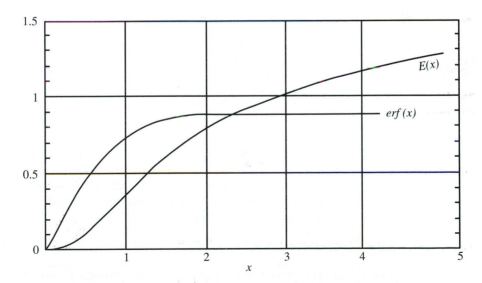

Figure 6.5 Values of the error function, *erf(x)*, as defined by equation (6.23c), and of Dawson's integral, *E(x)*, for a typical range of values of $x = \zeta H$. (Modified from Budd, 1969, Fig. 4.10. Reproduced with permission of the Australian Antarctic Division.)

At high latitudes and altitudes on polar ice sheets, the ice temperature can be slightly below the mean annual temperature because radiative cooling during the clear winter night is more effective than heating during the summer day.

Further discussion of these effects and additional references can be found in Hooke and others (1983).

Melting and Freezing at the Bed

In our discussion so far, we have tacitly assumed that the temperature at the base of the glacier is below the melting point. However, this assumption has not been incorporated into equations (6.24) or (6.25). To be specific, if the bed is at the pressure melting point and melting is occurring there, some of the geothermal heat is clearly being used for that purpose and is not being conducted upward into the ice. Thus, our estimate of β_o is likely to be too high. If we inadvertently insert such a value of β_o into equations (6.24) or (6.25), the calculated temperature at the bed, θ_0, will turn out to be greater than the pressure melting temperature.

To obtain a correct solution for the temperature profile in this case, β_o must be adjusted downward. The procedure is straightforward. Because $erf(0) = 0$ and $\theta_o = \theta_{pmp}$, the temperature at the pressure melting point, equation (6.24) can be solved for β_o; thus,

$$\beta_o = \frac{2\zeta}{\sqrt{\pi}} \frac{\theta_H - \theta_{pmp}}{erf(\zeta H)}. \tag{6.26}$$

The melting point is depressed approximately 0.098 K Mpa^{-1} (if the water produced is saturated with air); so, for example, θ_{pmp} under 500 m of ice would be ~ -0.4°C. Inserting the value of β_o obtained from equation (6.26) into equation (6.24) and solving for temperatures at other depths in the glacier will give the desired temperature profile. [Note that this approach is equivalent to solving equation (6.16) with a temperature boundary condition at the bed.]

The basal melt rate, dm/dt, can also be calculated. The heat available for melting is the difference between the geothermal heat flux and the heat flux into the ice, or $K(\beta_G - \beta_o)$, where β_G is the geothermal gradient. Thus, we obtain

$$\frac{dm}{dt} = K\frac{(\beta_G - \beta_o)}{L} \simeq 220(\beta_G - \beta_o) \quad \text{mm a}^{-1}, \tag{6.27}$$

where L is the latent heat of fusion, and the result is in mm a^{-1} if the gradients are in K m^{-1}.

It is also possible that water formed by basal melting at some distant locality has moved along the bed to the site at which the temperature profile is to be calculated. Until all such water is refrozen, perhaps incorporating sediment into the ice in the process, it will keep the basal temperature at the pressure melting point. Again, equation (6.24) does not know about this water, so the intelligent scientist must intervene. Presumably, he or she has calculated basal melt and freeze rates further upglacier, and has kept track of how much of the water produced has not refrozen. In any case, the procedure is similar to that above, except that now the value of β_o calculated from equation (6.26) will be greater than that necessary to conduct the geothermal heat upward into the ice, and dm/dt in equation (6.27) will be negative, indicating freezing.

Character of the Temperature Profile

Several temperature profiles calculated from equation (6.24) are shown in Figure 6.6a. For the conditions assumed, the ice is nearly isothermal in the upper few hundred meters, and then warms rapidly near the bed. Higher vertical velocities, resulting from higher accumulation rates at the surface, increase the thickness of the isothermal zone and decrease the basal temperature.

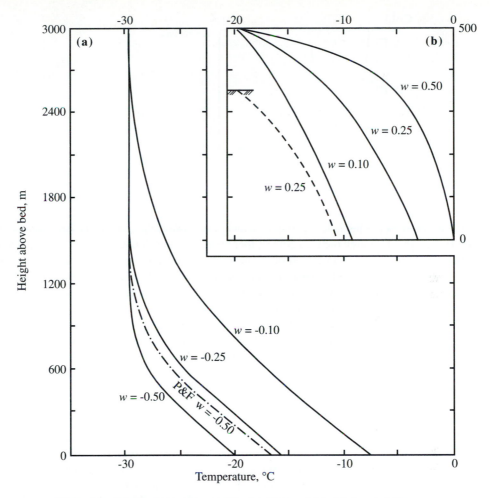

Figure 6.6 Calculated temperature profiles in polar ice sheets. (**a**) Accumulation zone. Vertical velocity is negative or downward. Profile labeled P & F is calculated using technique of Philberth and Federer (1971). (**b**) Ablation zone. Vertical component of velocity is positive or upward. (Modified from Hooke, 1977, Fig. 3. Reproduced with permission of The University of Washington.)

In essence, cold ice is advected downward from the surface, and the upward-moving geothermal heat warms this descending ice. With higher rates of advection (higher vertical velocities), the heat supplied can warm a smaller fraction of the descending ice, so the ice column as a whole is colder.

The shape of the temperature profile can be understood qualitatively in the following way. Consider the three elements of ice labeled A, B, and C in Figure 6.7. All three are moving downward, but because w decreases with depth [equation (6.15)], element A will be moving fastest and element C slowest. As element C moves down, it must warm up and this requires heat. Thus, the heat flux out of the top of this element will be less than that into the bottom, and the temperature gradient required to conduct this heat will be less at the top of the element than at the bottom. However, w is small in this part of the glacier, so despite the comparatively high temperature gradient here, this element does not have to warm up very much, and the change in temperature gradient through it is small, as shown. Element B has a higher velocity, and the temperature gradient is still comparatively high

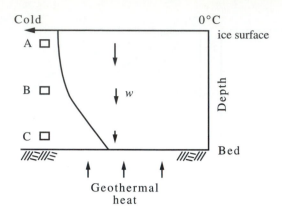

Figure 6.7　Qualitative illustration of effect of downward vertical velocity on a temperature profile.

here, at mid-depth in the glacier, so this element must warm up a lot. Thus, here the change in temperature gradient through the element is rather large. Element A has the highest vertical velocity, but at this level in the glacier nearly all of the heat introduced at the base has been consumed in warming deeper ice. Hence, the temperature gradient here is quite low, and despite its high velocity, element A does not have to warm up very much. Thus, again, the change in temperature gradient through the element is small, as shown.

Later we will examine temperature profiles in the ablation area, where the vertical velocity is upward. However, the reader may find it both challenging and instructive to try to deduce the character of the profile there, using the logic just presented.

Error Introduced by the Assumed Vertical Velocity Distribution

One of the most tenuous assumptions we made in deriving equation (6.24) was that the longitudinal strain rate, $\partial u/\partial x$, was independent of depth. This led to the use of a linear decrease in w with depth [equation (6.15)] in equation (6.16). We know that $\partial u/\partial x$ decreases with depth under most conditions; for example, if the ice is frozen to the bed, $\partial u/\partial x = 0$ at the bed.

Philberth and Federer (1971) recognized this problem and devised a complicated procedure for calculating temperature profiles when $\partial w/\partial z$ was higher near the surface than near the bed. Such a variation in $\partial w/\partial z$ is logical in view of the decrease in u, and hence usually also $\partial u/\partial x$, with depth. Philberth and Federer's procedure results in basal temperatures that are higher than those calculated using a constant $\partial w/\partial z$ (compare the profile labeled $P \& F$ in Figure 6.6a with that labeled $w = -0.50$). Using this technique, Philberth and Federer were able to model the measured profile in the Camp Century, Greenland, borehole much better than had been done previously.

As an alternative to the Philberth and Federer approach, Budd et al. (1971) suggested the *ad hoc* relation

$$w = \frac{h}{H}\frac{u(h)}{u_s}w_s, \tag{6.28}$$

which differs from our equation (6.15) by the addition of the ratio between the horizontal velocity at depth h and that at the surface, $u(h)/u_s$. As $u(h)$ is close to u_s near the surface, $\partial w/\partial z$ is then highest at the surface, as seems appropriate. An expression for $u(h)$ [equation (5.16)] could be inserted into equation (6.28), but use of the resulting expression in equation (6.14) would lead to a relation that no longer could be integrated. Thus, such approaches are only viable in numerical models.

TEMPERATURE PROFILES IN THE ABLATION ZONE

In the ablation area, except very near the equilibrium line, w is positive, or upward. In this case, equation (6.19) becomes

$$\frac{\partial \theta}{\partial z} = \beta_o e^{\zeta^2 z^2}. \tag{6.29}$$

The seemingly trivial change in sign in the exponent renders this equation unintegrable in closed form! However, it is relatively simple to integrate it numerically. Examples of the resulting profiles are shown in Figure 6.6b. Of particular interest is the exponential increase in gradient near the surface, a result that is apparent from equation (6.29). A consequence of this behavior is that basal temperatures must commonly reach the melting point in the ablation zone, even with modest ice thicknesses and vertical velocities. Furthermore, as a result of the small β_o, most of the geothermal and frictional heat is trapped at the bed, and basal melt rates are high.

TEMPERATURE PROFILES NEAR THE SURFACE OF AN ICE SHEET

Let us now consider the temperature profile in the firn area some distance from an ice divide, a problem studied by Robin (1970). We will restrict the problem to two dimensions: assume that strain heating is negligible; and ignore conduction as K is low in firn, while the advective terms are significant. Equation (6.13) then becomes

$$\frac{\partial \theta}{\partial t} = -u \frac{\partial \theta}{\partial x} - w \frac{\partial \theta}{\partial z}. \tag{6.30}$$

$\partial \theta / \partial t$ may be thought of as being composed of two parts: a thickening or thinning of the ice sheet with time, and climatic change; thus,

$$\frac{\partial \theta}{\partial t} = \lambda(\dot{\varepsilon}_{zz} H - b_n) + \frac{\partial \theta_o}{\partial t}.$$

Here, $\dot{\varepsilon}_{zz} H$ represents thinning of the ice sheet by flow (or vertical strain), b_n represents thickening by accumulation, and the difference between them is the net change in surface elevation. (Compressive $\dot{\varepsilon}_{zz}$ is taken as positive here.) This is multiplied by the lapse rate, or rate of decrease in temperature with increasing elevation, λ, to obtain the resulting change in temperature at the glacier surface. To this is added any change in temperature due to secular climatic change, $\partial \theta_o / \partial t$.

$u \cdot \partial \theta / \partial x$ represents the change in temperature at the glacier surface as the ice flows to lower elevations. $\partial \theta / \partial x$ can thus be replaced with $\alpha \lambda$, where α is the surface slope of the glacier. Finally, w is equated with the accumulation rate, b_n. Making these substitutions yields

$$\lambda(\dot{\varepsilon}_{zz} H - b_n) + \frac{\partial \theta_o}{\partial t} = -u \alpha \lambda - b_n \frac{\partial \theta}{\partial z}. \tag{6.31}$$

The meaning of the terms in equation (6.31) can be clarified by reference to Figure 6.8. A particle of snow deposited at "A" has moved to "B" after n years, and is buried under nb_n meters of new accumulation. In the absence of conduction, it is still at the temperature at which it was deposited at "A." The surface above "B" was at "C" when the snow was deposited at "A" and, owing to the lapse rate, snow now accumulating at "C" is $nu\alpha\lambda$ degrees warmer than that which was accumulating at "A." In addition, the ice sheet has thinned by an amount $n(\dot{\varepsilon}_{zz} H - b_n)$ over the intervening years, and the surface is now at "D," which is $n(\dot{\varepsilon}_{zz} H - b_n)\lambda$ degrees warmer than "C." Finally, there may have been secular climatic warming at a rate $\partial \theta_o / \partial t$, so snow at "D" is $n\,\partial \theta_o / \partial t$ warmer than it would be

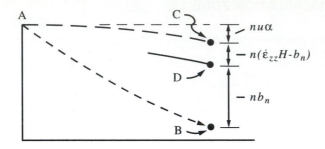

Figure 6.8 Interpretation of terms in equation (6.31).

otherwise. Thus the surface at "D" is

$$nu\alpha\lambda + n(\dot{\varepsilon}_{zz}H - b_n)\lambda + n\frac{\partial\theta_o}{\partial t}$$

warmer than the firn at "B." To obtain the temperature gradient from "B" to "D," divide by $n\,b_n$ and cancel the n's; thus,

$$\frac{\partial\theta}{\partial z} = \frac{u\alpha\lambda + (\dot{\varepsilon}_{zz}H - b_n)\lambda + \partial\theta_o/\partial t}{b_n}, \tag{6.32}$$

which, with minor manipulation, can be shown to be the same as equation (6.31).

When one is far from the edge of an ice sheet, it is very difficult to determine whether the ice sheet is thickening or thinning; that is, whether $(\dot{\varepsilon}_{zz}H - b_n)$ is positive or negative. Although b_n can be measured in snow pits, the problem is to measure $\dot{\varepsilon}_{zz}$ without an immovable base upon which to establish a survey point. Furthermore, such observations would span only a short time interval. However, suppose we can measure u, α, λ, and b_n, and have reason to believe that $\partial\theta_o/\partial t$ is negligible. Then, equations (6.31) or (6.32) can be solved for $(\dot{\varepsilon}_{zz}H - b_n)$. Two examples are shown in Table 6.2. The results, a 0.031 m a^{-1} thickening rate at Site 2 in Greenland and a 0.018 m a^{-1} thinning rate at Byrd Station in Antarctica, are surprisingly reasonable.

While potentially providing a sensitive measure of the state of health of an ice sheet, this technique is probably not especially useful because moderately deep boreholes are needed to obtain $\partial\theta/\partial z$, and $\partial\theta_o/\partial t$ is not known well enough. However, the derivation of equations (6.31) and (6.32) serves to emphasize that, in general, as one moves away from the divide, temperature gradients near the surface of an ice sheet become positive; that is, the temperature decreases with depth (decreasing z). We now turn our attention to more sophisticated models that enable us to investigate such temperature distributions deep in the ice and far from a divide.

TABLE 6.2 Values of parameters in equation (6.31) for Site 2 in Greenland and Byrd Station in Antarctica

	Site 2	Byrd Station
α	−0.00427	−0.00161
u_s, m a^{-1}	17	15
λ, K m^{-1}	−0.011	−0.008
b_n, m a^{-1}	0.4	0.2
$\partial\theta/\partial z$[a], K m^{-1}	0.00115	0.00024
$(\dot{\varepsilon}_{zz}H - b_n)$, m a^{-1}	0.031	−0.018

[a]$\partial\theta/\partial z$ is measured below 150 m depth.

TEMPERATURE DISTRIBUTIONS FAR FROM A DIVIDE

Budd and others (1971) have used two techniques for solving equation (6.13) in a more general form than those we have considered so far. The first is called the *Column model*, and the second is the *Flowline model*. Calculations using the Column model can be done by hand, but those using the Flowline model require a fast digital computer. We will consider the Column model in some detail, and we will use the Flowline model only to illustrate some of the limitations of the Column model.

The Column Model

The coordinate system we will use is shown in Figure 6.9. The temperature profile is to be calculated at a point a distance χ from the divide. Starting again with equation (6.13), we restrict the model to two dimensions, thus eliminating derivatives in the *y*-direction; we assume that temperature gradients in the *x*-direction are sufficiently small that their derivative is negligible; and we assume a steady state. With these assumptions, equation (6.13) becomes

$$0 = \kappa \frac{\partial^2 \theta}{\partial z^2} - u \frac{\partial \theta}{\partial x} - w \frac{\partial \theta}{\partial z} + \frac{Q}{\rho C}. \tag{6.33}$$

Let us now consider the strain-heating term, $Q/\rho C$. From equation (6.9) it will be seen that this term increases approximately as d^4, where d is the depth below the surface. In other words, because the strain rate increases rapidly near the bed, most of the strain heating occurs in the basal few metres of ice. [The student may find it interesting to study this effect by solving equation (6.13) with the assumption that all advection terms and the horizontal conduction terms are negligible, and that a steady state exists. Equation (6.9) is used for $Q/\rho C$. The problem is most easily tackled by using a coordinate system in which the *z*-axis points vertically downward.] Recognizing that significant strain heating occurs only *near* the bed, Budd (1969) assumed that it occurs only *at* the bed, and that this heat could thus be added to the geothermal flux. The basal boundary condition, β_b, thus becomes

$$\beta_b = \beta_G + \frac{\tau_b \bar{u}}{K}$$

$$\frac{K}{m} \quad \frac{K}{m} \quad \frac{\dfrac{N\ m}{m^2\ a}}{\dfrac{J}{m\,a\,K}} = \frac{K}{m} \qquad 1\,N - m = 1\,J. \tag{6.34}$$

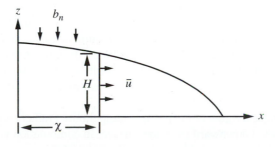

Figure 6.9 Coordinate system and parameters involved in Column model calculations.

Here, β_G is the gradient required to conduct the geothermal flux upward into the ice (previously, our β_o); τ_b is the basal drag, approximated by $\rho g d\alpha$; and $\bar{u}$ is the mean horizontal velocity. For $\bar{u}$ at χ we use the balance velocity

$$\bar{u} = \frac{1}{H}\int_0^\chi b_n(x)\,dx \tag{6.35}$$

[equation (5.1)]. In calculations, care must be taken to ensure that the sign of the $\tau\bar{u}/K$ term is the same as that of β_G; this sign is determined by the choice of coordinate axes.

We turn now to the term $u \cdot \partial\theta/\partial x$ in equation (6.33). In equation (6.31) we set $\partial\theta/\partial x = \alpha\lambda$, as this is the rate at which the atmospheric temperature increases as we move to lower elevations along the ice surface. This is, therefore, the rate at which near-surface ice must warm as a result of horizontal advection. If the glacier surface slope is sufficiently low, the deeper ice will warm at the same rate, with negligible lag. This led Budd (1969) to suggest that, to a reasonable first approximation, $u \cdot \partial\theta/\partial x$ can be replaced with $u\alpha\lambda$ in equation (6.33). The consequences of this assumption are discussed below.

With the additional substitution of $(w_s - w_b)\,z/H$ for w, equation (6.33) becomes

$$\kappa\frac{\partial^2\theta}{\partial z^2} - (w_s - w_b)\frac{z}{H}\frac{\partial\theta}{\partial z} = u\alpha\lambda, \tag{6.36}$$

which is to be solved using the boundary condition of equation (6.34). Here, w_s and w_b are the vertical velocities at the surface and bed, respectively. w_s can be calculated from the submergence velocity [equation (5.22)] when b_n, u_s, and α are known. If the velocity at the bed is assumed to be parallel to the bed, w_b can be estimated from knowledge of u_b and the bed slope. Assuming that $u_b = \bar{u} = u_s$ is probably a reasonable approximation in this calculation, but knowing $\bar{u}$, u_s and u_b could also be calculated from equations (5.18) and (5.19).

The solution to equation (6.36) is (Budd, 1969; Budd et al., 1971):

$$\theta(h) = \theta_H - \frac{\beta_b}{\zeta}[erf(\zeta H) - erf(\zeta h)] - \frac{2\bar{u}\alpha\lambda H}{(w_s - w_b)}[E(\zeta H) - E(\zeta h)] \tag{6.37}$$

where:

$$erf(x) = \int_0^x e^{-t^2}\,dt$$

$$E(x) = \int_0^x \left[e^{-y^2}\int_0^y e^{t^2}\,dt \right] dy$$

$$\zeta = \left[\frac{w_s - w_b}{2H\kappa}\right]^{1/2}.$$

Note that this solution uses Budd's definition of the error function, $erf(x)$. $E(x)$ is known as Dawson's integral, and it too has been tabulated. A plot of it for a reasonable range of ζ is shown in Figure 6.5.

Temperature profiles can be calculated readily using equation (6.37) and Figure (6.5). A typical one is shown in Figure 6.10 (profile a). The minimum temperature occurs at some depth below the glacier surface. This represents, as in Figure 6.8, cold ice that is advected downward and laterally from some point further upglacier where the surface is at a higher elevation and hence colder. However, the Column model does not include this longitudinal advection rigorously, but simply specifies a warming rate. Thus, the temperature at depth is only an approximation that becomes better as the warming rate decreases. This approximation is best, therefore, where surface slopes (α) and lapse rates (λ) are lowest.

Both the magnitude and the curvature of the positive temperature gradient near the surface are adjusted so that the heat conducted downward from the surface, in combination with heat advected downward, is sufficient to warm the ice everywhere above the point of minimum temperature at the

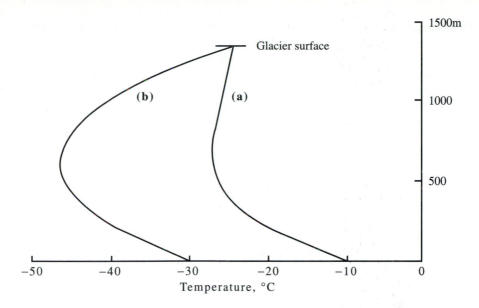

Figure 6.10 Temperature profiles calculated from (**a**) equation (6.37) and (**b**) equation (6.38). The following values of the parameters, approximately appropriate for Camp Century, Greenland, were used in the calculations: $\alpha = -0.01$; $u = 15$ m/a; $\lambda = -0.01$ K/m; $\kappa = 37.2$ m^2/a; $H = 1368$ m; $\theta_H = -24°$C; and $\beta_b = -0.0508$ K/m.

rate $u\alpha\lambda$. Ice below this point is warmed at this rate by heat from the bed–both geothermal and frictional. When the warming rate at depth that is specified (by representing it by $u\alpha\lambda$) is larger than that in the natural situation being modeled, the positive temperature gradient at the surface becomes too high, and the temperatures at depth, thus, too cold.

Further insight into the mathematical properties of the solution can be gained by comparing the profile calculated from equation (6.37) with one calculated from

$$\kappa\frac{\partial^2\theta}{\partial z^2} = u\alpha\lambda, \tag{6.38}$$

together, again, with the boundary condition of equation (6.34). [The integration of equation (6.38) is left as an exercise for the reader.] This is profile (b) in Figure 6.10. Equations (6.36) and (6.38) differ in that the vertical advection term is omitted in equation (6.38). By analogy with our discussion of equation (6.24) and Figure 6.7, one might think that because vertical advection moves cold ice downward from the surface, omission of this term would make profile (b) warmer than profile (a) at depth. However, in this case the ice is colder at depth than at the surface, and because the vertical advection term operates on ice at the surface at the point where the profile is being calculated, not at some point upglacier therefrom, the ice advected downward is warmer than the ice at depth. As a result of this downward advection of heat, included in equation (6.37), the $u\alpha\lambda$ warming rate does not need to be satisfied entirely by conduction from the surface in profile (a).

The Flowline Model

The most important distinction between the Flowline and Column models is that longitudinal advection is handled more accurately in the Flowline model. The model starts with the Robin solution [equation (6.24)] at the divide, where longitudinal advection is negligible. Then, at each successive point along the flowline at which a temperature profile is to be calculated, the horizontal velocity at each depth is estimated—for example, from equations like (5.6), using equation (5.18) to obtain u_s, knowing $\bar{u}$—and, after a trial solution for the temperature at that depth, a preliminary value of the longitudinal temperature gradient, $\partial\theta/\partial x$, is obtained from the difference between this profile and the previous one. This value can be used in the longitudinal advection term to refine the temperature calculation. This approach thus avoids the need for the constant $u\alpha\lambda$ term, and instead allows the warming rate to vary with depth.

Clearly, with an iterative approach such as this, it is not practical to solve the problem by "hand." A high-speed digital computer was used for the solutions that Budd and others (1971) obtained. This, in turn, allowed them to add other refinements, which we will not elaborate upon herein.

BASAL TEMPERATURES IN ANTARCTICA: COMPARISON OF SOLUTIONS USING THE COLUMN AND FLOWLINE MODELS

The reliability and weaknesses of the Column and Flowline models can be illustrated by comparing the results of calculations using the two approaches. Budd and others (1971) used the models to study the temperature distribution, and in particular basal temperatures, in the Antarctic Ice Sheet. They displayed the input to the models and their results in a series of maps, some of which are reproduced in Figure 6.11. Before discussing the basal temperatures, it is worth mentioning some characteristics of the ice sheet that are based on field observations, and some others that were derived from the models.

The surface elevation, accumulation rate, and near-surface temperature used as input to the models are shown in Figure 6.11a–c. Of particular interest is the accumulation rate. Maximum

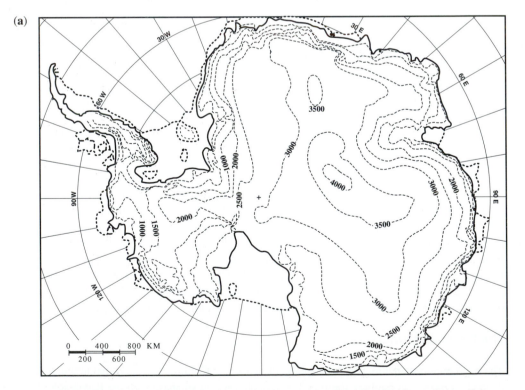

Figure 6.11 Maps of Antarctica. (a) Surface elevation, m. (Reproduced with permission of the Australian Antarctic Division.)

(b)

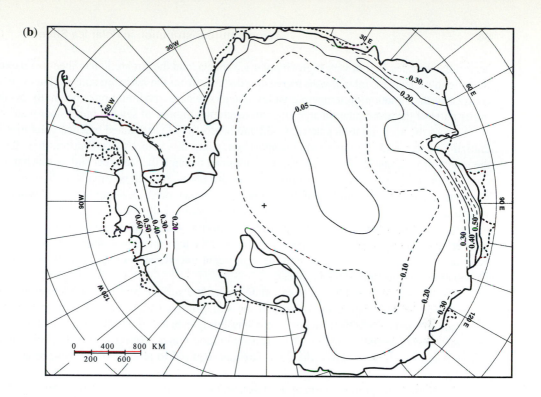

(c)

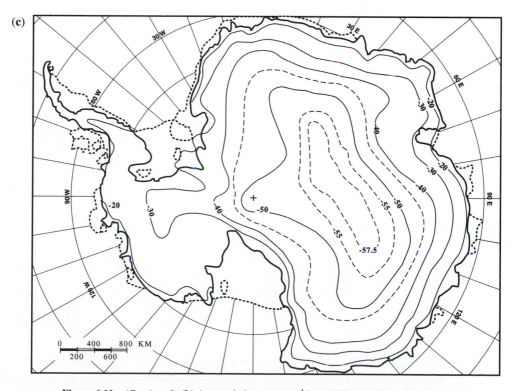

Figure 6.11 *(Continued).* (**b**) Accumulation rate, m a^{-1} ice; (**c**) Near-surface temperature, °C.

accumulation rates are not very high on polar ice sheets, and the Antarctic Ice Sheet is no exception. Rates tend to decrease inland, with increasing distance from the moisture source. This is in distinct contrast to the situation on temperate valley glaciers, where maximum rates normally occur in the upper parts of the accumulation area and can reach magnitudes of several meters per year.

From the accumulation pattern, Budd and others (1971) calculated balance velocities by generalizing equation (5.1) to two dimensions. Then, using this velocity distribution, they determined the length of time needed for a particle of ice to move through the ice sheet and be discharged at the coast, or the *residence time* (Fig. 6.11d). Some years ago a proposal was made to dispose of nuclear waste by leaving it in the middle of the Antarctic Ice Sheet, on the assumption that it would melt its way to the bottom and there remain isolated from the biological environment. Design criteria called for this isolation to last for at least a quarter of a million years. Calculations such as those summarized in Figure 6.11d revealed relatively few areas of the ice sheet where such residence times might be expected. This, in conjunction with the significant uncertainty in the calculations, was, in part, responsible for rejection of the proposal.

Basal drags (Fig. 6.11e) were calculated from the surface profile and thicknesses measured using radioecho techniques. In contrast to the situation on valley glaciers, where τ_b is typically between 0.05 and 0.15 MPa (Chapter 5), the basal drag is below 0.05 MPa over most of Antarctica. Note also that τ_b decreases inland. The low accumulation rates near the center of the ice sheet result in low balance velocities. Thus, the shear stresses required to provide those balance velocities are also low.

The basal temperature gradient used as a boundary condition in the modeling (equation 6.34) is shown in Figure 6.11f. As noted (Table 6.1), the gradient required to conduct the geothermal flux

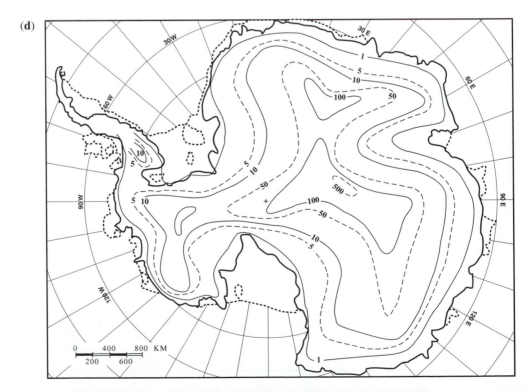

(d)

Figure 6.11   *(Continued).* **(d)** Residence times, ka.

(e)

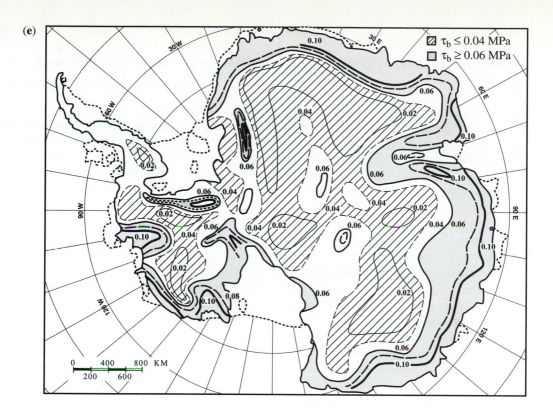

(f)

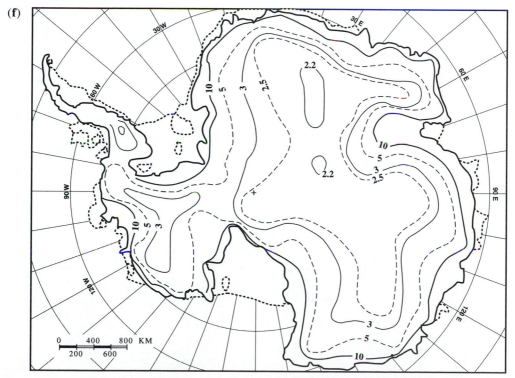

Figure 6.11 *(Continued).* (**e**) Basal drag, MPa; (**f**) Basal temperature gradient, β_b, K/100 m.

upward into the ice is estimated to be only about 0.02 K m^{-1} in East Antarctica. The increase in β_b toward the coast (Fig. 6.11f) is thus a result of the $\dfrac{\tau_b \bar{u}}{K}$ term in equation (6.34). Strain heating is nearly four times as important as geothermal heating near the coast.

   Basal temperatures calculated from the Column and Flowline models are shown in Figures 6.11g, h, respectively. The patterns from the two models are generally similar. Inland from the coast, the pattern from both models reflects the bed topography, as temperatures are lower where the ice is thinner. At ice divides, both models should give the same result, as there will be no horizontal advection. However, downglacier from divides basal temperatures are systematically lower in the flowline model. This, presumably, reflects the lag between the time that warming occurs at the surface and the time this warming is felt at depth. The Flowline model recognizes this lag, but the Column model does not. The colder temperatures at depth in the Flowline model mean that more heat is being conducted downward from the surface, so despite the fact that temperatures are colder in the Flowline model, the warming rate is higher.

   The effect of halving the velocity in the Column model is shown in Figure 6.11i (compare with Fig. 6.11g). In general, owing to the reduced warming rate required by the $u\alpha\lambda$ term, lower velocities result in basal temperatures that are warmer inland. However, because of the reduced strain heating, temperatures near the coast are colder.

   Differences among Figures 6.11g–i emphasize the need for caution in using the predictions of such steady-state numerical models. Indeed, recent field data suggesting that there is a lake beneath the Antarctic Ice Sheet at Vostok Station (Fig. 6.11h) suggest that the calculations of Budd and others (1971) may yield temperatures that are too low there. However, their flowline

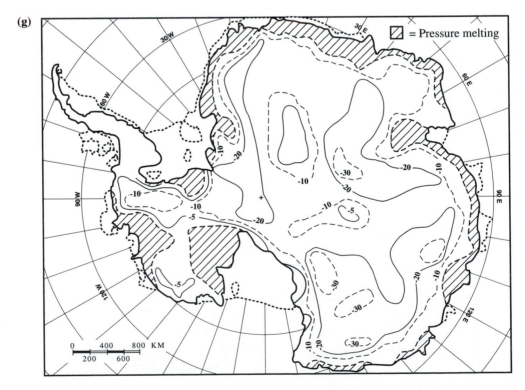

Figure 6.11   *(Continued).* **(g)** Basal temperatures, °C, from the Column model.

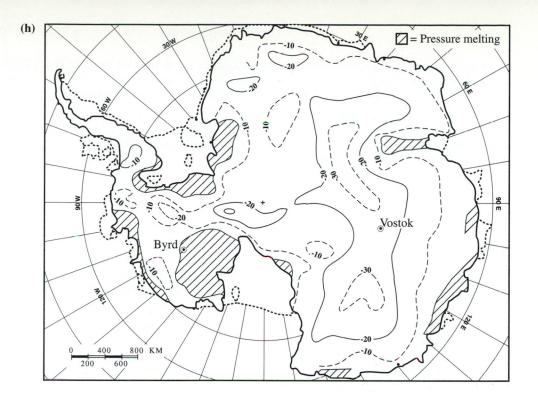

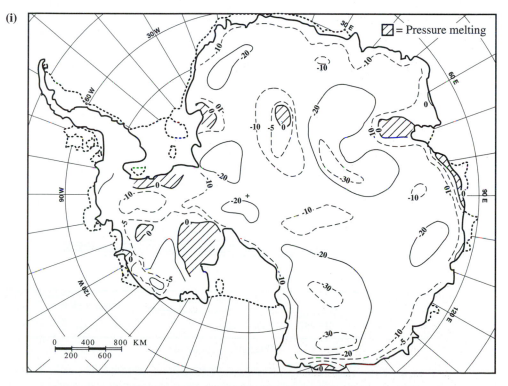

Figure 6.11 *(Continued).* (**h**) Basal temperatures, °C, from the Flowline model; (**i**) Basal temperatures, °C, from the Column model with velocities halved.

model did correctly predict pressure-melting conditions at the bed beneath Byrd Station. The need to assume a steady state and a basal temperature gradient are probably the two most serious limitations in these models, although the reader will readily think of others.

ENGLACIAL AND BASAL TEMPERATURES ALONG A FLOWLINE USING THE COLUMN MODEL

Let us now consider, in greater detail, the temperature distribution along a flowline calculated with the use of the Column model (Fig. 6.12). The original objective of the modeling shown in Figure 6.12 was to investigate the possibility that, along the margin of the Laurentide Ice Sheet in North Dakota, there could have been a ~2-km-wide zone in which the ice was frozen to the bed. Such a temperature distribution is implied by glacial landforms, as discussed further below (Moran et al., 1980). Thus, the flowline modeled was assumed to extend from Hudson Bay to North Dakota.

In the model, the accumulation rate was assumed to be 0.20 m a^{-1} 65 km upglacier from the equilibrium line, and to decrease linearly to 0.05 m a^{-1} at the divide, and to 0 at the equilibrium line. In the ablation area, the ablation rate increased linearly downglacier from the equilibrium line, and the rate of increase was adjusted to provide a balanced mass budget. The horizontal velocity was approximated by the balance velocity [equation (5.1)] modified to allow for divergence of the flow-lines. The ice-sheet profile was adjusted to provide the shear stress necessary to yield this horizontal velocity, using a relation similar to equation (5.19) with a sliding law to estimate u_b. Isostatic depression of Earth's crust was included. The vertical velocity was calculated from the submergence or emergence velocity relation [equation (5.22)], and was assumed to decrease linearly with depth [equation (6.15)]. The temperature at the margin was –7.5°C. The temperature along the surface was calculated assuming a lapse rate of –0.01 K m^{-1}, and making an empirical correction for warming effects of percolating meltwater. The geothermal fluxes used were appropriate to the geologic terrane along the flowline. To circumvent certain problems, discussed later, it was assumed that the warming rate was $\frac{1}{2}u\alpha\lambda$ instead of $u\alpha\lambda$.

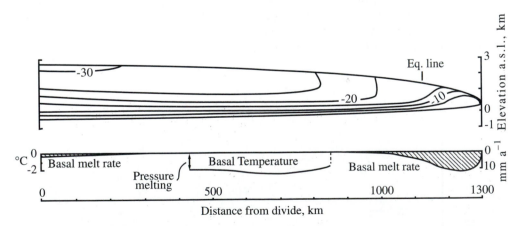

Figure 6.12 Temperature distribution along a flowline calculated with the use of the Column model. The double arrow labeled "pressure melting" shows the depression of the melting point due to the weight of the overlying ice. (From Moran et al., 1980, Fig. 6. Reproduced with permission of the International Glaciological Society.)

Several features of the temperature distribution in Figure 6.12 merit comment:

- The downward and outward advection of cold ice is represented by the reversal in slope of the −20°C and −25°C isotherms ~900 km from the divide.
- The progressive compression of the isotherms near the bed downglacier from the divide reflects the outward increase in basal temperature gradient as strain heating increases.
- Basal melting occurs over the first 250 km of the flowline because the accumulation rate here is low, and downward advection of cold ice is, thus, less important than it is further downglacier.
- Between ~250 and ~420 km from the divide, half of the meltwater formed in the first 250 km is refrozen to the base. The rest is assumed to have drained away into the bedrock. (Had it been assumed, instead, that more of the meltwater stayed at the ice/bed interface, the zone of sub-freezing temperatures between ~420 and ~840 km from the divide would be smaller or absent.)
- The zone of subfreezing basal temperatures between ~420 and ~840 km owes its existence to increased downward advection of cold ice as the accumulation rate increases outward.
- Basal melting resumes downglacier from ~840 km as strain heating warms the basal ice. It becomes particularly important in the ablation area where upward vertical velocities decrease the basal temperature gradient, thus trapping more heat at the bed.
- The basal frozen zone at the margin, not visible at the scale of Figure 6.12, is a result of cold atmospheric temperatures at the margin and decreasing vertical velocity as the margin is approached. The vertical velocity decreases because, as the surface slope steepens, a greater fraction of the ablation rate is balanced by the $u_s \tan \alpha$ term in the emergence velocity [equation (5.22)]. In addition, it is assumed that most of the meltwater formed in the outer ~500 km of the glacier leaves the system as groundwater or by way of localized subglacial conduits.

The fact that water from melting basal ice flows downglacier along the bed and refreezes is consistent with observations of layers of dirty ice, several meters thick, that were encountered at the bottoms of both the Byrd Station, Antarctica, and the Camp Century, Greenland, ice cores. In both cases, the dirt was dispersed throughout the ice, and the dirty ice had fewer air bubbles than did the overlying clean ice. In the Camp Century core, the oxygen isotope ratios indicated that the basal dirty ice was formed from water that originally condensed at lower temperatures than did the overlying ice. All of these observations are consistent with melting of ice that originally formed at a higher altitude than the overlying ice, downglacier flow of that water along the bed, and refreezing of the water incorporating dispersed dirt in the process. It is difficult to account for meters-thick layers of basal ice with dispersed dirt in any other way, although regelation of ice downward into till is a possible way of entraining layers of dirt with higher debris content (Iverson, 1993).

Problems with High $u\alpha\lambda$ Warming Rates in the Column Model

As the warming rate required by the $u\alpha\lambda$ term in the Column model increases, the curvature of the temperature profile increases, and the minimum temperature decreases (see Fig. 6.10 and discussion on pp. 80–81 above). Near the equilibrium line, the downglacier warming rate is high because melt-water percolating into the firn raises near-surface temperatures. (In the modeling for Fig. 6.12, as noted, this effect is included by using an effective value of λ that is higher than the atmospheric lapse rate.) If the ice at depth is assumed to be warming at the same rate, calculated minimum temperatures in profiles near the equilibrium line are often lower than the minima in profiles just upglacier. In plots such as Figure 6.12, this appears as a pocket of cold ice beneath the equilibrium line that is surrounded by warmer ice (Hooke, 1977, Fig. 4c). Such a temperature distribution is physically impossible; to have cooled off, this ice would had to have lost heat to colder ice, yet it is surrounded by warmer ice.

To circumvent this problem, as noted, a warming rate of $\frac{1}{2}u\alpha\lambda$ was used.

GEOMORPHIC IMPLICATIONS

Temperature distributions such as that in Figure 6.12 have implications for glacial erosion and deposition. First, erosion rates are likely to be highest where basal melt rates are low or where meltwater is refreezing to the glacier sole. Thus, we might expect to find that erosion was most intense some distance from the divide. Conversely, the formation of lodgement till by subglacial melting should be most prevalent beneath the ablation zone. Both are consistent with observation.

Second, zones of frozen bed, a couple of kilometers wide, should develop along ice-sheet margins in regions where mean annual temperatures are sufficiently low. This, indeed, seems to have been the case in North Dakota and adjacent areas of Alberta and Saskatchewan. Here, blocks of bedrock, tens to hundreds of meters on a side, became frozen to the base of the glacier and were moved outward a kilometer or so (Fig. 6.13). Detachment may have been facilitated by high pore water pressures in the unfrozen rock beneath the frozen zone. Upon deposition, these blocks formed hills. When the ice eventually receded, the basins from which the blocks were plucked became lakes (Moran et al., 1980).

SUMMARY

We began this chapter by deriving the energy balance equation. Given boundary conditions appropriate for a polar ice sheet, solutions to this equation yield the temperature distribution in the ice sheet. The boundary conditions most commonly used are (1) the temperature at the surface, which is approximated by the mean annual temperature, perhaps with a correction for heating by percolating melt water; and (2) the temperature gradient at the bed. The latter is based on an estimate of the geothermal flux. If calculations suggest that the bed is at the pressure melting point, the temperature gradient is adjusted to ensure that basal temperatures do not exceed the melting point.

Using appropriate simplifications, we studied solutions to the energy balance equation for the situation at an ice divide, for the situation near the glacier surface but some distance from the divide, and for a column of ice extending through an ice sheet some distance from the divide. A key

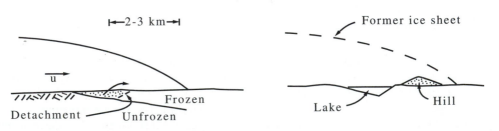

Figure 6.13   Hill-lake pair formed by thrusting at a frozen margin.

assumption in the latter, the so-called Column model, is that longitudinal advection can be approximated by assuming a warming rate at depth that equals that at the surface. By comparing solutions using the Column model with those using the Flowline model, a model that handles longitudinal advection more rigorously, we studied the effects of this and other assumptions on calculated temperature distributions at the base of the Antarctic Ice Sheet.

Finally, we investigated the way in which several physical processes are reflected in calculated temperatures along a flowline of the Laurentide Ice Sheet, and we noted some geomorphic implications of the theoretical basal temperature distributions.

7

The Coupling Between a Glacier and Its Bed

In Chapter 4 we found that the rate of deformation of ice, $\dot{\varepsilon}$, could be related to applied stress, σ, by: $\dot{\varepsilon} = (\sigma/B)^n$. The rigorous basis for this flow law will not be developed until Chapter 9, but some indications of the complexities involved in applying it have already been mentioned. Despite these complexities, calculations using this flow law are reasonably accurate. Computed deformation profiles are an example. This is, in large part, because ice is a crystalline solid with relatively uniform properties. The principal causes of inaccuracy in such calculations are a consequence of impurities in the ice, including water, of anisotropy associated with the development of preferred orientations of crystals, and of incomplete knowledge of the temperature and boundary conditions.

As mentioned briefly in Chapter 5 (Fig. 5.4), glaciers also move over their beds readily when the basal temperature is at the pressure melting point. However, the rate at which this movement occurs is far more difficult to analyze. This is again, in part, because the boundary conditions, principally the water pressure and the morphology of the bed, are not known. However, a more fundamental problem is the fact that granular rock debris is usually present, either in the ice or between the ice and the bed, or both. The processes involved in deformation of such material and the appropriate constitutive relation describing its behavior during deformation, or that of ice containing it, are subjects of considerable uncertainty. Furthermore, unlike the situation with pure ice, the properties of the rock debris vary, not only from glacier to glacier, but also from point to point beneath a single glacier.

Although it may be impossible to know the boundary conditions well enough to predict the rate of movement of a glacier over its bed with accuracy, it is nevertheless important to understand the processes so as to place limits on the rate. Thus, a significant effort has been made to analyze these processes, using judicious assumptions where necessary as a substitute for detailed data. This analysis has led to relations between the rate of movement and measurable quantities such as water pressure and basal drag that can be tested with field data.

We start this discussion by looking at the movement of clean ice over an irregular hard undeforming bed—the traditional sliding problem. Some of the principal shortcomings of the analysis are then discussed. Finally, we take up the problem of deformation of the granular materials over which many glaciers move.

SLIDING

The basic processes by which ice moves past an obstacle on a hard bed, *regelation* and *plastic flow*, were first discussed by Deeley and Parr (1914) and later quantified by Weertman (1957, 1964). *Regelation* involves melting of ice in the region of high pressure on the upglacier or stoss side of the obstacle and refreezing of that water in the region of lower pressure on its lee face, while *plastic flow* is simply deformation of ice in a three-dimensional flow field around the obstacle. In his analysis, Weertman used a simplified model of the bed geometry, sometimes called the tombstone model, consisting of uniformly spaced rectangular blocks on a flat surface (Fig. 7.1). This model has been roundly criticized as being unrealistic, and inappropriately defended by arguing that fudge factors can be inserted to make it applicable to actual situations. The real value of the model is that the physical principles involved in the sliding process are illustrated without the necessity of invoking sophisticated mathematical techniques.

Consider the bed shown in Figure 7.1. For simplicity we will use cubical obstacles of mean dimension, ℓ, instead of Weertman's rectangular ones. The mean spacing between obstacles is $\mathcal{L}$, and we therefore define $r = \ell/\mathcal{L}$ as the roughness of the bed. The mean drag on the bed is τ. As the ice is separated from the bed by a thin film of water, we assume that faces parallel to the flow cannot support a shear stress. Thus the entire drag over an area $\mathcal{L}^2$ must be supported by the obstacle. The total force on the obstacle is then $F = \tau\mathcal{L}^2$, so the stress difference across the obstacle, $\sigma_s - \sigma_l$, is

$$\sigma_s - \sigma_l = \frac{\tau\mathcal{L}^2}{\ell^2} = \frac{\tau}{r^2} \tag{7.1}$$

where the subscripts s and l refer to *stoss* and *lee*, respectively.

The average temperature at the bed is constrained to be at the pressure melting point determined by the average pressure in the water layer. The average pressure is a function of the local glacier thickness. The pressure in the water film on the stoss face of the obstacle is higher than the average, and the pressure on the lee face is lower (Fig. 7.2). Thus, there is a pressure difference across the obstacle, $\Delta P = \sigma_s - \sigma_l$, and this results in a temperature difference, $\Delta T = C\,\Delta P$, where C is a physical constant relating the change in pressure melting temperature to the pressure. It has the value 0.074 K MPa^{-1} for pure water and 0.098 K MPa^{-1} for air-saturated water.

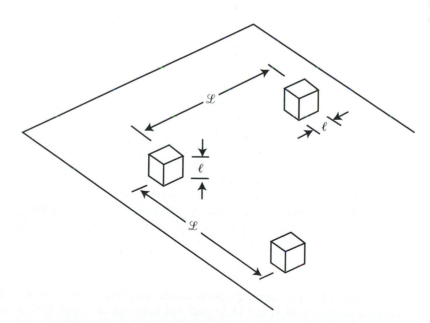

Figure 7.1   Bed geometry used in Weertman's (1964) analysis of basal sliding.

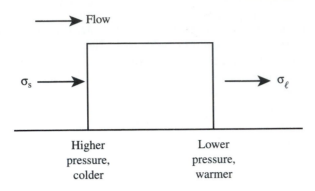

Figure 7.2   Pressure and temperature on stoss and lee sides of a rectangular bump on a glacier bed.

Using the equation (7.1), the temperature difference across the obstacle is

$$\Delta T = C\left(\frac{\tau}{r^2}\right) \tag{7.2}$$

and the temperature gradient through it is $\Delta T/\ell$. The heat flow through the obstacle is thus:

$$Q = \frac{\Delta T}{\ell} K_r \ell^2 = \Delta T K_r \ell \tag{7.3}$$

where the ℓ^2 in the middle term is the cross-sectional area of the obstacle and K_r is the thermal conductivity of rock. K_r has the dimensions J m^{-2} s^{-1} (K/m)$^{-1}$, or J m^{-1} s^{-1} K^{-1}. A typical value for rock or ice is 2.2. Q has the dimensions J s^{-1}.

This heat flow can melt ice at the rate $Q/H\rho$, where H is the heat of fusion, $3.3 \cdot 10^5$ J kg^{-1}, and ρ is the density of ice, ~900 kg m^{-3}. Thus, the melt rate is expressed in m^3 s^{-1}. Dividing this rate by the cross-sectional area of the obstacle, ℓ^2 gives the speed with which ice can move past the obstacle by regelation, S_r. Thus, using equations (7.2) and (7.3),

$$S_r = \frac{Q}{\ell^2 H\rho} = \frac{C\tau K_r}{\ell H\rho r^2}. \tag{7.4}$$

In reality, some heat also flows from the low-pressure region to the high-pressure region through the ice above the obstacle and through the rock beneath it, so this relation slightly underestimates S_r.

The water formed by melting ice on the stoss side of the obstacle flows from the high-pressure area either upglacier or downglacier to areas of lower pressure. The area of low pressure in the lee of the obstacle that we have been analyzing is one such sink. Because heat is conducted away from this area by the temperature gradient through the obstacle, this water freezes. Thus, to complete the regelation cycle, the water flux to the lee of the obstacle must equal the melt on the stoss side. We will examine the consequences of a failure of this condition later.

As a result of the high stresses on the obstacle, Weertman suggests that creep or plastic flow of ice around the obstacle is enhanced, and that this contributes to flow of the glacier past the obstacle. He assumes that the speed with which ice moves past the obstacle by this process, S_p, is proportional to ℓ and to the creep rate obtained by using the stress difference from equation (7.1) in the flow law; thus,

$$S_p = b\dot{\varepsilon}\ell = b\left(\frac{\tau}{Br^2}\right)^n \ell \tag{7.5}$$

where b is a dimensionless constant of proportionality. Note that dimensionally $\dot{\varepsilon}\ell$ is a speed.

Comparing equations (7.4) and (7.5), you will note that as ℓ increases, S_r decreases but S_p increases (Fig. 7.3). S_r decreases because the path that heat follows back through the obstacle

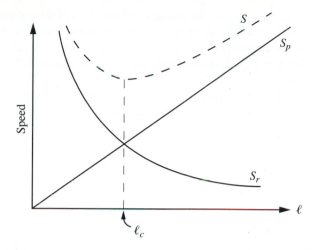

Figure 7.3 Relation between sliding speed and obstacle size for regelation, S_r, plastic flow, S_p; and their sum, S.

increases with ℓ, so the temperature gradient decreases, thus decreasing the heat flow. The physical reasons for the increase in S_p are less obvious; dimensionally, it results from the fact that to obtain a speed from a strain rate, one must multiply by a length scale, and the obvious length scale in the present situation is ℓ.

The total sliding speed, S, is generally considered to be the sum of the contributions from regelation and plastic flow (Fig. 7.3); thus,

$$S = S_r + S_p = A_r \frac{\tau}{\ell r^2} + A_p \left(\frac{\tau}{r^2}\right)^n \ell \tag{7.6}$$

where, for simplicity, the factors that are physical constants for any given situation have been lumped into the parameters A_r and A_p. We will also consider the contributions to be additive, but note that Nye (1969, p. 455) finds that this is not strictly correct on real beds consisting of roughness elements of many different sizes. This is because the pressure distribution resulting from regelation is then not the same as that resulting from plastic deformation.

Let us seek the obstacle size, ℓ_c, for which, as shown in Figure 7.3, S is a minimum. To do this we take the derivative of S with respect to ℓ; thus,

$$\frac{dS}{d\ell} = -A_r \frac{\tau}{\ell^2 r^2} + A_p \left(\frac{\tau}{r^2}\right)^n \tag{7.7}$$

set the result to 0, and solve for ℓ_c:

$$\ell_c = \sqrt{\frac{A_r}{A_p} \left(\frac{\tau}{r^2}\right)^{\frac{1-n}{2}}} \tag{7.8}$$

Inserting this back into the expressions for S_r and S_p [equations (7.4) and (7.5)] yields

$$S_r = \sqrt{A_r A_p} \left(\frac{\tau}{r^2}\right)^{\frac{n+1}{2}} \quad \text{and} \quad S_p = \sqrt{A_r A_p} \left(\frac{\tau}{r^2}\right)^{\frac{n+1}{2}}. \tag{7.9}$$

Thus when S is a minimum, $S_r = S_p$ and $S = 2S_r = 2S_p$.

A glacier bed has irregularities of many sizes, but the sliding speed in any one area of the bed must be the same for all of them. Suppose that r is also the same for all obstacle sizes in this area. Then, in the size range where plastic flow dominates, it is clear from equation (7.5) (or Fig. 7.3) that the drag exerted on the base of the glacier by obstacles of size ℓ_c will be greater than the drag exerted by larger obstacles [with S_p, b, B, and r all constant , τ varies inversely with ℓ]. Similarly, in the size range where regulation dominates, the drag exerted by obstacles of size ℓ_c will also be greater than that exerted by smaller obstacles. In other words, obstacles of size ℓ_c exert more drag on the base of the glacier than do obstacles of any other discrete size, and this has thus come to be called the *controlling* obstacle size. As implied by $S_r = S_p$, regulation and plastic flow contribute equally to motion of ice past these roughness elements.

When one considers a bed composed of a continuous spectrum of obstacle sizes, and particularly of roughnesses, the concept of a *controlling* size is no longer as relevant (Nye, 1969, p. 459). Nevertheless, an obstacle size for which $S_r = S_p$ normally appears when bed geometry is simplified in order to make theoretical studies of sliding mathematically tractable. The term *controlling obstacle size* for this size is probably irrevocably ingrained in the literature.

Making use of the fact that $S = S_r + S_p$ and the relations in equation (7.9), assuming that $n = 3$, and combining the constant factors into a single constant, A, yields

$$S = A\frac{\tau^2}{r^4}. \tag{7.10}$$

This is only true for beds composed of uniform obstacles of size ℓ_c or for beds with a homogeneous distribution of roughness elements—so-called white roughness. For beds composed of much smaller obstacles, $S \to A_r\tau/(\ell r^2)$ (equation 7.4), whereas for beds of much larger obstacles, $S \to A_p\ell\tau^3/r^6$ [equation (7.5)]. Of particular interest in equation (7.10) is the quadratic dependence of S on τ, and the strong inverse dependence on r.

Nye (1969) and Kamb (1970) have analyzed sliding of glaciers over a bed with a more realistic geometry consisting of superimposed sine waves. As in Weertman's development, the two processes by which ice was assumed to move past roughness elements on the bed were regulation and plastic flow. The mathematical techniques that Nye and Kamb employ are elegant (and beyond the scope of this book). A price paid for this realism was that to obtain exact solutions, both Nye and Kamb had to assume a linear rheology ($n = 1$) for ice. Kamb also obtained an approximate solution for a nonlinear rheology. Both Kamb [equation (45)] and Nye [equation (34)] concluded that $S \propto \tau/r^2$ in the linear theory. This is consistent with our equation (7.9) with $n = 1$. However, in Kamb's [equation (79)] nonlinear theory, the dependence of τ on ℓ_c [equation (7.8)] leads to $S \propto \tau^2$, at least for certain roughness spectra; this is of interest in light of some laboratory and field studies discussed below. The dependence of S on r in Kamb's nonlinear theory is more complicated owing to the way in which roughness is defined. Let us now examine this in more detail.

Roughness in the Nye-Kamb Theory

A different definition of bed roughness is needed in the Nye-Kamb models in which the bed topography is modeled by superimposed sine waves. Thus, Kamb introduces a relative roughness parameter, $\zeta = a/\lambda$ where a and λ are the amplitude and length of a sine wave (Fig. 7.4). Note that ζ is not so much a measure of the heights of bumps, a, but rather of the steepness of the adverse slope that they present to the glacier. As before, we can define a controlling wavelength, λ_c, for which $S_r = S_p$.

Lliboutry (1975) has suggested a modification of the Kamb-Nye approach. If $z(x,y)$ describes the elevation of the bed above some reference plane and $z_*(x,y)$ weights the contribution of various roughness elements to the total roughness in such a way that wavelengths bigger and smaller than the controlling size contribute less resistance to flow, then

$$r = \frac{1}{A}\int_{-\infty}^{\infty}\int_{-\infty}^{\infty}\left(\frac{\partial z_*}{\partial x}\right)^2 dx\, dy$$

where A is the area of the bed. Thus, the roughness is defined in terms of the square of the bed slope in the direction of flow, averaged over the bed and weighted as just described.

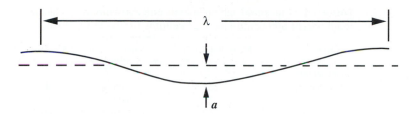

Figure 7.4   Wavelength and amplitude of a sinusoidal wave on a glacier bed.

When ζ is constant for all wavelengths, the spectrum is called *white*. This is one of the spectra for which $S \propto \tau^2$ in Kamb's nonlinear theory. From casual observations of glacier beds, however, it is quickly clear that ζ is not constant; there is commonly a distinct absence of short wavelengths. In his studies, Kamb found almost no obstacles with wavelengths less than 0.5 m in the direction of flow. He further observed that ζ was commonly ~0.05. On bedrock exposed in front of Glacier de Saint-Sorlin in France, on the other hand, roughness elements with wavelengths shorter than 0.5 m are common (Benoist, 1979).

The frequent absence of short wavelengths is usually attributed to preferential abrasion of these features by regelating ice. As noted, regelation is most effective over small obstacles. During regelation, any rock particles that become incorporated into the regelation layer, say by entrainment during refreezing in the lee of a previous obstacle, are forced into strong contact with the next obstacle upon the stoss side of which this ice melts. Thus, small obstacles are abraded away while larger ones, accommodated principally by plastic flow, are not.

Tests of Sliding Theories

The only sliding theory that can be reasonably tested with field data is Kamb's approximate nonlinear one. The sliding speed and other data used for the test were collected on Blue and Athabasca glaciers, using boreholes to the bed and tunnels along the bed. In neither of these techniques was a large enough area of the bed exposed to permit direct measurement of the roughness. Thus, instead, Kamb calculated ζ and the controlling wavelength, λ_c, from the measured sliding speeds and known glacier geometry (Table 7.1).

When Kamb used a full white roughness spectrum in his calculations, the values of ζ were about one-third those in Table 7.1. Thus, in accord with his observations, Kamb assumed that obstacles with short wavelengths had been abraded away, and instead of the full white roughness spectrum he used a truncated spectrum that did not have obstacles with those wavelengths. This yielded values of ζ (Table 7.1) that are consistent with observations on exposed bedrock outcrops, thus providing support for the theory. {It is noteworthy that in the absence of these shorter wavelengths, $S \propto \tau^3$ [our equation (7.5); Kamb's equation (90)].}

Another test of the theory comes from observations of the thickness of the regelation layer at the base of a glacier. Regelation ice can be distinguished from more highly deformed ice by grain size and crystal orientation. Thin sections of the ice viewed through crossed polarizers are used for this purpose. Kamb and LaChapelle (1964) measured thicknesses of the regelation layer in ice tunnels beneath Blue Glacier. They judged the average thickness to be about 5 mm while the maximum was 29 mm. These values can be compared with those calculated from Kamb's theory. The calculation is based on the fact that the thickness of the regelation layer in a depression in the lee of a bump is proportional to the degree to which the bump was accommodated by regelation. For example, for obstacles of the controlling size, accommodated half by regelation and half by plastic flow, regelation ice should half fill the depression between bumps (Fig. 7.5). The predicted thicknesses were 1 to 10 mm.

TABLE 7.1   Measured sliding speeds and corresponding calculated roughnesses and controlling wavelengths (from Kamb, 1970)

Location	Measured S, m/a	Calculated ζ, m	Calculated λ_c, m
Blue Glacier			
Borehole K	22	0.05	0.32–0.45
Borehole V	4	0.09	0.47–0.67
Western ice fall	6	0.02–0.04	0.62–1.12
Central ice fall			
On ridge	128	0.03	0.15–0.28
In trough	4	0.13	0.37–0.53
Athabasca Glacier			
Hole 1B	41	0.02	0.50–0.70
Hole 1A	42	0.02	0.33–0.47
Hole 209	3	0.06	0.59–0.84
Means		0.054	0.53

The fact that these thicknesses were less than those observed suggests that regulation may be more important than predicted by the theory.

Weaknesses of Present Sliding Theory

A number of processes are involved in sliding of ice over a hard bed that are not adequately described in the above theoretical models. An obvious example is the failure to consider frictional forces between rock particles in the basal ice and the underlying bedrock.

"Frictional" drag may also be provided by cold patches in which ice temporarily becomes frozen to the bed. Robin (1976) proposed two types of cold patch. In the first, which he termed the "heat pump effect" (Fig. 7.6a), water that is formed in the zone of high pressure on the stoss side of a bump, where the melting point is depressed, is squeezed out of the ice through veins along three-grain intersections (see Fig. 8.1). When this "cold" ice is transported to the top of the bump where the pressure is less, any water remaining in the ice and along the ice–rock interface refreezes, releasing the heat of fusion and thus warming the ice. The water within the ice is likely to freeze first, followed by that at the interface. If the amount of water present is sufficient, enough heat will be released to warm the ice to the new pressure melting point without freezing all of the water at the interface. However, if some of the meltwater escaped around the bump, as shown in Figure 7.6a, *all* of the water at the interface might freeze, thus cementing the glacier to the bed.

The second mechanism discussed by Robin involves local increases in water pressure in areas between bumps. Because the weight of the glacier is constant, any such increase will decrease the

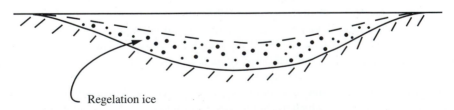

Figure 7.5   In the lee of a bump of the controlling size, regelation ice should fill the lower half of the space.

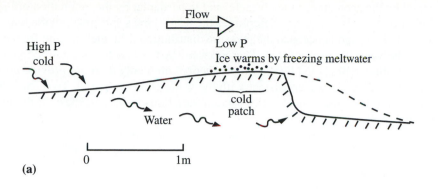

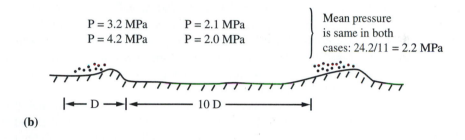

Figure 7.6   Formation of cold patches (after Robin, 1976). (a) Water that is squeezed out
of the ice on the stoss side of an obstacle may drain away and thus not be available to
refreeze in the low-pressure zone at the top of the obstacle. (b) Small changes in pressure
between obstacles result in large changes on tops of obstacles.

pressure on the stoss sides of bumps, where the pressure is already higher than average. In the example shown in Figure 7.6b, the area between bumps is 10 times the area of the bumps. Thus, a 0.1 MPa
increase in pressure between bumps reduces the pressure over the bumps by 1 MPa, resulting in a
~0.07°C increase in the pressure melting point. The ice, being at the pressure melting point, was
colder while the pressure was high. Thus, the decrease in pressure leads to freezing of any water present, potentially including any at the ice–rock interface.

It is worth noting that in addition to increasing the drag between the glacier and the bed, such
cold patches may be an effective erosional mechanism. Rock fragments that have been loosened
from the bed but do not project appreciably above it are separated from the ice by a melt film. As
long as the melt film exists, they may be held in the bed by rock-to-rock frictional forces that exceed
the drag exerted by the ice through the film. However, such fragments may be entrained if the melt
film becomes frozen.

There are also a number of problems surrounding the use of the simple regelation theory
presented above. Nye (1973) notes, for example, that at any point on an obstacle, the melt rate
(or freezing rate) required for movement of ice past that obstacle by regelation is completely

determined by the geometry of the obstacle, and in particular by the inclination of the face to the direction of motion. The melt rate determines the heat sources and sinks, so the temperature distribution is known, and hence also the pressure distribution. The melting and freezing rates also determine the water fluxes required. The awkward fact is that for normal bed geometries, the pressure distribution predicted by the simple theory commonly does not provide pressure gradients in the melt film that are consistent with the water fluxes required. To resolve this discrepancy, one has to take into consideration spatial variations in the thickness of the melt film and temperature gradients across it.

Impurities provide a second problem for regelation theory. Water moving in a melt film over an obstacle on the bed may absorb ions from the bed or from rock flour between the bed and the ice. Such impurities lower the freezing point. Thus, the temperature in the lee of the obstacle is lower than would be the case with pure water, and the temperature gradient through the obstacle is correspondingly reduced (see Fig. 7.2). This reduces the heat flux through the obstacle, and thus reduces S_r.

When impurities collect in the freezing water film in the lee of a bump, fractionation occurs; some of the impurities are carried away by the ice that forms, while the rest remain in the melt film. The steady-state situation is one in which the concentration of impurities in the film is such that the rate of removal of ions from the lee side during freezing equals the influx of ions in water coming from the stoss side of the bump. The impure ice thus formed will melt on the next suitable bump downglacier around which regelation is occurring, and the resulting impure meltwater will acquire more impurities. After several such cycles, the concentration of ions in water on the lee sides of obstacles becomes high enough to induce precipitation. The most common such precipitates are $CaCO_3$, but Fe/Mn coatings are also observed. Hallet (1976a, 1979) and Hallet and others (1978) have made detailed studies of the calcium carbonate precipitates, and Hallet (1976b) has undertaken quantitative estimates of the degree to which basal sliding over a hypothetical bed composed of sinusoidal waves of a single wavelength would be reduced by various concentrations of $CaCO_3$ in the melt film (Fig. 7.7). Note in Figure 7.7 that the wavelength for which S is a minimum, that is λ_c, is reduced from 0.6 m for the case of no solutes to 0.2 m for the highest solute concentration. This is because solutes reduce the efficacy of the regelation process, effectively shifting the S_r curve in Figure 7.3 downward.

A further effect of solutes has been observed in regelation experiments with wires (Drake & Shreve, 1973). As the stress driving the regelation increases, the pressure in the lee of the wire decreases, and it may reach the triple-point pressure. At this point a vapor pocket forms and the temperature cannot be raised further. Because the temperature on the stoss side can continue to decrease as the pressure increases, the mean temperature around the wire is less than the far-field temperature, and heat will flow from the surroundings toward the wire. This increases the rate of melting, but also means that some of the meltwater formed on the high-pressure side of the wire will not refreeze. This water collects in pockets that are then left behind in the ice, as a sort of wake, as the wire advances. Such a process might occur beneath glaciers in areas of relatively high basal shear stress.

The rheology of basal ice may also be rather different from that of ice well above the bed, thus altering the role of plastic flow. Basal ice may have fewer bubbles, may have a different solute content, and is quite likely to have more interstitial water because strain heating will be significant here, and there is no way to remove this heat other than by melting ice. In addition, the constant changes in stress field as the ice flows around successive bumps may result in zones of transient creep as the crystal structure adjusts to the changes.

Finally, cavities may form in the lee of obstacles. This effect is discussed further below.

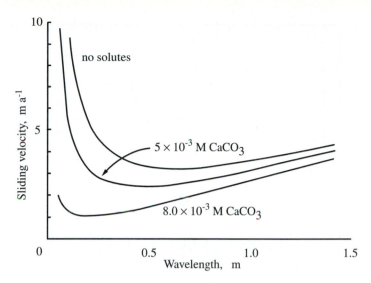

Figure 7.7   Effect of solutes on sliding speed. (After Hallet, 1976b. Reproduced with permission of the author and the International Glaciological Society.)

The Role of Normal Pressure

Another effect that is overlooked in the sliding theories discussed above is that of normal stresses. Budd and others (1979) carried out laboratory experiments in which ice blocks upon which a normal load, N, had been placed were dragged across rough rock surfaces. Temperature control was achieved by immersing the ice and rock surfaces in an ice-water bath. They found that $S \propto \tau^3/N$. The cubic dependence on τ might suggest that plastic flow was the dominant sliding process, and this may very well have been the case as it was possible for meltwater formed at the interface to escape to the surrounding bath. Thus, the only heat available for melting would be frictional heat and heat conducted from the bath to the interface. (The interface would be colder than the bath owing to depression of the melting point.)

More puzzling is the inverse dependence on N; this is what one expects in a purely frictional system, such as would be provided by a rock being dragged across a bedrock surface. To the extent that the ice–rock interface in the experiments was perfectly lubricated by a melt film, however, we would presume that no shear stresses could have been supported parallel to the surface. In this case, the sliding speed should have been independent of the normal pressure. However, some erosion of the rock surface occurred during the experiments, so rock particles in the basal ice and in contact with the bed may have been responsible for the observed increase in drag with increased normal load.

Budd and others suggested that in studies of real glaciers, N should be replaced by the effective normal pressure, N_e, or the normal pressure minus the water pressure, a factor first vigorously emphasized by Lliboutry (1968 and earlier). The importance of water pressure on sliding speed is now widely recognized (e.g., Fig. 7.8), but as we will discuss next, some of the mechanisms involved are not frictional as first suggested.

Cavities and the Effect of Water Pressure

Elevated water pressures increase the sliding speed in two ways: (1) by increasing the degree of separation of ice from the bed, thereby increasing the shear stress on parts of the bed still in contact with the ice; and (2) by exerting a net downglacier force on ice in cavities. In addition, they weaken any

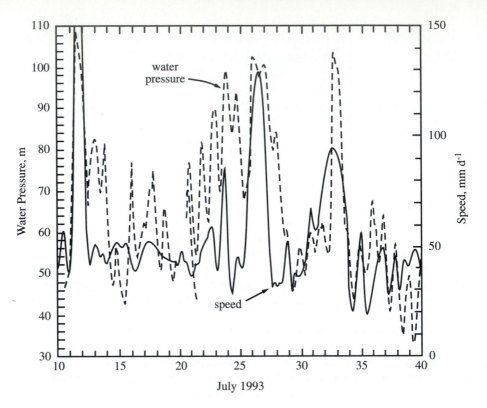

Figure 7.8   Diurnal variations in surface speed on Storglaciären, Sweden, measured with the use of a computer-controlled laser distance meter. The distance from a point off the glacier to a stake on the glacier was determined every 10 minutes. The dashed line shows corresponding water pressures, in meters of water, measured in nearby boreholes. Only the major peaks in speed are clearly related to water pressure peaks. (Hansen & Hooke, unpublished data.)

deforming subglacial till over which the glacier is moving, thus increasing u_d (Fig. 5.4). We consider here the first two of these. The third will be addressed in connection with our discussion of subglacial till deformation.

The degree of separation of ice from the bed in the lee of obstacles is increased when water pressures remain elevated for periods of a few days or weeks. Let us briefly examine the conditions required for such separation in an idealized situation (Iken & Bindschadler, 1986). The pressure at the bed is

$$P(x, y) = \sigma_o + P_o(x, y)$$

where σ_o is the ice overburden pressure and $P_o(x,y)$ is a fluctuating contribution that is high on the stoss sides of bumps and low on lee sides. The basal drag due to this effect can be expressed in terms of P_o; thus,

$$\tau_b = \frac{1}{A_r} \int_{A_r} P_o(x, y) \frac{\partial z}{\partial x} \, dx \, dy. \tag{7.11}$$

Here, the x-axis is directed downglacier and z is normal to the mean bed and positive upward, so $\partial z/\partial x$ is the slope of the bed in the direction of flow. Negative values of $\partial z/\partial x$ indicate downglacier-sloping surfaces. A_r is an area of the bed large enough to be representative of average conditions.

If we consider a sinusoidal bed of amplitude a and wavelength λ, P_o will vary sinusoidally; Iken and Bindschadler (1986) find that its maximum amplitude is

$$|P_o|_{\max} = \sqrt{2}\,\langle P_o^2 \rangle^{\frac{1}{2}} = \frac{\lambda \tau_b}{a\pi} \tag{7.12}$$

where $\langle P_o^2 \rangle^{\frac{1}{2}}$ is the root-mean-square of P_o. The minimum pressure will occur at inflection points on the downglacier faces of the undulations, and is

$$P_{\min} = \sigma_o - \frac{\lambda \tau_b}{a\pi}. \tag{7.13}$$

Note that $P_{\min}$ decreases as τ_b, and hence the sliding speed, S, increase. If the water pressure exceeds this minimum value, separation occurs and cavities will grow to a size determined by the degree to which the water pressure exceeds $P_{\min}$. Roughness elements on the bed that are bridged by such cavities no longer exert any drag on the ice. The task of balancing the downslope component of the body force, $\rho g h \alpha$, is thus shifted to places where the ice is still in contact with the bed, and the shear stress on these areas increases, thus increasing S.

The second mechanism by which elevated water pressures lead to acceleration of a glacier is a type of hydraulic jacking. If the subglacial drainage system is reasonably well connected to cavities in the lee of bumps, increasing water pressures in the drainage system result in high water pressures in the cavities. [Pressures in the water film on the stoss sides of the bumps are always in excess of the overburden pressure, and thus are not affected appreciably by changes in the (lower) pressure in the cavities.] The water in the cavity thus pushes upglacier against the bedrock and downglacier against ice. The result is a downglacier force that is added to the downglacier component of the body force. Drag forces on the bed must then increase to balance this additional down glacier force. An increase in the sliding speed results in the required increase in drag.

Over time spans of the order of hours, cavity sizes cannot change appreciably because such changes require ice flow. Thus, on bedrock beds, diurnal changes in speed resulting from input of meltwater or from storms (Fig. 7.8) must be a result, principally, of hydraulic jacking. Of course, cavity size increases during hydraulic jacking as a result of the increased flow rate, so if high speeds are sustained the degree of separation will increase sufficiently to result in a significant further increase in speed.

The effect of changing water pressures in a lee-side cavity is nicely illustrated by a numerical modeling study conducted by Röthlisberger and Iken (1981) (Fig. 7.9). When the cavity pressure was 0.05 MPa lower than necessary to support the cavity, velocity vectors were toward the cavity (Fig. 7.9a), tending to close it. An increase in cavity pressure of only 0.07 MPa was sufficient to cause the cavity to begin to enlarge at a rate of about 10 mm d^{-1} (Fig. 7.9b). Note, in particular, the substantial reductions in normal pressure, indicated by the bold numbers in Figure 7.9b; the decrease of over 1.2 MPa at the crest of the bump could easily have resulted in freezing there in a real situation, as suggested by Robin (1976) (Fig. 7.6b).

Iken (1981) notes that for a given adverse bed slope, β, measured with respect to the average slope of the bed, there is a critical pressure, P_{crit} above which the glacier may accelerate without bound. Such accelerations would occur if (Fig. 7.10)

$$P_w \lambda \sin \beta > \rho g \lambda \sin(\beta - \alpha) \tag{7.14}$$

The right side of this equation is the component of the body force acting parallel to the back slope of the bump and in the upglacier direction, and $\lambda \sin \beta$ is the projected area of the cavity face, normal to the back slope of the bump, against which P_w acts. Using the expressions for σ_o and τ_b shown in Figure 7.10 and the trigonometric identity, $\sin(\beta - \alpha) = \sin \beta \cos \alpha - \cos \beta \sin \alpha$, we obtain

$$P_{crit} = \sigma_o - \frac{\tau_b}{\tan \beta}. \tag{7.15}$$

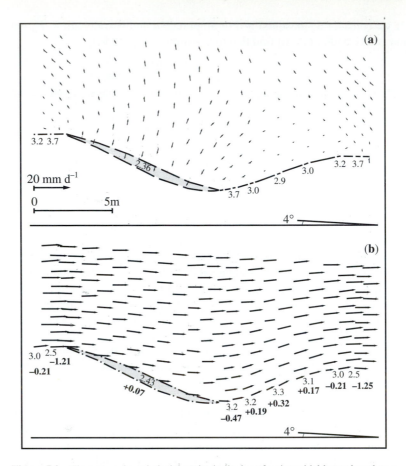

Figure 7.9   Flow around a subglacial cavity in the lee of a sinusoidal bump based on a numerical model using the finite element method. (**a**) The water pressure in the cavity, 2.36 MPa, is too low so the cavity is shrinking. (**b**) The water pressure in the cavity is too large, so the cavity is expanding. Light-face numbers show pressure at bed. Bold numbers show change in pressure following a 0.07-MPa change in water pressure in the cavity. The basal shear stress was 0.103 MPa in both experiments, and the mean pressure at the bed was 2.72 MPa. The cavity would be stable at a pressure of 2.41 MPa. (After Röthlisberger & Iken, 1981, Fig. 3. Reproduced with permission of the authors and the International Glaciological Society.)

On an actual glacier bed consisting of a variety of sizes and shapes of obstacles, P_{crit} would not be exceeded everywhere simultaneously. Thus, for most situations, $P_{crit} \approx \sigma_o$ is probably more realistic.

Iken and Bindschadler (1986), working on Findelengletscher (Findelen glacier), have collected an outstanding set of field data on the relation between water pressure and surface speed (Fig. 7.11). Here, the expected exponential increase in speed with increased water pressure, with water pressure asymptotically approaching the limit σ_o, is clearly demonstrated.

Iken and Bindschadler suppose that the character of the bed in front of Findelengletscher is similar to that beneath the glacier, and thus are able to calculate sliding speeds using Kamb's (1970) theory. For wavelengths and roughnesses that they believe to be appropriate, the theory gives sliding speeds that are too large, compared with the surface speed, to be realistic. They attribute the discrepancy largely to failure of the theory to take rock-to-rock friction into consideration.

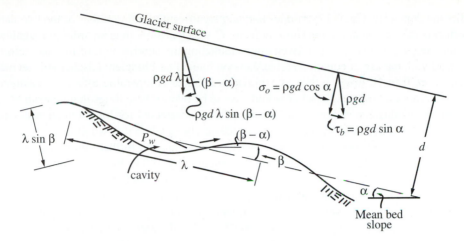

Figure 7.10 Diagram illustrating calculation of P_{crit}.

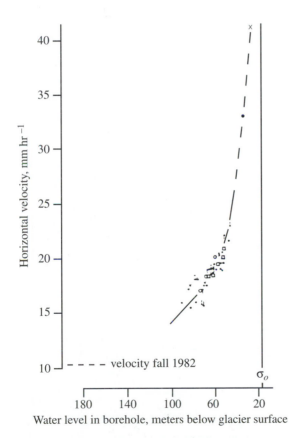

Figure 7.11 Speed of a stake on the surface of Findelengletscher as a function of water pressure, represented here by the water level in a borehole. Were the water level to rise to 18 m below the surface, the glacier would float. (After Iken & Bindschadler, 1986, Fig. 6. Reproduced with permission of the authors and the International Glaciological Society.)

Jansson (1995) has studied the relation between effective pressure, N_e, and surface speed, u_s, on Findelengletscher and Storglaciären, using Iken and Bindschadler's (1986) data (Fig. 7.11) for Findelengletscher, and finds that a relation of the form

$$u_s = CN_e^{-0.4} \tag{7.16}$$

fits the data well (Fig. 7.12). τ_b does not vary significantly within either of the two data sets, so its effect is incorporated into the constant factor C, which is more than an order of magnitude higher on Findelengletscher (Fig. 7.12). Even after subtracting the contribution of internal deformation, estimated with the use of equation (5.16), Jansson found that Findelengletscher still seemed to be sliding more than 10 times as fast as Storglaciären under comparable effective pressures. The basal shear stress on Findelengletscher would have to be two to three times that beneath Storglaciären to explain this difference, but shear stresses (albeit uncorrected for longitudinal stress gradients; see Fig. 11.7) at the sites of the measurements are nearly equal on the two glaciers.

More recent data from Findelengletscher serve only to further emphasize our lack of understanding of the effect of water pressure on sliding. By 1985, three years after the measurements shown in Figures 7.11 and 7.12, the surface speed had decreased 25%, and by 1994, it had decreased an additional 35% for comparable water pressures (Iken & Truffer, in press), yet there have not been any changes in the geometry of the glacier, and hence in driving stress, that could explain this deceleration. Iken and Truffer (in press) suggest that the basal water system was better connected in 1982, so that high water pressures reached more subglacial cavities. Thus, in effect, there may have been more subglacial hydraulic jacks urging the glacier forward in earlier years.

DEFORMATION OF SUBGLACIAL TILL

We have known for decades that ice moving over granular subglacial materials can deform these materials. (Herein, the term "granular material" should be understood to include materials with significant amounts of clay, although a distinction between granular materials and clays is usually made

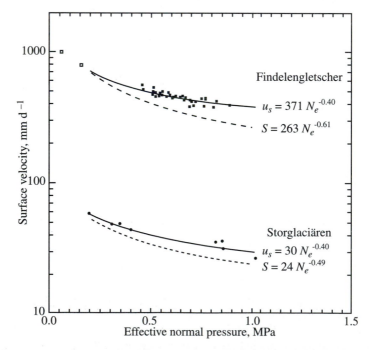

Figure 7.12   Relation between surface speed, u_s, sliding speed, S, and effective pressure, N_e, on Findelengletscher and Storglaciären. Dashed lines show sliding speed estimated by subtracting internal deformation from u_s. (After Jansson, 1995. Reproduced with permission of the author and the International Glaciological Society.)

in the soil mechanics literature.) Commonly, the granular material is till, either formed by erosion during the present glacial cycle, or left from a previous one. Recently it has become clear that a large fraction of the surface velocity of a glacier may be a result of deformation of such till (Fig. 5.4).

Intense interest in the rheology of till dates from work on Ice Stream B in Antarctica where studies of seismic velocities suggested that a layer with high porosity, saturated with water, was present beneath the ice (Blankenship et al., 1986). These characteristics suggest active deformation; thus, the high speed of the ice stream, about 450 m a^{-1} at the site of the study, was attributed to this process. Subsequent drilling revealed that the ice stream was, indeed, underlain by till (Engelhardt et al., 1990), but it is not yet known whether the till is deforming. It is possible, instead, that high water pressures have simply decoupled the ice stream from the till.

More recently, some scientists studying the Quaternary period have suggested that the large volumes of material found in till sheets in the midwestern United States and the large volumes of glacigenic material found in some submarine fans surrounding the Barents Sea could only have been transported to their present locations in deforming subglacial till layers (e.g., Alley, 1991; Hooke & Elverhøi, 1996). It is estimated that the amount of material that could be transported in basal ice or by subglacial melt streams is too low to account for the volumes of these deposits in the time inferred to be available for their formation. In the Barents Sea case, calculated basal melt rates are so high that little material is likely to have been entrained by basal ice, and yet they are too low to provide the water volumes required for significant fluvial transport.

Because glacial till is a granular material, its rheology is quite different from that of ice. Granular materials normally have a yield strength below which they do not deform. This yield strength, s, is related to two physical properties of the material, the cohesion, c, and the angle of internal friction, φ, by the classical Mohr-Coulomb relation:

$$s = c + N_e \tan \varphi \qquad (7.17)$$

where N_e is the effective normal pressure. To determine c and φ, laboratory tests are conducted in which the stress needed to initiate deformation of a material is measured at various effective normal pressures. When s varies linearly with N_e (Fig. 7.13), the slope of the line is φ, and the intercept is the apparent cohesion.

The term *apparent* cohesion is used because detailed measurements often show that the variation of s with N_e is not linear at low effective normal pressures, but rather is as shown by the dashed line in the inset in Figure 7.13. The true cohesion is the value of s at the intercept of this dashed line with the ordinate. Because the apparent cohesion varies directly with the true cohesion, however, we normally will not draw a distinction between the two quantities.

Let us now examine the physics involved in cohesion, and the physical significance of φ.

Cohesion

True cohesion in soils is a consequence of cementation, of electromagnetic forces between clay particles, and of electrostatic forces resulting from charge imbalances among ions absorbed on clay minerals (Mitchell, 1993, pp. 125, 373). Cementation is the major source of cohesion in subaerial soils, but would not be significant in continuously deforming subglacial tills. Thus, the magnitude of c in such tills is determined primarily by the amount and species of clay minerals present.

In situ deforming subglacial tills formed by erosion in the current cycle of glaciation do not seem to have much clay-sized material unless the glacier has moved over a bed containing such material. Furthermore, most of the clay-sized particles that are present in such clay-poor tills are not clay minerals. Thus, c may be small in such tills. For example, records from tiltmeters emplaced in till beneath Storglaciären demonstrate that this till is deforming. However, only ~5.4% (by weight) of the particles in samples collected through boreholes are less than 2 μm, and these particles are largely

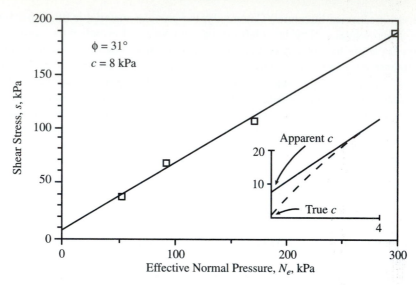

Figure 7.13 Relation between s and N_e obtained from laboratory tests on a sample of till from beneath Storglaciären (Iverson, unpublished data). Inset shows, schematically, how s may actually vary with N_e at low effective pressures.

quartz and hornblende (Iverson, unpublished data). Laboratory tests yield $c = 8$ kPa for this till (Fig. 7.13). Values for silty and clayey sands are typically between 20 and 75 kPa (Hausmann, 1990).

The absence of clay-sized material in deforming tills is likely to be largely a consequence of flushing by subglacial streams. In addition, however, it is noteworthy that the deviatoric stress required to fracture a grain increases as the particle size decreases, and that in the limit very fine grains deform plastically rather than fracture into still smaller particles (Kendall, 1978). That the clay-sized particles present in such tills tend not to be clay minerals is due to the absence of subaerial weathering processes. Higher concentrations of clay-sized particles and of clay minerals in Pleistocene tills may be a consequence either of subaerial weathering after retreat of the ice, or of incorporation of previously-weathered material over which the ice moved.

Cohesion is not increased by saturation by water unless clay minerals are present. The well-known fact that the walls of wet sand castles stand up better than dry ones is, rather, a result of surface tension. Surface tension effects are present when the sand is wet but pore spaces still contain air. This is because surface tension is a result of stresses associated with the air–water interface.

Consolidation

When a granular material accumulates gradually, it compacts under it own weight. Such a material is called *normally consolidated*. If an additional load, such as a shear stress, is then placed on the material, it becomes *overconsolidated*. The term overconsoldated is also used to describe a granular material that, after being normally consolidated, experiences a reduction in overburden pressure due to erosion.

Angle of Internal Friction

When an overconsolidated granular material begins to deform, it must *dilate* so that individual grains can move over one another (Fig. 7.14a). A normally consolidated material may either dilate or compact slightly, depending on the granulometry (size distribution of particles) and the conditions under which it accumulated.

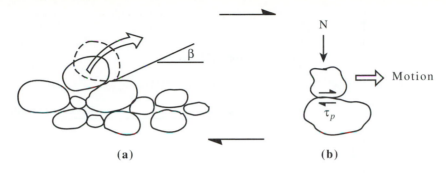

Figure 7.14 Deformation of a granular medium involves both (**a**) dilation as grains move apart in order to pass over one another, and (**b**) friction between grains that are constrained to slide past one another.

Grains in such a deforming material must also *slide* past one another locally (Fig. 7.14b). The forces resisting this sliding motion are frictional. Frictional forces are a consequence of the interlocking of microscopic asperities on the surfaces of the materials (Mitchell, 1993, p. 362). The maximum shearing stress that can be supported by friction between two particles, τ_p, is proportional to the effective normal pressure, N_e: $\tau_p = \mu N_e$. The constant of proportionality, μ, is called the coefficient of friction.

Let us define β as the angle, relative to the shear plane, that particles must ascend during dilation from an overconsolidated state (Fig. 7.14a) and also define $\omega = \tan^{-1} \mu$. Then $\varphi = \beta + \omega$ (Iverson et al., 1996). In granular materials that do not have much clay, ω is typically 20 to 25 degrees and φ is typically between 25 and 40 degrees (Mitchell, 1993, pp. 343, 366). Thus, more than half of the resistance to deformation of such a material is a consequence of frictional forces, while the remainder is due to processes such as dilation and crushing (Mitchell, 1993, p. 401).

As a result of the dependence of φ on β, φ also depends on the granulometry of the material. For example, if spaces between particles in Figure 7.14a were filled with finer material, a particle could not settle down into the gap between subjacent particles, and thus would not have to rise so much to move over its neighbor. β would then be lower, and hence so would φ.

Normal pressures suppress dilation and also force particles into firmer contact, thus increasing τ_p. These two factors account for the dependence of s on N_e.

When a granular material is deformed at a constant strain rate (with the shear stress, τ, being measured as a function of time or displacement), τ first increases rapidly and nearly linearly to a peak—the yield strength. If the material was initially overconsolidated, τ then declines slightly before reaching a constant value. The final value of τ, normally reached after a shear strain of the order of only 10%, is called the *residual strength* (Fig. 7.15) (Skempton, 1985). The difference between the peak and the residual strength reflects the additional stress needed to induce dilation. Once dilated, the material remains dilated as long as deformation continues. Thus, the stress required for deformation remains constant. [In materials in which clay-sized particles are abundant (>20%) and are predominantly clay minerals, a further decline in strength may occur as the platy clay particles become aligned parallel to the direction of shear.]

Grain Fracture and the Granulometry of Deforming Subglacial Till

If a granular medium is sheared at a constant rate between moving platens, in one of which there is a pressure sensor that is many times the diameter of individual grains but much smaller than the platen itself, the pressure recorded by this sensor varies with time (Fig. 7.16) (Iverson et al., 1996;

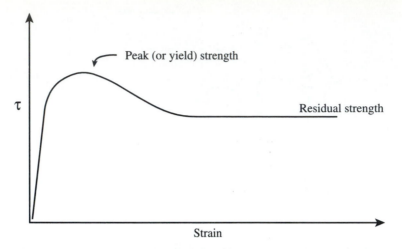

Figure 7.15   Schematic illustration of the variation of mean shear stress as a function of time (or displacement) in a granular medium that is sheared at a constant rate.

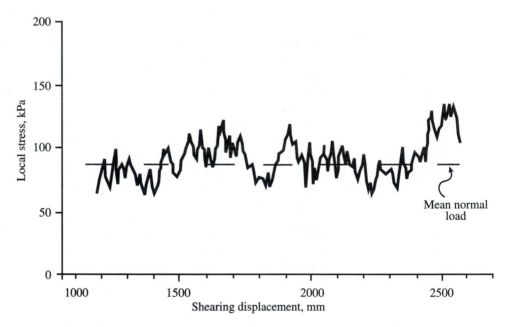

Figure 7.16   Variation in local pressure with time in a granular medium, ~55 mm thick, as it is deformed in a ring-shear experiment (After Iverson et al., 1996, Fig. 6.)

Mandl et al., 1977). Sometimes it exceeds the mean normal load on the sample by as much as a factor of 2, while at other times it is significantly less than the mean. One logical explanation for this is that grains in the medium become aligned to form bridges such as that shown in Figure 7.17a. When traced through a granular material of significant thickness, these bridges are much more complicated than suggested by the simple sketches in Figure 7.17a; high contact stresses are probably distributed along a three-dimensional array of routes, forming what might be called a grain-bridge network (Iverson et al., 1996).

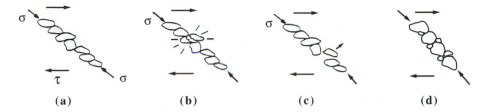

Figure 7.17 (**a**) A grain bridge, formed by nearly coaxial alignment of several grains in a deforming granular medium. The bridge may fail by (**b**) fracture of a grain or (**c**) slip between grains. (**d**) Stresses at contacts between grains are reduced when additional particles occupy pore space. Heavy arrows show shear stress applied to material, τ, and component of this stress along grain bridge, σ. (Modified from Hooke & Iverson, 1995, Fig. 1.)

For deformation to occur, grain bridges must fail. Failure can be a consequence either of fracture of a grain (Fig. 7.17b) or of slippage between grains (Fig. 7.17c). Fracture is most likely when two adjacent grains are of roughly equal size and when the space between them is not filled with smaller grains that absorb some of the stress. Slip between grains occurs when resolved stresses parallel to contacts between particles are greater than τ_p. Because the deviatoric stress required to fracture a grain varies with particle size, and because contacts between grains may have different orientations leading to different resolved stresses, there must be a wide range of bridge strengths.

Grain fracture alters the granulometry of a material. Biegel and others (1989) argue that the end product of this process is a granulometry that maximizes the support that each particle receives, and thus minimizes stress concentrations capable of causing fracture. For example, forces between particles in Figure 7.17d are distributed over several contact points, so local stresses are less likely to reach a level that will cause fracture. The granulometry that provides maximum support, according to Biegel and co-workers (1989), is one in which no two particles of the same size are in contact. Such a material has a fractal particle-size distribution with a fractal dimension of ~2.6. That is, if $\mathcal{N}_o$ is the number of particles of a reference size, d_o, then the number, $\mathcal{N}(d)$, of particles of size d is

$$\mathcal{N}(d) = \mathcal{N}_o \left(\frac{d}{d_o} \right)^{-m},$$ (7.18)

where m is the fractal dimension. [As is evident from equation (7.18), fractal size distributions appear to be the same at all scales. Thus, if there is one particle of unit size in a field of view, there will be 10^m particles that are 1/10th this size, regardless of the units used in making the measurement.] Sammis and others (1987) have shown that gouge from the Lopez Canyon Fault in California has such a particle-size distribution.

Deforming subglacial tills also have a fractal granulometry, with a fractal dimension close to 2.9 (Fig. 7.18). This suggests that grain fracture may play an important role in till deformation. That the fractal dimension is larger than the ideal of 2.6 is attributed to the production of fine material by abrasion, a process that would be inhibited by the higher effective normal pressures characteristic of deformation in active faults (Hooke & Iverson, 1995). However, more work is needed to understand fully the processes that give rise to fractal distributions.

Rheology of Deforming Subglacial Till

Our discussion thus far has focused on the strength of granular materials. A considerable volume of literature exists on this topic because of the interest in conditions under which slopes fail, leading, for example, to landslides or the collapse of highway embankments. In contrast, studies of the

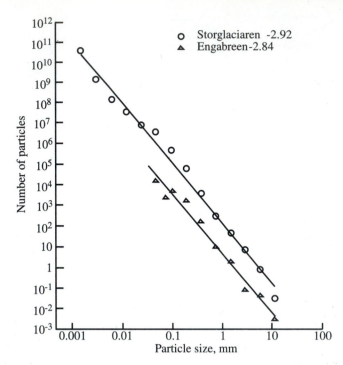

Figure 7.18   Grain-size distributions in two subglacial tills that were deforming. Fractal dimension of each till is shown. (Modified from Hooke & Iverson, 1995, Fig. 2.)

time-dependent behavior of deforming granular materials are less common. Furthermore, they often deal with materials containing an abundance of clay minerals, as clays are an important constituent of many landslides.

As noted, failure occurs when the peak or yield strength of a material is exceeded (Fig. 7.15). We, however, are interested in the rate of steady deformation some time after failure. Thus, the relevant measure of strength for studies of rheology is the residual strength. More specifically, we need to know whether the residual strength increases with strain rate, other factors such as effective pressure, granulometry, mineralogy, and so forth, remaining constant (Kamb, 1991). If such an increase occurs, the strain rate may be a unique function of the applied stress, and a "flow law" for till may exist.

In Chapter 4 we discussed possible mechanisms that might control the rate of deformation of ice. Let us now do the same for till. The principal processes we have discussed are dilation and failure of grain bridges. Dilation occurs early in the deformation process, and once the medium is dilated it remains so. Thus, dilation should not be rate controlling, and in the absence of repeated formation of grain bridges, we might expect the material to deform steadily and homogeneously, once a yield stress is exceeded. However, grain bridges do form, and deformation proceeds only when a bridge fails. This suggests that failure of grain bridges may be the rate-controlling process in till deformation. If this is the case, and if the formation of grain bridges is stochastic in time and space, then a mechanistic rheological model for till deformation should be based on these processes. Analysis should focus on the frequency of failure of grain bridges and on the amount of deformation resulting from each failure.

Studies of processes that are thermally activated, such as the creep of ice [equation (4.6)], provide a conceptual framework for such a model. In thermally activated processes, the process operates or proceeds when a certain energy barrier is exceeded. In the creep of ice, the barrier is the energy needed to break an atomic bond, thus allowing movement of a kink in a dislocation (Fig. 4.6). Fundamental to the theory of thermally activated processes is the premise, based on principles of statistical mechanics, that the probability distribution, $p(f)$, of energy levels, f, in atomic bonds is given by

$$p(f) = Ae^{-\alpha f} \tag{7.19}$$

where A and α are constants (Glasstone et al., 1941, p. 159).

In containers filled with beads and subjected to a normal load, the distribution of force levels at intergranular contacts is, indeed, given by equation (7.19) with f now defined as the force at such contacts (Liu et al., 1995). Thus, it seems plausible that the theory of thermally activated processes can be adapted to the analysis of deformation of granular materials. Mitchell and others (1968) and Mitchell (1993, pp. 349–361) have used this approach, and they conclude that a relation of the form

$$\dot{\varepsilon} = \Gamma e^{\gamma\tau} \tag{7.20}$$

should describe the steady strain rate in a granular material. Here, τ is a mean shear stress sufficient to cause deformation and thus maintain dilation, and Γ and γ are constants presumably dependent upon the strength and granulometry of the material.

A flow law for till that is of this form is consistent with laboratory data, which commonly can be described by an equation of the form

$$\frac{\tau}{\tau_o} = 1 + b\ln\left(\frac{\dot{\varepsilon}}{\dot{\varepsilon}_o}\right) \tag{7.21}$$

(e.g., Mitchell, 1993, Fig. 14.15). Here, τ_o is normally taken to be the stress at a reference strain rate, $\dot{\varepsilon}_o$; τ_o must be greater than s [equation (7.17)] as the material is deforming. For sandy materials, $b \approx 0.015$, whereas for materials with significant quantities of clay, it ranges from about 0.015 at strain rates of $\sim10^2$ a^{-1} to 0.07 at strain rates of $\sim10^6$ a^{-1} (Nakase & Kamei, 1986, Figs. 1 and 14). Inferred strain rates in till are at the lower end of this range (Table 7.2). Treating $\dot{\varepsilon}_o$ and τ_o as constants, equation (7.21) may be written

$$\dot{\varepsilon} = k_1\dot{\varepsilon}_o e^{\frac{\tau}{b\tau_o}} \tag{7.22}$$

where $k_1 = e^{-1/b}$. Comparing this with equation (7.20) we see that $k_1\dot{\varepsilon}_o = \Gamma$ and $1/b\tau_o = \gamma$. Based on the experimental values of b, and noting that strain rates in subglacial till tend to be at the lower end of the range given above, it is clear that strain rates should increase substantially with only a small increase in τ, as Kamb (1991) recognized.

Such a dependence of $\dot{\varepsilon}$ on τ in till is also consistent with field measurements on Storglaciären (Hooke et al., in press). The measurements were made by inserting instruments into till beneath about 120 m of ice, using boreholes through the glacier to gain access to the till. One instrument consisted

TABLE 7.2 Sliding speed, till thickness, and strain rate in the till beneath various glaciers

Glacier	Sliding speed, u_b, m a^{-1}	Till thickness, h_t, m	$\dot{\varepsilon}_t = u_b/h_t$ a^{-1}	Reference
Blue Glacier[a]	4	0.1	40	Engelhardt et al. (1978)
Breidamerkurjökull	24	0.5	48	Boulton & Hindmarsh (1987)
Ice Stream B	450	6[b]	75	Alley et al. (1987a)
Storglaciären[a]	10	0.2	50	Hooke et al. (1992)

[a] Values of $\dot{\varepsilon}_t$ for Blue Glacier and Storglaciären may be maximum estimates of the critical strain rate as the deformation is inferred to extend to the till/bedrock interface.

[b] Thickness of deforming till layer beneath Ice Stream B is inferred on the basis of geophysical data, whereas other values of h_t are based on more direct measurements.

of a cylinder with conical ends (the "fish") that was dragged through the till by a wire connected to a load cell. The force required to drag the fish varied between about 407.4 N and 408.6 N during a period of several days when the speed of the glacier varied diurnally (Fig. 7.19). The variations were basically in phase with those in N_e, which is consistent with equation (7.17).

However, the force was not related to the surface speed. As the speed with which the "fish" was pulled through the till should have varied in phase with the surface speed and the variations should have been of similar amplitude, it appears that the force on the "fish" did not depend significantly on its speed through the till, and hence did not depend on the strain rate in the till surrounding it. This is consistent with a constitutive relation of the form of equation (7.22).

In a number of papers (e.g., Alley et al., 1987b; Boulton & Hindmarsh, 1987; MacAyeal, 1989), a constitutive relation of the form

$$\dot{\varepsilon} = k_2 \frac{(\tau - \tau_c)^m}{N_e{}^p} \tag{7.23}$$

has been used to describe till rheology. Here, k_2 is a constant and τ_c is a critical shear stress below which no deformation occurs. Some authors assume that $\tau_c = 0$, and others let $\tau_c = s$ [equation (7.17)]. This relation is entirely intuitive; there are no reliable field or laboratory data supporting it.

In equation (7.23), it is normally assumed that $1 < m < 2$, so the sensitivity of $\dot{\varepsilon}$ to τ is far less than suggested by equation (7.22). This is illustrated in Figure 7.20. The other major difference between equations (7.22) and (7.23) is the way in which the effective pressure is incorporated. As τ_o [equation (7.22)] must be greater than s by an amount sufficient to deform the material at a rate $\dot{\varepsilon}_o$, τ_o must vary directly with N_e, albeit in a poorly constrained and probably nonlinear way. Thus, lower effective pressures increase $\dot{\varepsilon}$. In equation (7.23), lower effective pressures also increase $\dot{\varepsilon}$, but in this case the

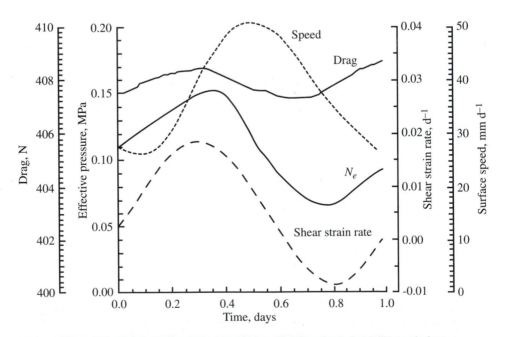

Figure 7.19 Relation among force on a cylinder pulled through subglacial till beneath about 120 m of ice, surface speed, water pressure in a nearby borehole, and shear strain rate in the till. Data are from a period of about 10 d in August 1992, at a time when all parameters were varying diurnally, and are "stacked" by averaging the values obtained at the same time of day each day.

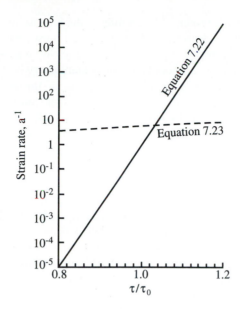

Figure 7.20 Experimentally determined relation between stress and strain rate in granular media (solid line) compared with a commonly used constitutive relation for till rheology (dashed line).

influence is through both the explicit effect in the denominator and the implicit effect on τ_c in the numerator.

Variation of Strain Rate with Depth in a Deforming Subglacial Till

Let us now address two questions, to which there are currently no firm answers: (1) What controls the thickness of the layer of subglacial till that can be mobilized by an overriding glacier; and (2) what is the shape of the velocity profile through this layer? We start by considering the variation in the relevant variables, τ and N_e, with depth.

Beneath a glacier of thickness h, with surface slope α and a horizontal bed:

$$\tau = \tau_b + \rho_t g z \alpha$$
$$N_e = N_{eb} + (\rho_t - \rho_w)gz.$$

Here ρ_t and ρ_w are the bulk density of the till and the density of water, respectively, subscript b refers to the conditions at the ice–till interface, and z is measured downward from the interface. Assuming $\rho_b \approx 2000 \ kg \, m^{-3}$, and taking derivatives with respect to z yields

$$\frac{d\tau}{dz} = \rho_b g \alpha \approx 20\alpha \ kPa \ m^{-1}$$
$$\frac{dN_e}{dz} = (\rho_b - \rho_w)g \approx 10 \ kPa \ m^{-1} \tag{7.24}$$

For typical surface slopes, the rate of increase of N_e with depth clearly exceeds that of τ. Thus, it seems likely that the decrease in $\dot{\varepsilon}$ in the till is nonlinear. For example, Alley (1989) has suggested that $\dot{\varepsilon}$ decreases rapidly near the ice–till interface and more slowly at depth. Furthermore, if the till layer is thick enough, then at some depth the strength of the till must exceed the applied stress and deformation will cease. However, to predict the actual velocity profile and the depth at which deformation ceases requires a better understanding of the dependence of τ_o on N_e in equation (7.22).

In situations in which water is produced by melting at the ice–till interface and is lost downward by flow through a permeable substrate, the hydraulic head must decrease downward through the till. Then dN_e/dz will be higher than in the purely hydrostatic case represented by the second of equations (7.24), and the deforming layer should be thinner than otherwise (Alley, 1989).

Relevant to this discussion is the observation in a tunnel beneath the glacier Breidamerkurjokull in Iceland, that within about 0.5 m of the glacier sole deformation of the till was pervasive, while at greater depths it was localized in shear zones (Boulton & Hindmarsh, 1987). Such localization of deformation is characteristic of virtually all laboratory experiments on granular materials. Whether the shear zones beneath Breidamerkurjokull are similar in origin to those in the laboratory samples is not known, however. Nor do we know the extent to which the thickness of a zone of deformation expands with time, or to which its thickness depends upon the granulometry of the material. Both the laboratory and the field observations suggest, however, that there may be a critical strain rate below which till does not remain dilated (Alley, 1989b). For example, estimates of strain rates in till based on field data fall within a relatively narrow range, from 40 to 75 a^{-1} (Table 7.2). Perhaps the thickness of the deforming layer is limited by the length of grain-bridge networks, and networks are longer when stresses build up rapidly. If stress buildup is slow, multiple small adjustments between particles could limit the length of the network.

PLOUGHING AND DECOUPLING

At sufficiently high water pressures, or low effective pressures, a glacier on a deformable bed may become partially decoupled from the underlying till. For example, shear strain rates in till beneath Storglaciären, measured by inserting tiltmeters into the till through boreholes, decrease as the effective pressure decreases and surface speed increases (Figs.7.19 and 7.21) (Hooke et al., in press; Iverson et al., 1995). In Figure 7.21, this is particularly true during the three periods of low effective pressure on about July 26, August 3, and August 12, but is also evident during some of the less pronounced low N_e events.

The extent of such decoupling is not known, nor is it clear whether diurnally varying water pressures, typical of valley glaciers, are essential for such decoupling. These questions are of interest, however, in view of recent suggestions, mentioned earlier, that during the Pleistocene huge amounts of sediment may have been moved long distances in deforming layers of subglacial till (Alley, 1991; Hooke & Elverhøi, 1996). If extensive decoupling is common beneath ice sheets, such sediment transfer would be less plausible.

The process of decoupling may be rather complicated. It seems likely that as the water pressure rises, a glacier may begin to slide over the till in some places while in others, clasts that are gripped in the ice and project down into the till continue to plough through it, causing local bed deformation. Evidence for such ploughing is occasionally preserved in tills in deglaciated areas (Clark & Hansel, 1989)

The ploughing process has been studied by Brown and others (1987). They considered spherical clasts of radius R embedded half in the ice and half in the till, and suggested that the force required to push such a clast through the till scales with the cross-sectional area of the clast, or with R^2. As this force must be provided by the ice, and as the ice is at the pressure melting point, regelation and plastic flow must be occurring around the clast. As with obstacles on a glacier bed, the stress that the ice exerts on the clast will be low for both small clasts accommodated largely by regelation and for large clasts accommodated largely by plastic deformation (Fig. 7.22). Whether a clast ploughs or not will be dependent upon the strength of the till, which is controlled by the effective pressure. However, based on Figure 7.22, it appears that clasts in the 10- to 40-mm size range will be the first to move.

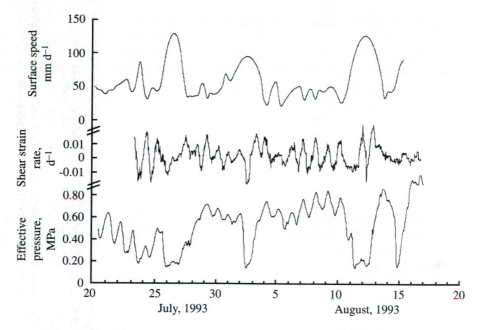

Figure 7.21   Relation among shear strain rate in till beneath Storglaciären, water pressure in a nearby borehole and surface velocity. (Modified from Iverson et al., 1995.)

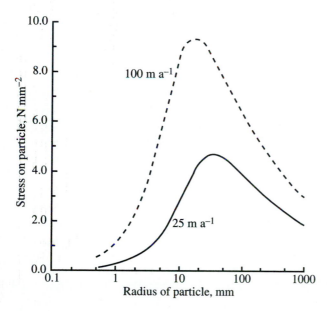

Figure 7.22   Stress exerted by ice on a spherical clast half embedded in till. Curves were calculated with the use of equation (2) of Brown et al. (1987).

SUMMARY

In this chapter we have explored the coupling between glaciers and both rigid and deformable beds. In the former, the dominant processes by which ice moves past irregularities on the bed are regelation and plastic flow. As small obstacles are accommodated more readily by regelation and larger obstacles by plastic flow, there are, theoretically, obstacles of intermediate size that exert more drag on a glacier than do larger or smaller ones, at least if the roughness is constant. This intermediate size has come to be called the controlling obstacle size, although on a glacier bed with a continuum of obstacle sizes and roughnesses, the concept of a "controlling" size becomes less relevant.

Theoretically, the speed with which ice moves past obstacles by regelation is proportional to τ_b, whereas for plastic flow it is proportional to τ_b^3. When both processes are involved, $S \propto \tau_b^2$.

The theory of sliding over rigid beds is imperfect because it does not take into account friction resulting from rock fragments entrained in the basal ice and dragged over the bed, local freezing of the ice to the bed, certain complications in the regelation theory for obstacles of irregular geometry, impurities in the meltwater that are acquired during regelation, and effects of changing water pressure. A good deal of recent research effort has focused on the last of these effects. Increases in water pressure can increase sliding speed by hydraulic jacking and, over somewhat longer time spans, by increasing the degree of separation of the glacier from the bed.

Our understanding of the movement of glaciers over soft beds is still modest. It is well known that the strength of granular materials depends on cohesion, on friction between individual grains, and on the need for such materials to dilate before appreciable deformation can occur [equation (7.17)]. Once the strength is exceeded, however, we wish to know the relation between the stress and the strain rate. Theoretical considerations, geotechnical studies, and field measurements suggest that $\dot{\varepsilon} \propto e^{\gamma\tau}$, although an alternative relation in which is $\dot{\varepsilon}$ far less sensitive to τ has often been used. Equally crucial, however, is developing an understanding of the conditions under which a glacier becomes decoupled from such a soft bed so that it glides over it without deforming it significantly.

Water Flow in and under Glaciers: Geomorphic Implications

A great deal has been learned about water flow through glaciers in the past two decades. Much of the progress has been theoretical, as experimental techniques for studying the englacial and subglacial hydraulic systems are few and not yet fully exploited, and observational evidence is difficult to obtain for obvious reasons. An added benefit of the recent progress is that we have gained a much better understanding of glacial erosional processes and of the origin of certain glacial landforms that owe their existence to the interaction between water and ice.

We begin this chapter with a discussion of the development of englacial water conduits and their geometry. We then examine the subglacial part of the system. Finally, we consider geomorphic implications of some of the recent research.

THE UPPER PART OF THE ENGLACIAL HYDRAULIC SYSTEM

Veins and the Initial Development of Passages

Nye and Frank (1973) argued that veins should be present along lines where three ice crystals intersect, and that these veins should join together at four-grain intersections to form a network of capillary-sized tubes through which water can move. They concluded that temperate ice should thus be permeable.

Such capillary passages have been observed in ice cores obtained from depths of up to 60 m on Blue Glacier, Washington (Fig. 8.1a) (Raymond & Harrison, 1975). The veins are triangular in shape (Fig. 8.1b) and roughly 25 μm across.

Estimates of the permeability of the ice resulting from this vein system vary widely. Expressed in terms of the thickness of a water layer that would be transmitted downward into the ice, values range from ~1 mm a^{-1} in coarse-grained ice with relatively few crystal boundaries per unit volume (Raymond & Harrison, 1975) to 1 m a^{-1} in fine-grained ice (Nye & Frank, 1973). Lliboutry (1971) noted that the existence of supraglacial streams precludes the possibility of significantly higher permeabilities. He further argued that, at permeabilities near the upper end of this range, the potential energy released by the descending water would rapidly enlarge the conduits to the point of completely melting the glacier.

Lliboutry (1971) concluded that deformation and recrystallization of ice would constrict the veins, rendering the ice essentially impermeable. Alternatively, air bubbles located along the veins

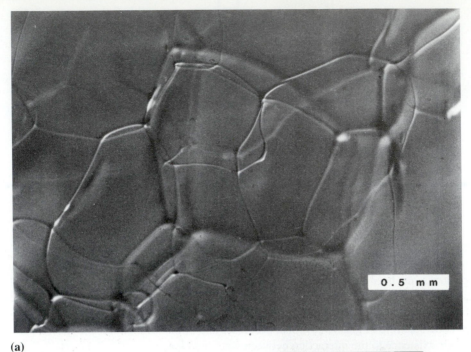

(a)

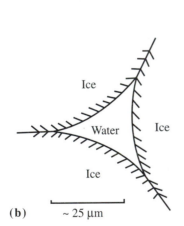

(b)   ~ 25 μm

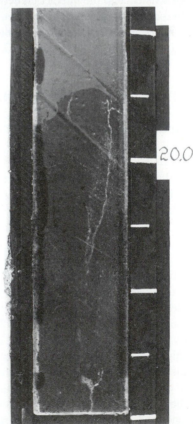

Figure 8.1   (a) Veins in ice from Blue Glacier.
(b) Cross section of a vein with approximate scale.
(c) Millimeter-sized tubes from a depth of 20 m in
Blue Glacier. (a and c from Raymond & Harrison,
1975. Reproduced with permission of the authors
and the International Glaciological Society.)   (c)

might block water movement. Lliboutry considered and rejected the latter idea, but Raymond and Harrison (1975) thought that it might have merit in coarse ice with few veins.

When water moves through such a vein system, viscous energy is dissipated in the form of heat. The amount of heat produced is proportional to the water flux. As the ice is already isothermal and at the pressure melting point, to a first approximation, this heat cannot be conducted away from the veins, but instead must be consumed by melting ice. In this way, passages are enlarged. Shreve (1972) and Röthlisberger (1972) argued that when two such passages of unequal size separate and rejoin, the larger passage carries more flow per unit of wall area, and it is thus enlarged at the expense of the smaller passage. Thus, we may anticipate that some of the capillary passages will become enlarged to millimeter-scale tubes a short distance below the surface.

Raymond and Harrison observed such tubes in a slab of ice from a depth of 20 m in Blue Glacier (Fig. 8.1c). The tubes formed an upward-branching arborescent network, as expected. Because the Shreve-Röthlisberger argument applies equally well to larger anastomosing passages, we may imagine that at greater depths, the arborescent network continues to evolve, with ever larger conduits developing.

Connections to the Surface

In the accumulation area, one can visualize continuous connections between the vein system and the overlying porous firn. As these veins do not always transmit downward all of the percolating meltwater, a local water table sometimes forms in the firn (Fountain, 1989; Vallon et al., 1976).

In the ablation area, however, there may be a surface layer of cold ice, several meters in thickness, in which the veins are frozen. This cold layer forms on glaciers in more continental climates where snowfall is low enough to allow appreciable cooling of the ice by conduction during the winter (Hooke et al., 1983). It is less likely to form in maritime climates where larger snowfalls form an effective insulating layer. When present, it is likely to persist well into the melt season, if not entirely through it, and thus forms an effective barrier to penetration of surface meltwater.

Because of this cold layer, and because the vein system, even on glaciers without such a cold layer, is relatively ineffective in transmitting water downward, it is principally by way of crevasses that surface water is able to reach the interior of the glacier. When a crevasse first forms, it may fill with water and overflow. In larger crevasses, however, this situation normally does not persist for long. It seems probable that once a crevasse penetrates deep enough to intersect the millimeter-scale conduit system, increasing the water supply to these conduits dramatically, the conduits are quickly enlarged until they can transmit all of the incoming water downward into the glacier.

Crevasses may close as they are moved into areas with lower tensile stresses. However, where melt streams pour into such a crevasse, the viscous energy dissipated maintains a connection to the englacial conduit system. The hole thus formed in the glacier surface is called a *moulin*.

When a crevasse opens across a melt stream upglacier from a moulin, it cuts off the water supply to the moulin. In the absence of further dissipation of viscous heat, the moulin's connection to the deeper drainage system is then constricted by inward flow of ice, and during the winter the upper part of the moulin fills with snow. In due course, the snow becomes saturated with water, which eventually freezes. These processes result in distinctive structures in the ice.

Holmlund (1988) has descended into moulins on Storglaciären during the winter. He also, over a period of several years, carefully mapped these distinctive structures as ablation exposed ever deeper levels in the glacier. He found that moulins are typically 30 to 40 m deep, although deeper ones occur in other glaciers, that channels leading from the bottoms of moulins are typically meandering and trend in the direction of the initiating crevasse, and that after some distance the meandering channel ends in a vertical conduit leading deeper into the glacier.

Shreve (1972) has compared the drainage system we have just described with one developed in a permeable limestone in which karst has developed. The anastomosing vein system provides the basic permeability, while the moulins and larger arborescent network of conduits are the analog of the karst system. Our task now is to consider the geometry of the system of larger conduits deeper in the glacier, below the level of Holmlund's mapping. One possibility is that these conduits are not vertical, but rather slope steeply downglacier, normal to equipotential surfaces in the glacier. We develop the theory behind this idea next, following closely the analysis of Shreve (1972).

EQUIPOTENTIAL SURFACES IN A GLACIER

In a permeable porous medium, water flows in the direction of the negative of the maximum gradient of the potential, Φ, where Φ is defined by

$$\Phi = \Phi_o + P_w + \rho_w gz. \tag{8.1}$$

Here, Φ_o is a reference potential, P_w is the pressure in the water, ρ_w is the density of water, g is the acceleration of gravity, and z is the elevation above some datum level such as sea level.

To gain some appreciation for this concept, consider the situation in a lake (Fig. 8.2). Let $\Phi = \Phi_1$ at point 1 on the lake surface. Moving down a distance Δz to point 2 increases P_w by $\rho_w g \Delta z$ but decreases the third term on the right in equation (8.1) by the same amount. Therefore $\Phi_2 = \Phi_1$ and there is no flow. However, if the lake surface slopes gently toward the outlet, moving horizontally at constant $z = z_2$ from point 2 to point 3 will result in a decrease in P_w. Flow will then be toward the position of lower P_w, which also is a position of lower Φ. In other words, it is not simply the gradient in P_w that controls the direction of flow, but the gradient in Φ.

To determine the potential field in a glacier from equation (8.1), we must determine P_w everywhere. P_w is not hydrostatic because the water is moving, and most of it is a long way from the surface through many small passages.

In general, the pressure in the ice, P_i, is different from that in the water, and the ice deforms as a result of this pressure difference. P_w rarely exceeds P_i significantly, but it can be much less than P_i. Passages may thus increase in size slightly at very high water pressures, and they decrease in size rapidly at low pressures. In addition, as noted, heat generated by viscous dissipation melts conduit walls, enlarging passages (Fig. 8.3). In a steady state, the rate of closure of passages by creep of ice, u, is equal to the melt rate, $\dot{m}$, so the net rate of increase in size of the passages, $\dot{r} = \dot{m} - u$, is 0. (Although mathematically untidy, note that we have defined positive u as being inward, while positive $\dot{m}$ is outward. This simplifies some of the later equations.)

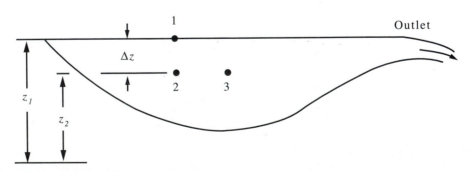

Figure 8.2 Illustration of difference between pressure field and potential field.

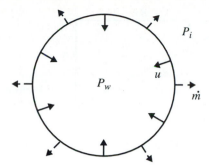

Figure 8.3 Closure of a cylindrical conduit, u, is balanced by melt, $\dot{m}$, in the steady state.

Let us assume that the flow of ice can be represented by $\dot{\varepsilon}_e = (\sigma_e/B)^n$, that ice is incompressible and isotropic, and that the passages are circular in cross section. We further define P_c, the pressure causing creep closure, by

$$P_c = P_i - P_w \qquad (8.2)$$

(Fig. 8.3). To a good approximation, $P_i = \rho_i g(H - z)$, where H is the elevation of the ice surface above the datum level (Fig. 8.4). Then

$$\frac{u}{r} = \left(\frac{P_c}{nB}\right)^n \qquad (8.3)$$

(Nye, 1953). This relation will be derived in Chapter 11 [equation (11.22)]. In the derivation it is assumed that $\sigma_e = \frac{1}{\sqrt{2}}\sigma_{rr}$, the radial stress. In other words, other components of the stress tensor, and hence of the strain rate tensor, are assumed to be negligible [see equation (2.10)]. Thus, there can be no deformation of the ice other than that resulting from the presence of the passage. In the present application, in which this assumption is clearly violated, we add a multiplying factor, K, which is approximately 1. K equals 1 if $\sigma_e = \frac{1}{\sqrt{2}}\sigma_{rr}$. Rearranging and substituting for P_c and P_i, we can rewrite equation (8.2) as:

$$P_w = \rho_i g(H - z) - KnB\left(\frac{u}{r}\right)^{\frac{1}{n}} \qquad u \geq 0. \qquad (8.4)$$

(If $u < 0$, implying that the passage is opening as a result of water pressures in excess of the ice pressure, $|u|$ must be used, and the sign of the second term adjusted accordingly, but $u < 0$ is rare in nature.) In excess of a couple of kilometers from the glacier terminus, changes in u along a tunnel will be relatively

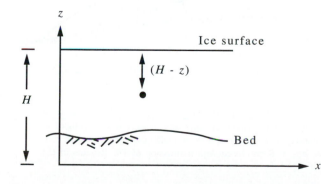

Figure 8.4 Coordinate axes used in discussion of conduit closure.

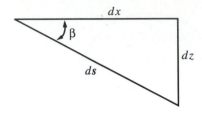

Figure 8.5 Sloping conduit showing how ds and β are defined.

small. Thus, combining equations (8.1) and (8.4) and taking the derivative with respect to an arbitrary direction, s, yields:

$$\frac{\partial \Phi}{\partial s} = \rho_i g \frac{\partial (H - z)}{\partial s} + \rho_w g \frac{\partial z}{\partial s}. \tag{8.5}$$

To determine the orientations of equipotential planes in the glacier, we make use of the fact that if s lies in such a plane, $\partial \Phi / \partial s = 0$; so,

$$-(\rho_w - \rho_i)\frac{\partial z}{\partial s} = \rho_i \frac{\partial H}{\partial s}. \tag{8.6}$$

Our objective is to define the dip of the plane, β. The dip in some horizontal direction, x, will be dz/dx because z is the vertical coordinate of the plane (Fig. 8.5). Therefore, multiply equation (8.6) by ds/dx and rearrange; thus,

$$\frac{\partial z}{\partial s}\frac{\partial s}{\partial x} = -\frac{\rho_i}{\rho_w - \rho_i}\frac{\partial H}{\partial s}\frac{\partial s}{\partial x}$$

or, inserting numerical values for the respective densities, letting $\alpha = dH/dx$, the slope of the glacier surface, and noting that $\tan \beta = dz/dx$:

$$\beta \approx -\tan^{-1}(11\alpha). \tag{8.7}$$

Thus, the equipotential planes dip upglacier (note the minus sign) with a slope of about 11 times the slope of the glacier surface (Fig. 8.6), a result first obtained by Shreve (1972).

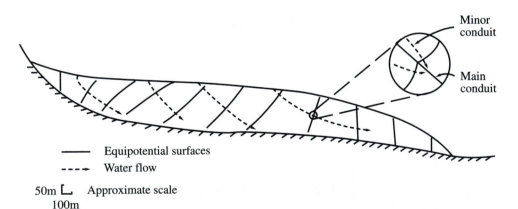

—— Equipotential surfaces

- - -▸ Water flow

50m ⌐ Approximate scale
100m

Figure 8.6 Longitudinal section of a glacier showing upglacier-dipping equipotential surfaces and the theoretical directions of englacial water flow. Inset shows dimpling of an equipotential surface and consequent diversion of flow in smaller passages toward the conduit. (After Hooke, 1989, Fig. 1. Reproduced with permission of The Regents of the University of Colorado.)

Rigorously, the equipotential surfaces are defined only within the water passages, but with suitable caution they can be treated as though they were defined throughout the glacier. Because $P_w \neq P_i$ under normal conditions, the equipotential surfaces are dimpled in the vicinity of the conduits. As P_w is usually less than P_i, the dimples point upglacier, as shown in Figure 8.6. With the use of the theory presented below, it can be shown that the difference between P_i and P_w increases with increasing conduit size, so dimples around larger conduits are larger. Hence, water in smaller conduits flowing normal to equipotential surfaces will be deflected toward the larger ones. This strengthens the tendency of the conduit system to evolve toward an arborescent pattern.

Alternative Derivation of Equipotential-Plane Dip

Consider the situation in Figure 8.7. We wish to determine under what conditions water will be forced outward over the subglacial hill, or phrased differently, under what conditions the upglacier slope of the hill will be parallel to an equipotential plane so that water will not flow up it. The ice pressure at (1) is $P_{i1} = \rho_i g (h_1 + h_2 + \Delta h)$, and that at (2) is $P_{i2} = \rho_i g h_2$. In the absence of water flow and conduit closure, the pressure in the water at (1), P_w, would be the sum of P_{i2} plus the hydrostatic head $\rho_w g h_1$. If $P_{i1} > P_w$, the conduit will begin to close and water will be forced out over the hill. Thus, the condition we seek is $P_{i1} = P_w$, or:

$$\rho_i g(h_1 + h_2 + \Delta h) = \rho_i g h_2 + \rho_w g h_1 . \tag{8.8}$$

Solving this for h_1, dividing by Δx, noting that $\alpha = -\Delta h/\Delta x$ and $\tan \beta = h_1/\Delta x$, and inserting numerical values for the densities leads directly to equation (8.7) Q.E.D.

MELT RATES IN CONDUITS

Let us now consider the rate of melting of conduit walls, following Shreve, 1972). The total amount of energy available per unit length of conduit, Δs, is

$$Q \frac{\partial \Phi}{\partial s} \Delta s$$

$$\frac{m^3}{s} \frac{N/m^2}{m} m = \frac{N-m}{s} = \frac{J}{s}. \tag{8.9}$$

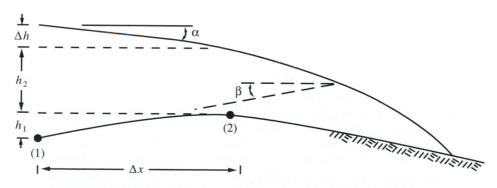

Figure 8.7 Sketch illustrating alternative derivation of dip of equipotential planes in a glacier.

Some of this energy must be used to warm the water to keep it at the pressure melting point as ice thins in the downglacier direction. The rest is available to melt ice; thus,

$$\dot{m}\,\Delta s(2\pi\ r\ \rho_i\ L) + \rho_w\ C_w\ \mathbb{C}\ \frac{\partial(H-z)}{\partial s}\rho_i\ g\,\Delta s\,Q = Q\frac{\partial\Phi}{\partial s}\,\Delta s$$

$$\frac{m}{s}\,m \qquad m\frac{kg}{m^3}\frac{J}{kg}\ \frac{kg}{m^3}\frac{J}{kgK}\ \frac{K}{N/m^2}\ \frac{m}{m}\ \frac{kg}{m^3}\frac{m}{s^2}\,m\ \frac{m^3}{s}.$$

(8.10)

Here, r is the radius of the conduit, L is the latent heat of fusion, C_w is the heat capacity of water, and $\mathbb{C}$ is the change in the melting point per unit of pressure. As you will see from inspection of the terms and the dimensions of the various quantities in them, the first term on the right is the energy used to melt tunnel walls, and the second is the energy needed to warm the water to keep it at the pressure melting point. Here, we have implicitly taken the positive s-direction to be upglacier, in the direction opposite to that of the water flow. Thus, both $\partial\Phi/\partial s$ and $\partial(H-z)/\partial s$ are positive.

It is common to define $k = \rho_w C_w \mathbb{C}$. Inserting numerical values [$\rho_w = 1000$ kg m^{-3}, $C_w = 4180$ J kg^{-1} K^{-1}, and $\mathbb{C} = 0.074 \times 10^{-6}$ K Pa^{-1}] we find that $k = 0.309$ and that it is dimensionless. If we assume that the water is saturated with air, and adjust $\mathbb{C}$ accordingly, $k = 0.410$.

Then, using equation (8.5) and dividing by Δs yields:

$$\dot{m}(2\pi r\rho_i L) + k\left(\frac{\partial\Phi}{\partial s} - \rho_w g\frac{\partial z}{\partial s}\right)Q = Q\frac{\partial\Phi}{\partial s}$$

(8.11)

or, solving for $\dot{m}$:

$$\dot{m} = \frac{Q\left[(1-k)\dfrac{\partial\Phi}{\partial s} + k\rho_w g\dfrac{\partial z}{\partial s}\right]}{2\pi r\rho_i L}.$$

(8.12)

It is interesting to insert some numbers into this equation to get a sense of the magnitude of $\dot{m}$. Consider a horizontal tunnel so $\partial z/\partial s = 0$. Suppose the tunnel has a diameter of 0.5 m and that it is in a glacier with a surface slope of 0.1. We now need a relation between Q and the tunnel roughness. The Gaukler-Manning-Strickler equation is one of two that is commonly used for such calculations. It is:

$$\bar{v} = \frac{Q}{\pi r^2} = \frac{R^{2/3}S^{1/2}}{n'}.$$

(8.13)

Here, $\bar{v}$ is the mean velocity over the tunnel cross section, R is the hydraulic radius of the tunnel, or the cross-sectional area divided by the perimeter (so $R = r/2$ in circular tunnels), S is the nondimensional headloss

$$S = \frac{1}{\rho_w g}\frac{\partial\Phi}{\partial s},$$

(8.14)

which is approximately equal to the glacier surface slope, and n' is known as the Manning roughness coefficient. For smooth channels, n' may be as low as 0.005 m$^{-1/3}$s, but for conduits beneath Storglaciären we have found that values ~0.2 m$^{-1/3}$s seem to be consistent with flow velocities determined from dye-trace experiments (Hock & Hooke, 1993; Seaberg et al., 1988). [This is actually a rather high value, suggesting that the walls of these tunnels are quite irregular, presumably due to differential melting. Significantly lower values, ranging from 0.08 to 0.12 m$^{-1/3}$s have been obtained from studies of floods, called *jökulhlaups*, resulting from subglacial drainage of ice-dammed lakes (Björnsson, 1992). This may be because tunnels formed during such floods are larger compared with the size of roughness elements on the tunnel walls]. Equation (8.13) then gives a mean velocity of about 0.4 m s^{-1}, or $Q \approx 0.08$ m^3 s^{-1}, and equation (8.14) gives $\partial\Phi/\partial s \approx 88.2$ N m^{-3}. Whence $\dot{m} \approx 0.3$ m a^{-1}. This may not seem like a lot, but volumetrically, the amount of ice melted in a year would be nearly four times the size of the original conduit.

A consequence of this melting and the resulting inward flow of ice toward the conduit is that structures such as foliation in the ice are also bent inward. A beautiful example of this is shown in Figure 8.8.

Some heat may also be advected into the glacier in water originating at the glacier surface and entering the englacial conduit system by way of moulins. The melt rate from such water, $\dot{m}_s$, is

$$\dot{m}_s = \frac{Q\rho_w C_w \dfrac{\partial \theta}{\partial s}}{2\pi r \rho_i L} \tag{8.15}$$

(Shreve, 1972). Here, $\partial \theta / \partial s$ is the rate at which the water cools as it flows through the conduit. If we assume that $\partial \theta / \partial s \approx 0.01$ K km^{-1} and use the discharge in the previous example, then $\dot{m}_s \approx 0.22$ m a^{-1}. Thus, this is a heat source that must be considered. However, a possible mitigating factor is that ice crystals are often carried in streams on a glacier surface. Thus, some of the heat would be used to melt these crystals rather than the conduit walls. It is not clear how the energy will be partitioned in this situation.

WATER PRESSURES IN ENGLACIAL AND SUBGLACIAL CONDUITS

Our next task is to determine the water pressure in conduits. Our discussion focuses on subglacial conduits, but it is equally applicable to englacial ones that are deep enough so that P_w is greater than atmospheric pressure. The water pressure in subglacial conduits is of considerable interest owing to its effect on sliding speed.

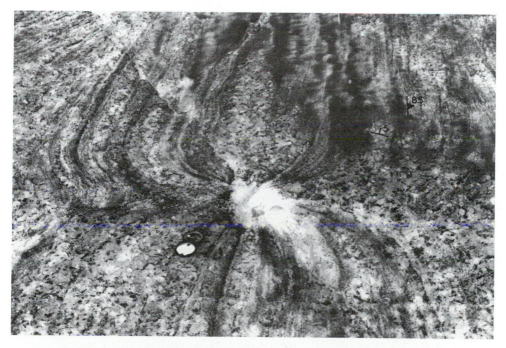

Figure 8.8 Foliation deflected into a conduit, now closed, by inward flow of ice to balance melting. (From Taylor, 1963, Fig. 11. Reproduced with permission of the author and the International Glaciological Society.)

Qualitatively, we expect P_w to increase upglacier because the ice thickness increases, and P_c must remain relatively constant so that $u = \dot{m}$. The increase in P_w, or more rigorously in Φ, provides the potential gradient that drives water toward the terminus.

Less obvious is the way in which P_w should vary with Q, yet this is quite important because if P_w decreases as Q increases, water will be drawn from smaller conduits toward larger ones, leading to the development of an arborescent drainage network. Conversely, if P_w increases as Q increases, the conduit system will tend to be anastomosing or braided. In the type of system that we have been discussing, consisting of conduits that may vary in size but not in shape in the longitudinal direction, it turns out that P_w decreases as Q increases. However, in subglacial drainage systems in which conduits locally lie in the lee of bedrock steps, so that conduit geometry is controlled by the steps, the reverse may be true. Qualitative explanations of these phenomena are difficult because the result depends upon the details of the way in which conduit size changes with discharge and in which u and $\dot{m}$ change with conduit size. We thus turn to a quantitative analysis, following closely the work of Röthlisberger (1972).

We start with the steady-state condition, $u = \dot{m}$, and obtain an expression representing this condition by combining equations (8.3) and (8.12), writing $P_i - P_w$ for P_c in the former; thus,

$$r\left(\frac{P_i - P_w}{nB}\right)^n = \frac{Q\left[(1-k)\dfrac{\partial \Phi}{\partial s} + k\rho_w g \dfrac{\partial z}{\partial s}\right]}{2\pi r \rho_i L}. \tag{8.16}$$

Noting that $R = r/2$ in circular conduits and solving equations (8.13) and (8.14) for r yields

$$r^2 = \frac{2^{\frac{1}{2}} n'^{\frac{3}{4}} Q^{\frac{3}{4}} (\rho_w g)^{\frac{3}{8}}}{\pi^{\frac{3}{4}}\left(\dfrac{\partial \Phi}{\partial s}\right)^{\frac{3}{8}}}. \tag{8.17}$$

Combining equations (8.16) and (8.17) to eliminate r, letting

$$D = 2^{\frac{3}{2}} \pi^{\frac{1}{4}} (\rho_w g)^{\frac{3}{8}} \rho_i L = 3.63 \times 10^{10} \left(\frac{N}{m^2}\right)^{\frac{11}{8}} m^{-\frac{3}{8}}.$$

and simplifying, we obtain

$$\left(\frac{\partial \Phi}{\partial s}\right)^{\frac{11}{8}} + \frac{k}{1-k}\rho_w g \frac{\partial z}{\partial s}\left(\frac{\partial \Phi}{\partial s}\right)^{\frac{3}{8}} = \frac{Dn'^{\frac{3}{4}}(P_i - P_w)^n}{(1-k)Q^{1/4}(nB)^n}. \tag{8.18}$$

We now need to relate $\partial \Phi / \partial s$ to P_w and the geometry of the tunnel system. Differentiating equation (8.1) with respect to s yields

$$\frac{\partial \Phi}{\partial s} = \frac{\partial P_w}{\partial s} + \rho_w g \frac{\partial z}{\partial s}. \tag{8.19}$$

Referring to Figure 8.9, note that

$$ds = \frac{dx}{\cos \beta} \tag{8.20}$$

so, noting again that $\tan \beta = dz/dx$, equation (8.19) becomes

$$\frac{\partial \Phi}{\partial s} = \left[\frac{dP_w}{dx} + \rho_w g \tan \beta\right]\cos \beta. \tag{8.21}$$

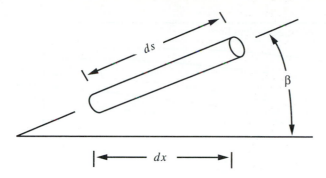

Figure 8.9 Definition of ds in terms of β and dx.

Letting

$$G = \left[\frac{dP_w}{dx} + \rho_w g \tan\beta\right] \tag{8.22}$$

and using equations (8.21) and (8.22) in equation (8.18) yields

$$G^{\frac{11}{8}} - kG^{\frac{11}{8}} + k\left(G - \frac{dP_w}{dx}\right)G^{\frac{3}{8}} = \frac{Dn'^{\frac{3}{4}}(P_i - P_w)^n}{Q^{\frac{1}{4}}(nB)^n \cos^{\frac{11}{8}}\beta} \tag{8.23}$$

Canceling the two terms in $kG^{11/8}$ and replacing G by its equivalent from equation (8.22) results in the relation we have been seeking:

$$\left[\frac{dP_w}{dx} + \rho_w g \tan\beta\right]^{\frac{11}{8}} - k\left[\frac{dP_w}{dx} + \rho_w g \tan\beta\right]^{\frac{3}{8}}\frac{dP_w}{dx} = \frac{Dn'^{\frac{3}{4}}(P_i - P_w)^n}{Q^{\frac{1}{4}}(nB)^n \cos^{\frac{11}{8}}\beta} \tag{8.24}$$

Equation (8.24) is a nonlinear differential equation that can be integrated numerically to obtain the water pressure, P_w, as a function of distance from the terminus, $x = 0$, subject to the boundary condition that $P_w = 0$ at the terminus. If the conduit is not full of water at the terminus, the boundary condition applies some distance upglacier from the terminus, where the conduit first becomes full, and the integration must start at this point. (Atmospheric pressure is ignored, as it is uniform over the area.) To carry out the integration, one uses the surface and bed topography along the course of the conduit to calculate P_i and β at each step, dx.

Equation (8.24) is clearly quite complicated, but we can gain insight into the general behavior of the water pressure field by studying some idealized cases. Consider, for example, a circular tunnel at the base of a slab of ice, 250 m thick, resting on a horizontal bed. Then, $\beta = 0$ and $P_i = $ constant. Röthlisberger presented some solutions for this case, which are shown in Figure 8.10. Water pressure is represented on the ordinate by the height to which water would rise in a vertical borehole that intersects the tunnel, or the *piezometric head*. A line connecting these water levels in a series of boreholes along a tunnel is called the *hydraulic grade line* or *energy grade line*, and its slope is $\partial\Phi/\partial s$. The *water equivalent line* in Figure 8.10 is the piezometric head at which the glacier would float. The values of n' and B used in the calculations are shown.

Aside from the obvious increase in P_w in the upglacier direction, thus providing the hydraulic head necessary to drive the flow on a horizontal bed, there are two characteristics of the patterns in Figure 8.10 that merit comment:

1. Water pressures increase as B decreases [compare curves (2), (3) and (4)]. This is because lower values of B imply softer ice and hence higher tunnel closure rates. Thus, higher pressures are necessary to reduce the closure rate so that $u = \dot{m}$.

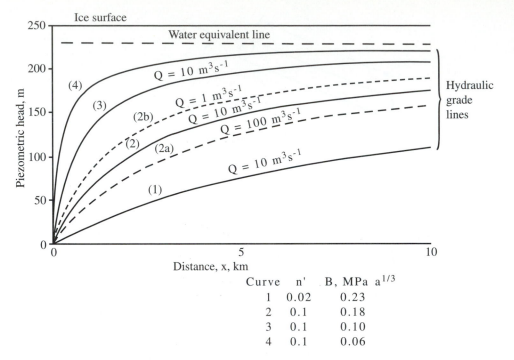

Curve	n'	B, MPa $a^{1/3}$
1	0.02	0.23
2	0.1	0.18
3	0.1	0.10
4	0.1	0.06

Figure 8.10 Hydraulic grade lines for a circular horizontal conduit under a slab of ice 250 m thick. (Modified from Röthlisberger, 1972, Fig. 2. Reproduced with permission of the International Glaciological Society.)

2. Water pressures increase as Q decreases. Although obvious from inspection of equation (8.24), this is somewhat counterintuitive. Consider the consequences of halving Q, holding, for the moment, P_w and hence $\partial\Phi/\partial s$ constant. Thus, $\dot{m}$ will be halved [equation (8.12)]. Clearly, halving Q will require a decrease in the cross-sectional area of the conduit $\mathbf{A}$. Now $\mathbf{A}$ varies as r^2 but the conduit closure rate, u, varies as r [equation (8.3)]. Thus, by the time r has decreased enough to halve u and thus match the new $\dot{m}$, $\mathbf{A}$ would have decreased so much that P_w would be forced to increase.

A simple numerical calculation may be useful. If subscripts "o" refer to the conditions before the change and subscripts "1" after the change, we find that $r_1 = 0.77\, r_o$ and, because $u \propto r$ [from equation (8.3)], $u_1 = 0.77\, u_o$. Now when $\partial\Phi/\partial s$ is constant, equation (8.12) can be written as $\dot{m}r = \Gamma Q$ where Γ is a constant. Therefore, $\dot{m}_1 r_1 = \tfrac{1}{2}\dot{m}_o r_o$. So, $\dot{m}_1 = \tfrac{1}{2}\dot{m}_o r_o / r_1 = 0.65\, \dot{m}_o$. Thus, $\dot{m}$ is reduced 35% while u is reduced only 23%, and they are now unequal. The tunnel will thus close, constricting the flow and raising the pressure.

Röthlisberger also presented some solutions for other idealized situations, and three of these are shown in Figure 8.11. Three interesting points merit discussion.

1. The negative P_w in Figure 8.11b implies that for $u = \dot{m}$, there must be suction. In other words, the u provided by the pressure of the overlying ice, alone, is not adequate to match $\dot{m}$, even with $P_w = 0$. Suction is necessary to increase u. Actually, the natural result is more likely to be that air will enter the channel from the terminus, resulting in open channel flow.

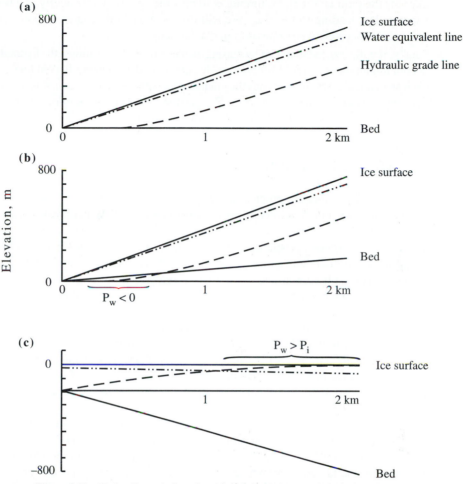

Figure 8.11 Hydraulic grade lines for some idealized situations with constant discharge of 10 m^3 s^{-1}. (Modified from Röthlisberger, 1972, Fig. 5. Reproduced with permission of the International Glaciological Society.)

In this case, u is low because the ice is thin, and $\dot{m}$ is high because the bed slopes downward in the direction of flow, thus increasing the contribution of the second term on the right in equation (8.5) to the energy dissipation. In fact, it is easy to show that unless the ice is more than a couple of hundred meters thick, even slight positive bed slopes will increase the energy dissipation sufficiently to lead to open channel flow in circular or semi-circular conduits (Hooke, 1984). The energy available to melt ice is then easily calculated from the decrease in potential energy, mgh.

2. In Figure 8.11c, $P_w > P_i$ some distance from the terminus. In this situation, the tunnel would be expanding, if it existed, and the tunnel size would be maintained by freezing of ice to the walls. Actually, in these situations it is more likely that water leaks out along the bed and the glacier, in effect, floats.

This condition arises when, as in Figure 8.11c, the bed has an adverse slope that is so steep that water flowing up it does not dissipate enough energy to remain warmed to the pressure melting point. Mathematically, the second term on the left in equation (8.10) exceeds the term on the right, forcing $\dot{m}$ to become negative. Physically, the water becomes supercooled, leading to freezing. We will discuss this further later, in connection with the origin and shape of overdeepenings in glacier beds.

3. Finally, it will be noted that with increasing distance from the terminus, the hydraulic grade lines in both Figures 8.10 and 8.11 become nearly parallel to the water equivalent line. As the slope of the water equivalent line represents the gradient in ice pressure at the bed, this means that $P_i - P_w$, and hence also u, are nearly constant. This justifies ignoring the $KnB(u/r)^{1/n}$ term in equation (8.4) when differentiating to obtain $\partial\Phi/\partial s$ (equation 8.5).

SHAPES OF SUBGLACIAL CONDUITS

Equation (8.24) makes specific predictions about basal water pressures. These predictions have been tested in the field, and the agreement with theory is not good. Water pressures are much higher than expected.

There are two obvious problems with application of equation (8.24) to subglacial conduits. First, it was derived for conduits with circular cross sections. Clearly, tunnels at the bed of a glacier are not circular. Thus, it is appropriate to see whether agreement can be improved by assuming some other tunnel shape. Second, stresses and the resulting strain rates parallel to the conduit axis are ignored in the derivation.

Let us deal first with the question of tunnel shape. Unfortunately, equation (8.3) is only valid when the conduit is circular. However, a reasonable expectation is that subglacial tunnels may be approximately semicircular (Fig. 8.12). Then, if $\tau_{\theta r}$ (Fig. 8.12) were vanishingly small, equation (8.3) should provide a reasonable approximation to the true closure rate. Unfortunately, when equation (8.24) is modified to apply to semicircular tunnels, the agreement of theory with field measurements is only marginally improved.

There are two reasons to suspect that subglacial conduits are not likely to be semicircular. First, $\tau_{\theta r}$ is unlikely to be negligible. Second, when a significant fraction of the water in the tunnels is derived from the glacier surface, water fluxes vary diurnally. During periods of low flow, the water may not fill the tunnel, in which case melting will be concentrated low on the walls. Thus, with melting concentrated near the bed and closure inhibited there, the tunnel is likely to become broad and low (Fig. 8.13).

Closure of a broad, low conduit will be faster than that of a semicircular one with the same cross-sectional area. This is because the conduit roof receives much less support from the walls. To a first approximation, closure of a conduit such as that in Figure 8.13 can be described by equation (8.3) if $\frac{1}{2}(H + W/2)$ is used in place of r. Numerical modeling with the use of the finite element

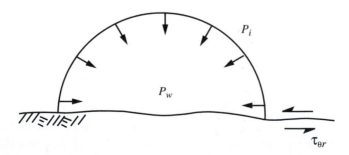

Figure 8.12   Some of the stresses around a semicircular subglacial tunnel.

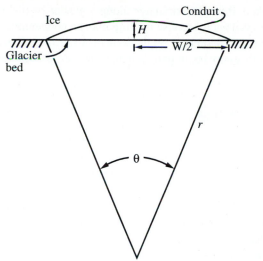

Figure 8.13 Geometry of an idealized broad, low subglacial conduit.

technique suggests that for conduits ~0.1 m high, this will overestimate closure rates by ~10% for $W \approx 5$ m, and up to 50% for $W \approx 25$ m.

When equation (8.24) is modified to apply to such broad, low tunnels, the agreement of theory with field measurements is greatly improved. However, this is, at least in part, because the angle θ (Fig. 8.13) can be varied. The results of a calculation for Storglaciären, in which good agreement between theory and observation was obtained with $\theta = 2°$, are shown in Figure 8.14. In other cases studied, agreement was obtained with θ between 2° and 36° (Hooke et al., 1990).

The second problem with equation (8.24) is that it ignores the effect that longitudinal and shear stresses parallel to the conduit axis have on the closure rate. As mentioned briefly in Chapter 2 [equation (2.10)], and as we will explore in detail in Chapter 9, such stresses increase the rate of tunnel closure even though they do not act in the direction of the closure. An investigation of the extent to which this effect can account for the discrepancy between measured and calculated water pressures is needed.

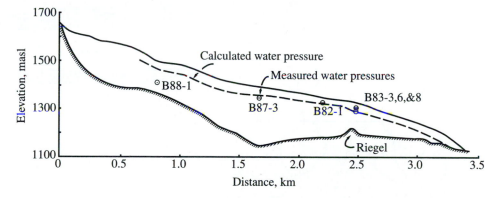

Figure 8.14 Longitudinal section of Storglaciären showing locations of boreholes, mean winter water levels in holes, and calculated hydraulic grade line. For several reasons, the water level in hole B88-1 is not thought to be representative of winter conditions. (From Hooke et al., 1990, Fig. 3. Reproduced with permission of the International Glaciological Society.)

In conclusion, it seems likely that when drainage along a glacier bed is in well-defined tunnels, the tunnels are broad and low. This is because melting may be concentrated low on the tunnel walls, and closure is inhibited there. Theoretical predictions of water pressures are also then more consistent with observation, but this may be, in part, because the theory does not include the effect of longitudinal stresses.

THE LINKED-CAVITY DRAINAGE SYSTEM

In some experiments on Variegated Glacier, Alaska, it was found that despite a water discharge, Q, of 5 m^3 s^{-1}, dye moved through the subglacial drainage system with a speed, v, of only 0.025 m s^{-1}. Suppose that this flow were all in a single conduit. Then, since $Q = v\mathbf{A}$, where $\mathbf{A}$ is the cross-sectional area of the conduit, $\mathbf{A}$ would have had to have been ~200 m^2. This presents a problem because, for any reasonable conduit roughness, n', equation (8.13) would then predict velocities that were one to two orders of magnitude higher than those observed.

Kamb (1987) suggested that the flow, rather than being in a single conduit, was in network of linked cavities (Fig. 8.15a). The cavities are believed to form in the lee of steps in the bed (Fig. 8.15b). Indeed, precipitates and the lack of striations in such locations on deglaciated bedrock surfaces argue strongly for their existence. The cavities are linked together by orifices that are much smaller in cross-sectional area than the cavities (Fig. 8.15b). The cavities provide the large $\mathbf{A}$ required, and the orifices throttle the flow, reducing the velocity.

An orifice formed in the lee of a step is shown in Figure 8.16. Diagram (a) shows the orifice geometry under certain basal-water-pressure and sliding-velocity conditions. Diagram (b) illustrates the geometry when heat released by viscous dissipation in the flowing water enlarges the orifice by melting the cavity roof. Kamb assumes that all of the heat is used to melt ice in the orifice in which

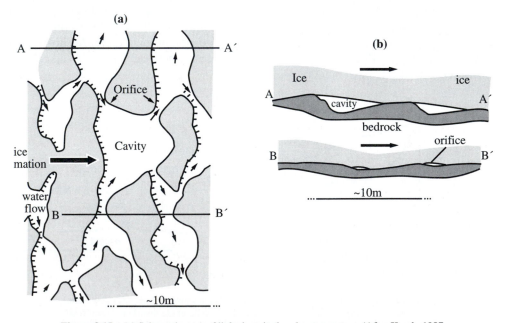

Figure 8.15   (a) Schematic map of linked-cavity basal water system. (After Kamb, 1987, Fig. 1.) (b) Vertical cross sections through linked-cavity system of *a*. Approximate scale is indicated by 10-m bar. (After Kamb, 1987, Fig. 2.)

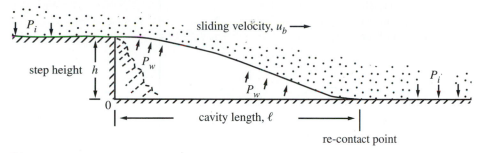

(a) Without Roof Melting

step height h

sliding velocity, u_b →

P_i

P_w

P_w

P_i

0

cavity length, ℓ

re-contact point

(b) With Roof Melting

h

melt back

cavity roof

u_b →

ℓ

Figure 8.16 Geometry of ideal orifice in lee of step in bed. **(a)** without roof melting. **(b)** with roof melting.

the heat is produced. As some of the heat will be advected into the next cavity, this will overestimate the orifice size. He also assumed that deformation of ice could be represented by a Newtonian flow law ($n = 1$) if the viscosity were chosen appropriately. (This assumption is commonly made when the problem is otherwise mathematically intractable.)

A critical parameter in Kamb's theoretical development is the orifice stability parameter, Ξ. For an orifice in the lee of a step of height h, Ξ is given by

$$\Xi = \frac{2^{\frac{1}{3}}}{\sqrt{\pi}} \frac{S^{\frac{1}{2}}}{\Lambda n'} \left(\frac{\mu}{u_b P_c} \right) h^{\frac{7}{6}} \tag{8.25}$$

where Λ is a constant factor involving the latent heat of fusion, μ is the equivalent Newtonian viscosity, u_b is the sliding speed, and the other factors are as defined previously. The hydraulic head, S, contains a correction for the sinuosity of the flow.

The cross-sectional shape of the orifice is governed by Ξ, as illustrated in Figure 8.17. With increasing Ξ, the orifice becomes increasingly arched until a point of instability is reached at a value of Ξ of about 1.0. In this figure, η is the height of the orifice, so the ordinate, η/h is the height of the orifice scaled to the height of the step. The orifice also becomes longer by about a factor of 3 as Ξ increases from 0 to 1.0, but this is not shown because the ordinate is scaled to the final length in each case.

Note that the cavity roof is less arched when u_b is high. This is because higher sliding speeds constantly replenish the ice. The cavity is also more arched when P_c is small ($\Rightarrow P_w$ large) because closure rates are reduced.

The instability of the orifice shape at $\Xi \approx 1$ is of considerable interest. Actually, the instability is of a mathematical nature only. It does not have a direct physical interpretation. However, it is a manifestation of a physical instability that Kamb has investigated further. He finds that if Ξ is high enough, say about 0.8, and that if P_w is increased enough to decrease P_c by about 20%, the length of the orifice,

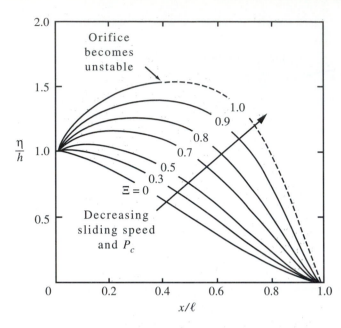

Figure 8.17 Steady-state configuration of orifice roof. (After Kamb, 1987, Fig. 8. Reproduced with permission of the author. Copyright by the American Geophysical Union.)

ℓ in Figure 8.16, will increase without bound. This may be the point at which a linked-cavity system is transformed into a tunnel system.

The way in which P_w changes with Q in a linked-cavity system is also of interest. We noted that decreasing Q in a tunnel system resulted in an increase in the steady-state pressure in the system (Fig. 8.18). However, in a linked-cavity system, decreasing Q has a relatively small effect on Ξ and an even smaller effect on the cross-sectional area of the orifices. This is because the orifice is, in effect, propped open by the step (Fig. 8.16). As a result, the lower discharge can be driven with a decreased head gradient (Fig. 8.18).

As noted, a consequence of the direct dependence of Q on P_w in the linked-cavity system is that a stable system of many interconnected cavities can exist. In a tunnel system, in contrast, the larger tunnels have lower pressures and thus capture the flow of the smaller ones.

Another important feature of Figure 8.18 is that for discharges in excess of ~0.1 m^3 s^{-1} and with orifices generated by step heights less than ~0.1 m, a much higher pressure is required to drive the flow in a linked-cavity system. This is because high pressures are required to open and maintain the orifices. High water pressures, of course, increase the speed of a glacier (Chapter 7).

TRANSITIONS BETWEEN CONDUIT AND LINKED-CAVITY SYSTEMS

Tunnels are commonly observed emerging from the margins of glaciers, and the rapidity with which dye poured into moulins often appears in outlet streams at the terminus (e.g., Hock & Hooke, 1993; Seaberg et al., 1988; among others) argues strongly for tunnel flow. However, as long as part of a glacier's movement is by sliding, linked cavities are certainly present, as ice must separate from the bed in the lee of at least some obstacles, and striations or joints in the rock will provide connections between resulting cavities. We thus need to investigate the conditions for stability of tunnels in the presence of cavities.

Fowler (1987, p. 265) and Raymond (unpublished, cited by Fowler, 1987) have studied this problem. Their approach is to consider what happens if, say, the pressure in the tunnel part of the

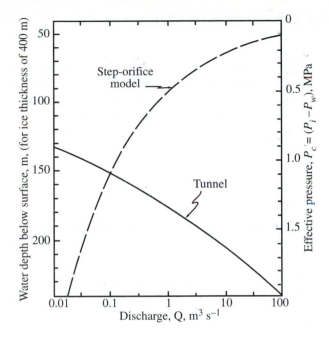

Figure 8.18 Relation between Q and P_c for tunnel and linked-cavity systems. (After Kamb, 1987, Fig. 12. Reproduced with permission of the author. Copyright by the American Geophysical Union.)

system is increased a small amount. This will lead to flow of water from the tunnels to the linked cavities, thus necessitating an increase in volume of the cavities, and hence in pressure in the cavities. If the increase in cavity pressure required is greater than the initial increase in pressure in the tunnel, the pressures will eventually equilibrate and the combined system will be stable.

Whether the change in pressure in the cavities satisfies this requirement for stability depends upon the sliding speed, bed geometry, water pressure, and water discharge in the conduits. Lower sliding speeds and lower water pressures tend to favor stability, whereas lower discharges favor collapse of the tunnels, with only the linked-cavity part of the system surviving.

Both the Kamb (1987) and the Fowler (1987) theories suggest that the limit of stability of the tunnel system is approached in winter when discharges are low, tunnels are thus shrinking, and water pressures are increasing. Locally, the tunnel system may collapse entirely, especially under thicker glaciers. Where remnants of the previous year's tunnel system are still present in the spring, however, they will increase in size rapidly as water inputs increase owing to the high melt rates on their walls and the high pressures that inhibit closure. If such remnant tunnels do not exist, or if they do not enlarge rapidly enough, the water pressure may rise to the level where Kamb's stability limit for the linked-cavity system is exceeded. New tunnels may then form by growth and coalescence of links. In either case, water pressures drop abruptly as the tunnel system is reestablished. Measurements of seasonal variations in water pressure (Hooke et al., 1989), of dye dispersion (Hock & Hooke, 1993; Seaberg et al., 1988), and of surface velocity (Hodge, 1974; Hooke et al., 1989) lend strong support to this model.

A reasonable supposition is that the linked-cavity and tunnel models represent end members of a continuum of drainage types. The drainage system beneath fast-moving glaciers, including those that are surging (discussed below), may be near the linked-cavity end member in this continuum. Conversely, ice sheets that are nearing stagnation, and thus beneath which there is little separation and cavity formation, may have drainage systems approximating the tunnel end member.

MULTIBRANCHED ARBORESCENT SYSTEM OF INDIVIDUALLY BRAIDED CONDUITS

A conduit system that has some of the characteristics of both the tunnel and the linked-cavity systems is one that consists of an arborescent network of broad, low conduits, each of which is individually braided (Fig. 8.19). Hock and Hooke (1993) called this a *multibranched arborescent system*. In contrast to the tunnel system, which often seems to be visualized as consisting of one or two main conduits, perhaps with tributaries normal to these conduits forming a trellis pattern, the multibranched arborescent system forms a dendritic pattern. However, as in the tunnel system, subglacial streams are inferred to flow in tunnels that are, in comparison with the linked-cavity system, relatively straight and uniform in size. In the multibranched arborescent system, as in the linked-cavity system, it is specifically recognized that no part of the glacier bed should be far from a conduit.

Hock and Hooke developed the idea of the multibranched arborescent system from tracer studies on Storglaciären. Dye was injected into a moulin about a kilometer from the terminus, and its arrival in one of the two streams draining the glacier, Sydjåkk (Fig. 8.19a), was recorded. As is commonly observed in tracer studies, the water velocities obtained from these tests were too low to be consistent with the existence of a single large tunnel from the moulin to the terminus. However, the velocities could be modeled rather well by assuming that the drainage system bifurcated several times between the terminus and the moulin, so that each of the highest tributaries of the system carries an average of only ~3% of the discharge at the terminus (Fig. 8.19a). This model is also consistent with the observation that many small streams and moulins are present on the surface at this location on the glacier, implying many sources.

That the individual channels are braided (Fig. 8.19b) is suggested by multiple peaks in some of the dye-return curves (Fig. 8.20). Such peaks occur when dye following one anabranch of a braided system moves more slowly than that in another anabranch. The two packages of dye then reach the terminus at different times.

The multibranched arborescent model is consistent with two basic observations: (1) that measured water velocities are much lower than they would be in a single tunnel; and (2) that water

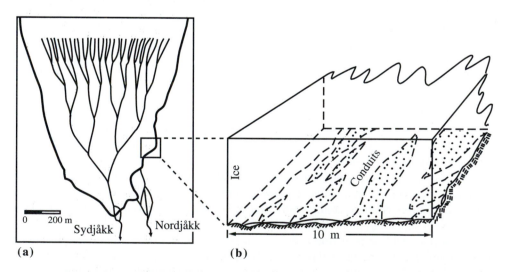

(a) **(b)**

Figure 8.19 Schematic sketches (**a**) of the multibranched arborescent conduit system postulated to exist beneath Storglaciären, and (**b**) of the individual channels in the system. (After Hock & Hooke, 1993, Figs. 4 and 6.)

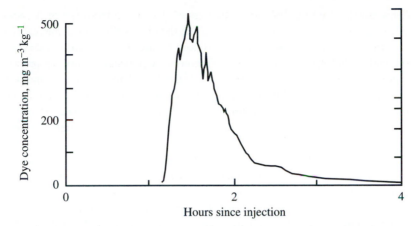

Figure 8.20 Typical dye-return curve from a test on Storglaciären. Ordinate is mg of dye per m^3 of water per kg of dye injected. (From Hock & Hooke, 1993, Fig. 3.)

pressures are relatively uniform over the bed, as observed by Iken and Bindschadler (1986) on Findelengletscher. The uniform drainage for all parts of the glacier bed provided by the multi-branched arborescent system should result in such relatively uniform water pressures. It is likely, however, that many of the smaller tributaries providing this uniform drainage are actually linked cavities.

In contrast to Kamb's (1987) model of the linked-cavity system, steady-state water pressures in the multibranched arborescent system probably decrease with increasing discharge. Winter water pressures beneath Storglaciären have, in fact, been modeled rather well (Fig. 8.14) with the use of Röthlisberger's theory (equation 8.24), modified to accommodate the upglacier decrease in conduit size. This test, however, is by no means definitive.

Even within the framework of the multibranched arborescent model, however, there can be appreciable differences among nearby glaciers. Iken and Bindschadler (1986, p. 110), for example, used electrical conductivity measurements in proglacial streams and other observations to study differences between the subglacial drainage of Findelengletscher and Gornergletscher. These two glaciers are both near Zermatt in the Swiss Alps and are of comparable size. They concluded that the drainage of Gornergletscher is in a few large tunnels, whereas that of Findelengletscher is in a larger number of smaller conduits. They did not propose an explanation for this difference.

SURGES

A surge is a rapid advance of a glacier, lasting a few months to a couple of years, that is unrelated to changes in mass balance. During a surge, the terminus might advance as much as a few kilometers at speeds of 10^1 to 10^2 m d^{-1}, and relatively stagnant ice in the terminus region may be overridden. As a result of the high strain rates, surges are accompanied by dramatic crevassing.

During a surge, a large amount of ice is transferred from a *reservoir area* which is usually, though not always, in the accumulation area, to a *receiving area* in the terminus region. Accordingly, the surface elevation in the reservoir area is drawn down and the receiving area thickens. Changes in thickness of tens of meters are common.

Surges are followed by a period of quiescence, lasting on the order of decades. During quiescence ice speeds are less than the balance velocity so the glacier thickens in the reservoir area and thins in the receiving area, thus becoming steeper. Before the resulting increase in driving stress can raise speeds to equal the balance velocity, however, another surge occurs. Thus, the process is periodic.

Surges can occur on glaciers resting on either hard beds composed mainly of bedrock, or on soft beds composed of till. Surges can also occur on glaciers that are, at least in part, frozen to their beds. Any complete theory of surging would have to accommodate all of these possibilities. While such a theory does not yet exist, it is likely that high basal water pressures play a role in all surges.

In the case of temperate glaciers on hard beds, the increase in thickness and speed during build up to a surge means that water pressures must rise higher to exceed the limits of stability of the linked-cavity system [equation (8.25)]. Surging may begin when this stability limit is not reached, so a tunnel system is not regenerated as water pressures rise in the winter or spring (Kamb, 1987). The resulting increase in sliding speed decreases the size of orifices (Fig. 8.17), thus further increasing P_w and hence u_b in a positive-feedback process.

According to this model, surging occurs when the glacier geometry is such that the linked-cavity system can persist for several weeks or months, the destabilizing effect of increases in water pressure being exceeded by the stabilizing effect of the increase in sliding speed. Eventually, however, owing either to further increases in water pressure or to changes in glacier geometry or both, the tunnel system is finally reestablished and the surge ends.

This model is consistent with observations leading up to and during the 1982/1983 surge of Variegated Glacier, Alaska, one of the best-studied examples of a surging glacier in the world (Kamb et al., 1985; Kamb, 1987). During the decade leading up to the surge, the glacier became thicker and steeper and the surface speed increased (Fig. 8.21a,b) (Raymond & Harrison, 1988). Calculated and measured rates of internal deformation indicate that the increase in u_s was largely a result of an increase in u_b. The acceleration to surge speeds started in the early winter (Fig. 8.21c) when decreasing water input is likely to have led to increasing water pressures, and the high sliding speed may have resulted in destruction of a tunnel system. Dye-trace data, mentioned earlier, suggest that the drainage system consisted largely of linked cavities at this time. The rate of acceleration gradually increased, presumably in the positive feedback process mentioned above. Eventually, large floods of dirty water emerged at the terminus and the surge ended. A dye-trace following the surge indicated that the drainage system had reverted to a tunnel configuration. Measurements of water pressure in a borehole confirm that P_w was within 0.5 MPa of the overburden pressure during the surge, with frequent fluctuations to within 0.15 MPa of overburden, and occasionally above it. Before and after the surge it was typically 0.8 to 1.6 MPa below the overburden pressure, which was 3.6 MPa at the site of the hole.

SUBGLACIAL CONDUITS ON DEFORMING TILL

Heretofore, our discussion of subglacial conduits has focused on situations in which the bed was comparatively rigid. Some glaciers, however, move over beds of deformable till. We found in Chapter 7 that the deformability of this material and the nature of the coupling between it and the glacier sole are both strongly dependent upon the effective pressure, $P_i - P_w$. In turn, the effective pressure depends, through P_w, on the nature of the drainage system at the ice–till interface.

As is the case with a conduit at the base of a glacier on a rigid bed, the roof of a conduit at an ice–till interface will tend to close when $P_w < P_i$, and in the steady state this tendency is balanced by

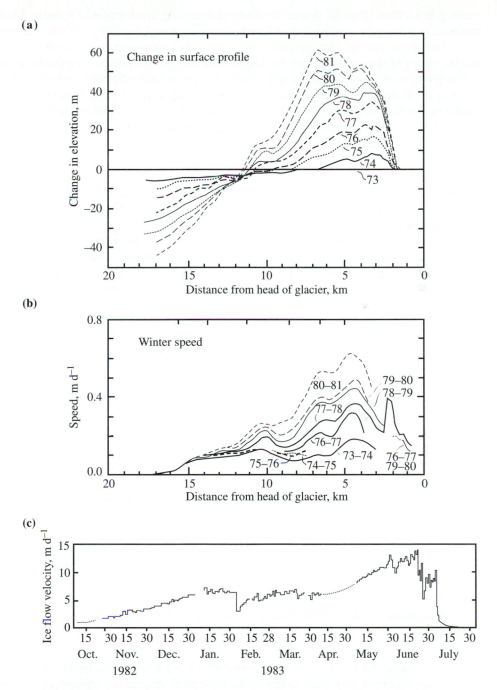

Figure 8.21 Evolution of the (**a**) surface profile and (**b**) speed of Variegated Glacier during its build-up to a surge (After Raymond and Harrison, 1988, Figs. 4 and 5b. Reproduced with permission of the authors and the International Glaciological Society). (**c**) Surface speed of the upper part of Variegated Glacier during its surge in 1982/83 (After Kamb et al., 1985, Fig. 2b. Reproduced with permission of the authors and the American Association for the Advancement of Science.)

melting. In addition, however, flow of till into the channel may tend to constrict it, and in the steady state this must be balanced by erosion of the till by the flowing water (Alley, 1989a; Walder & Fowler, 1994).

The physics of the latter processes are still poorly understood, as neither the rheology of till nor the theory of sediment transport are known well enough. Despite this, we can make some qualitative predictions about the nature of the drainage system under different conditions, based on an analysis by Walder and Fowler (1994). When P_w is high, flow of ice inward toward a conduit is inhibited, but flow of till is enhanced, and conversely. Walder and Fowler conclude that closure of the conduit by ice flow and till flow will be roughly equal when $P_i - P_w \approx 0.8$ MPa, but the uncertainty in this figure is quite large.

Ice sheets tend to have relatively low surface slopes. Potential gradients are thus low, so $\dot{m}$ is low. The conduit system then adjusts to provide a high P_w, thus inhibiting closure by ice flow. Deformation of till into conduits should then be enhanced. Conversely, valley glaciers normally have higher surface slopes, so $\dot{m}$ will be high and P_w lower, leading to conduit geometries controlled by the inward flux of ice.

We have already shown that when conduit closure by ice flow dominates in a tunnel system, P_w decreases as Q increases and water is thus diverted from smaller conduits to larger ones, leading to a dendritic drainage network. A key result of Walder and Fowler's analysis is that in conduits in which closure is predominantly by flow of till, P_w *increases* as Q increases. This is a direct result of their assumption that the depth of such a conduit is determined by the nature of the subglacial material, and is independent of the width. This assumption is open to question, however, as the depth scales with the width in alluvial rivers (Parker, 1979).

The consequences of a direct dependence of P_w on Q are quite interesting, however. Such a drainage system, like the linked-cavity system, will tend to be distributed or anastomosing as the higher water pressures should force water out of larger conduits into smaller ones. Landforms characteristic of dendritic drainage networks, such as eskers (see below), might not form; indeed, eskers are much less common in areas where ice sheets moved over thick till units (Clark & Walder, 1994). Instead, we might expect to find remnants of a distributed drainage system. In fact, gravel lenses are common in till sheets deposited by continental ice sheets, and these have been interpreted to be the beds of wide shallow subglacial conduits (Brown et al., 1987; Eyles et al., 1982).

In Chapter 5 we discussed ice streams, and we commented upon the fact that they appear to operate for periods of perhaps centuries, and then shut down, rather like an exceptionally long surge. We noted that ice streams rest on deformable beds and that their fast motion, despite very low driving stresses, must be due to high water pressures that either decouple the ice from the bed or allow the bed to deform. The reader will have a better appreciation for these concepts now that we have discussed water pressure, channels between ice and till, and till deformation in some detail.

SUBGLACIAL DRAINAGE PATHS AND THE FORMATION OF ESKERS

Traces of subglacial drainage paths are often preserved in the landscape. Eskers are one of the more common geomorphological features that define these paths. Eskers are sinuous ridges of gravel with undulating crests that were deposited by streams flowing in large subglacial tunnels. They are largest and best developed in areas that were covered by continental ice sheets during the late Pleistocene. Some are tens of meters in height and tens or hundreds of kilometers in length.

Eskers often appear to follow rather bizarre paths when viewed from the perspective of people accustomed to the courses of subaerial streams. Eskers climb hills, trend diagonally down valley sides, and run along valley sides instead of in valley bottoms. Shreve (1972, 1985a, 1985b) has shown that these characteristics can be readily understood from consideration of the hydraulic potential field beneath an ice sheet.

We noted (Fig. 8.6) that englacial water is expected to move in directions that are normal to equipotential planes in a glacier. Similarly, along the bed of a glacier water flow should be normal to the intersections of these equipotential planes with the bed. This is equivalent to saying that water flow down a hillslope should be normal to topographic contours, as topographic contours are the intersections of surfaces that are a constant height above sea level (≡ equipotential surfaces) with the topography.

Let us consider some examples of this. The solid lines in Figure 8.22a are topographic contours. They depict a gentle slope leading down to a valley that drains to the south. East of the valley there is a ridge that varies in elevation. Now, visualize the situation when an ice sheet covered the landscape, as shown in Figure 8.22b. The surface of the glacier sloped to the east, so the equipotential planes dipped westward. The dashed contours show the intersections of these planes with the landscape. These intersections are precisely analogous to the outcrop pattern that would be formed on the landscape by a westward-dipping sedimentary rock unit.

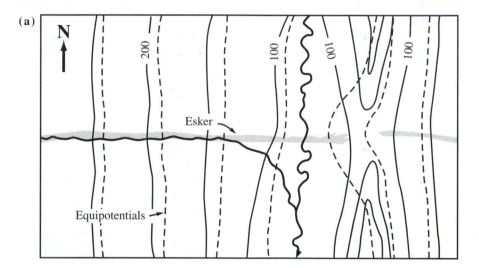

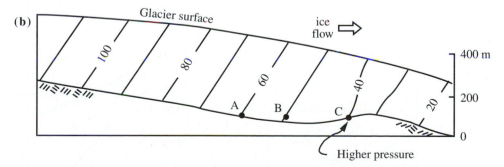

Figure 8.22 (a) Contour map of a landscape on which are superimposed contours of equipotential from a time when an ice sheet covered the landscape. (b) Topographic cross section from a time when ice was present showing the equipotential planes in the ice sheet.

Under subaerial conditions, creeks would run down the gentle slope on the west side of the area, and then turn south. However, when ice covered the area, subglacial water would not have turned south. Instead, because water would flow normal to the contours of equipotential, it would have been deflected toward the low point in the ridge. If such a subglacial stream could not carry all of the sediment delivered to it, we might now find an esker crossing the ridge at its lowest point. This is commonly observed in situations in which ridges cross the paths of eskers.

The equipotential planes in the vicinity of the ridge are distorted. This is because the ice is flowing and the pressure is thus higher on the stoss side of the ridge than on the lee side. To understand why the planes are distorted as shown, remember that in a situation in which z is constant, variations in potential are due to variations in P_w [equation (8.1)]. Thus, some distance away from the ridge a potential drop of 10 units might occur over the horizontal distance $\overline{AB}$, whereas near the stoss side of the ridge, where the pressure is elevated, a longer distance, $\overline{BC}$, would be required for the same drop. In the lee of the ridge, the distortion is in the opposite sense.

Because the velocity of water in the tunnel is proportional to $\partial\Phi/\partial s$ [equations (8.13) and (8.14)], this distortion of the potential field affects the velocity. In particular, where $\partial\Phi/\partial s$ is higher over the crest of the ridge, the velocity would be higher. This is consistent with the observation that eskers are commonly discontinuous across the crests of such ridges; the higher velocity flow there presumably inhibits deposition.

Another hypothetical situation is shown in Figure 8.23. Here, a topographic valley drains southeastward, diagonally across the direction of glacier flow. As a result, the trough in the equipotential contours is on the valley side rather than in the valley bottom, and this is where an esker would be found if conditions were otherwise suitable for its formation. Again, eskers are commonly found in such positions under these circumstances.

With an understanding of the physical processes that determine the locations of eskers in situations such as those in Figures 8.22 and 8.23, it is sometimes possible to determine the surface slope of the glacier beneath which the esker formed. As an example of this, consider the section of the Katahdin esker near the town of Medway, Maine, shown in Figure 8.24a. Ice flow was roughly from north to south in this area. In the northern part of the map, the two branches of the esker follow respective branches of the Penobscot River, but are slightly offset from the river, up onto the

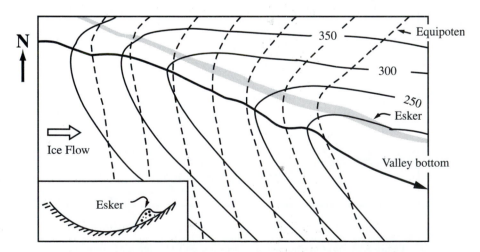

Figure 8.23 Topographic map of a valley trending diagonally across the direction of ice flow, showing how an esker formed in such a situation would be on the side of the valley.

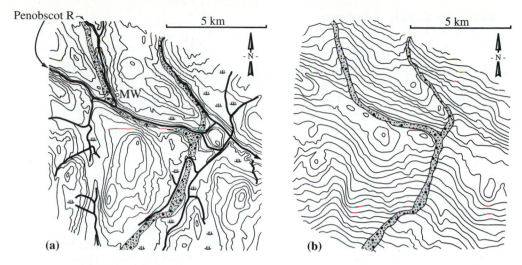

Penobscot R.

MW

5 km

5 km

-N-

-N-

(a)

(b)

Figure 8.24   (**a**) Map of the Penobscot River and Katahdin esker near Medway (MW), Maine. Near the middle of the map, the esker leaves the river valley and trends south-southwestward up a small tributary valley. (**b**) Map of equipotential contours beneath a glacier with a southward surface slope of 0.0048. The esker generally follows a trough in the potential surface. (After Shreve, 1985a. Reproduced with permission of the author and the Geological Society of America.)

valley sides, in the downglacier direction. However, south of the junction between the two branches, the esker departs from the valley of the Penobscot to run *up* the valley of a small tributary and then across the divide between this tributary and another small southward-flowing creek. To clarify the reasons for this, Shreve (1985a) constructed a series of maps of the potential field in the Medway area for different possible ice-surface slopes. The one that best explained the course of the esker (Fig. 8.24b) utilized a surface slope of 0.0048.

By determining ice-surface slopes in this way at a number of locations along a single esker system, one can reconstruct the surface profile of the ice sheet, and from this calculate the basal shear stress. Shreve (1985b) did this for the 150-km long Katahdin system (Fig. 8.25). His reconstruction, however, has been questioned because it is not certain that the entire system was active simultaneously. Nevertheless, in situations in which contemporaneity can be demonstrated, this is one of the few techniques available for determining the surface profiles of vanished ice masses.

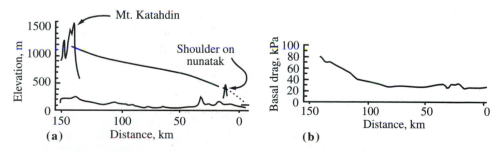

Figure 8.25   (**a**) Surface profile, and (**b**) basal shear stress of the part of the Laurentide Ice Sheet beneath which the Katahdin esker system formed. (After Shreve, 1985b. Reproduced with permission of the author and the University of Washington.)

Sediment Supply to Eskers

Eskers form where the sediment load delivered to a subglacial stream exceeds the transport capacity of the stream. The debris-laden basal ice of the glacier is one source of such sediment. As the energy dissipated by the flowing water melts this ice, the debris is released and an inward flow of ice toward the tunnel is induced.

This process can be seen in lithologic pebble counts from a section of the Great Pond esker in Maine, down-flow from a point where the esker crosses bedrock units of distinctive lithology (Van Beaver, 1971). The concentration of pebbles of these lithologies in the esker reaches a maximum about 3 km down-flow from the point where the esker crosses the units (Fig. 8.26). Had the stream been acquiring the pebbles directly from the bedrock—a difficult task at best once the esker began to develop on top of the rock—the concentration should have peaked at the down-flow edge of the unit. Rather, we infer that it was the glacier that eroded the pebbles from the bed and carried them along arcuate paths, as shown, until they were released into the stream.

Calculations suggest that this source of sediment is quite adequate to overload a subglacial stream, leading to deposition. For example, suppose the median (by weight) grain size in an esker is 0.05 m, and that the esker is forming in a tunnel beneath a glacier with a surface slope of 0.01. Then, based on equations for sediment transport in gravel-bedded streams (Parker, 1979), a conduit ~0.6 m high with a discharge of 0.9 $m^3 s^{-1}$ per meter of conduit width would be required to transport this material, and the sediment load would be at least 0.06 $m^3 d^{-1}$ per meter width. Under these conditions, the energy available for melting would be ~ 400 $J m^{-2} s^{-1}$ [equations (8.9) and (8.14)], and the melt rate on the conduit roof would be ~ 0.02 $m d^{-1}$. If the basal ice contained 10% debris by volume, the debris released by this melting would overload the stream after it flowed along the conduit for only 30 m ($0.02 \times 0.10 \times 30 = 0.06$ $m^3 d^{-1}$ per meter width).

In some eskers, however, some of the water and sediment load may have been derived from the glacier surface by way of moulins. For example, Mooers (1990) found eskers in central Minnesota that headed in conical hills of glaciofluvial gravel. He inferred that the hills were formed by sediment-

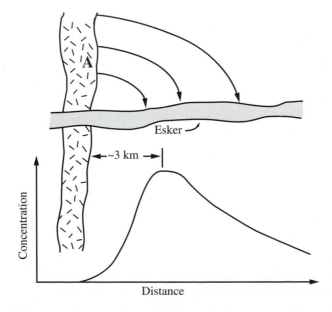

Figure 8.26 Schematic sketch showing variation in concentration of lithology **A** in an esker down-flow from the point where the esker crosses the outcrop of this lithology. Rocks are eroded by the glacier and carried along arcuate paths downglacier and inward toward tunnel.

laden supraglacial streams that reached the glacier bed through moulins and deposited a significant fraction of their load there before continuing to the margin through the subglacial conduits in which the esker formed.

Size and Location of Water Conduits on Eskers

It is natural to assume, as a first approximation, that the tunnel within which an esker formed was comparable in size to the esker (Fig. 8.27a). This is consistent with the observation that some eskers are composed of coarse gravel with a dearth of sedimentary structures. However, this may be a poor assumption in many instances, as the flux of water implied by a tunnel of this size would be horrendous. As basal melt rates are relatively low, such high fluxes would require either collection of water from a very large area of the bed or a surficial source. In some areas the former is improbable as the drainage area between the heads of eskers and inferred ice divides is insufficient. Likewise, surficial sources many tens of kilometers from the glacier margin are problematical, as near-surface ice temperatures are likely to be below 0°C at these elevations.

An alternative is that the tunnel was comparatively small (Fig. 8.27b). This might further suggest that the esker formed slowly over a period of many years.

If the tunnel is small compared with the size of the esker, it is also of interest to determine whether it is on top of the esker or low on the side (Fig. 8.27c). Lliboutry (1983) addressed this problem, using the following argument. If the height of the esker is Δh (Fig. 8.27), and there are two channels that are connected hydraulically, one on top of the esker and one on the side, then:

$$P_w^{side} = P_w^{top} + \rho_w g \Delta h$$
$$P_i^{side} = P_i^{top} + \rho_i g \Delta h \ . \tag{8.26}$$

The pressure causing tunnel closure is $P_c = P_i - P_w$, so subtracting the first of equations (8.26) from the second:

$$P_c^{side} = P_c^{top} - (\rho_w - \rho_i) g \Delta h \ .$$

Now, from equation (8.18), holding all other factors constant and using $n = 3$, we find that $Q \propto P_c^{12}$. Thus

$$\frac{Q^{top}}{Q^{side}} = \left[\frac{P_c^{top}}{P_c^{side}} \right]^{12} = \left[\frac{P_c^{side} + (\rho_w - \rho_i) g \Delta h}{P_c^{side}} \right]^{12} \ .$$

Thus, $Q^{top} > Q^{side}$ so the conduit on top of the esker will expand at the expense of that on the side. Phrased differently, owing to the nonlinearity of the flow law, the increase in potential closure rate as one moves down off the esker is not offset by the increase in P_w; hence, closure rates are higher in the conduit on the side, and water in it is forced up onto the top of the esker.

This analysis is valid as long as $u = \dot{m}$. However, near the terminus of a glacier where the ice is thin, $\dot{m}$ may exceed u, as we discussed in connection with Figure 8.11b. In this case, the conduit may

Figure 8.27   Esker of height Δh with (**a**) conduit comparable in size to esker; (**b**) conduit smaller than esker but on top of it; and (**c**) conduit low on side of esker.

slip off the side of the esker. In a valley called Atnedalen in Norway, there are several examples in which a small esker dropped down off the side of a larger one, paralleled it for a short distance, and then climbed back up onto the parent esker again (Fig. 8.28), presumably indicating that the ice thickness required to keep the conduit on top of the esker was becoming marginal. Closer to the ice margin, where the ice was thus thinner, these daughter eskers grew to nearly the same size as the parent. Then, still farther downstream, the esker ridges lost distinction and merged into a kettled outwash plain.

WATER PRESSURE FLUCTUATIONS AND GLACIER QUARRYING

Quarrying is an important process of glacier erosion. In quarrying, blocks of bedrock must first be loosened, either along preglacial joints or along fractures formed by subglacial processes. They then must be entrained by the glacier. Rapid water-pressure fluctuations within cavities in the lee of a bump on a glacier bed may play a role in both the fracture and entrainment processes (Iverson, 1989; Röthlisberger & Iken, 1981). Let us consider fracture first.

Water inputs to a glacier due to rain or melt can vary rapidly, causing subglacial cavities in the lees of bumps on the bed to fill and drain faster than they can adjust by flow of the ice. The resulting pressure fluctuations transfer the weight of the glacier first to and then from the tops of the bumps. Under 250 m of ice, for example, the pressure could vary from a relatively uniform 2.2 MPa on all faces of a bump to over 6 MPa, say, on the top, and nearly zero on the lee face. All rocks contain microcracks, fractions of a millimeter to a few millimeters in length, and such stress variations can lead to propagation of tensile fractures at the tips of these cracks when they are favorably oriented (Griffith, 1924). This can occur even at stresses well below the experimentally determined tensile strength of the rock (Atkinson, 1984; Atkinson & Rawlings, 1981; Segall, 1984). The likelihood of crack growth increases when the water pressure within cracks remains elevated while that in an adjacent cavity drops, or when stress corrosion resulting from repeated pressure changes reduces the strength of the rock (Iverson, 1991). Even higher and more concentrated stress differences can result when a cobble or boulder is dragged over a bump by the ice. Thus, it now seems safe to conclude that even sound crystalline rocks can be fractured subglacially through the action of ice and water, despite the fact that the ice is much weaker than the underlying rock.

Boulders with smooth stoss faces and plucked lee faces that are embedded in till, called *bullet boulders*, provide convincing field evidence for this process (e.g., Sharp, 1982). These boulders must have been transported by the ice and become lodged in the till as the basal ice melted. They

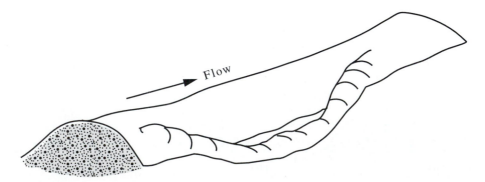

Figure 8.28   Sketch of small daughter esker diverging from and later rejoining parent esker. This may occur near the margin where the ice is thin and conduit closure rates are low.

would not have had their characteristic shape prior to lodgement, nor could they have been transported to their present location intact if they had preexisting fractures. Thus, their shape must have been produced by the overriding glacier after they became lodged.

Entrainment requires that bed-parallel forces tending to slide loosened blocks out of position exceed frictional forces tending to hold them in place (Iverson, 1989). Both are affected by fluctuations in water pressure. As noted earlier (Fig. 7.6), pressure-release freezing may occur on tops of bumps when increases in water pressure in cavities transfer part of the weight of a glacier away from the bumps (Robin, 1976), and similar cold patches can also develop owing to simple flow of the ice from the stoss side of a bump to its crest. Both processes increase the drag exerted by the ice on the block. The latter process is more effective at higher sliding velocities, so increases in subglacial water pressure that cause increases in sliding speed should increase its effectiveness.

Frictional forces resisting dislodgement of loosened blocks are reduced as water pressures rise (Iverson, 1989). This is because the normal pressure that ice exerts on a bedrock surface upglacier from a cavity is reduced, thus decreasing the friction along fractures that bound loosened blocks. In addition, once fractures are well-developed and in hydraulic communication with cavities, increases in water pressure in the fractures themselves reduce the effective pressure across fracture surfaces.

In summary, fluctuations in subglacial water pressure and associated transient changes in glacier sliding speed appear to be necessary for quarrying. Abrupt reductions in water pressure promote subglacial fracture, whereas increases, whether rapid or more gradual, promote the dislodgement of loosened blocks.

ORIGIN OF CIRQUES AND OVERDEEPENINGS

Cirques and overdeepened basins in glacier beds, such as those in Figure 8.29, are similar in form. Both have steep headwalls and both tend to have beds with adverse slopes. We will discuss the headwalls first.

Headwalls have jagged surfaces, apparently resulting from fracture and removal of blocks of rock. This morphology suggests that they are eroded by glacial quarrying. As we have just discussed, quarrying appears to be a result of water-pressure fluctuations on time scales of hours to days. These fluctuations seem to be most pronounced close to areas of water input (Hooke, 1991). In the case of cirques, the water input is localized by the bergschrund (the crevasse that forms between the headwall and the glacier), and in the case of overdeepenings, by crevasses that form over the convexities at their heads (Fig. 8.29). Thus, these water inputs and resulting pressure fluctuations occur at precisely the points where erosion is necessary to maintain the headwalls.

In the case of the headwall of an overdeepening, a positive feedback process appears to be operating. Crevassing over a minor convexity in the bed, an *initial perturbation* localizes water input and hence erosion. The erosion is concentrated on the downglacier side of the convexity. Thus, as erosion progresses, the convexity is amplified, resulting in further crevassing.

The other defining characteristic of cirques and overdeepenings is the gentle adverse slope of their beds. The steepness of this slope may be limited by the ability of water to flow along the bed. For example, if $k = 0.309$ and $\rho = 916$ kg m^{-3} in equation (8.12), it is easy to show [using equation (8.5)] that $\dot{m} = 0$ when $dz/ds = -1.7 \, dH/ds$. (This is left as an exercise for the reader.) In other words, when the adverse bed slope, dz/ds, is 1.7 times the surface slope, all of the energy dissipated in the flowing water is needed to warm the water to keep it at the pressure melting point as the ice thins in the downglacier direction [see equation (8.10) and discussion of Fig. 8.11c]. Where the adverse slope is steeper, equation (8.12) predicts, mathematically, that $\dot{m}$ should become negative, so water

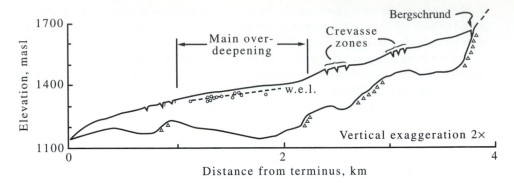

Figure 8.29 Longitudinal section of Storglaciären, Sweden, along an approximate flowline showing cirque, overdeepened basins, water-input points (crevasse zones and bergschrund), and inferred locations of quarrying (indicated by ΔΔΔ). w.e.l. = water equivalent line, circles (o) show heights of water in boreholes. (Modified from Hooke, 1991, Fig. 2.)

is freezing in the conduit, and equation (8.24) then predicts that $P_w > P_i$. Indeed, measured water pressures in the main overdeepening of Storglaciären are quite close to the overburden pressure (Fig. 8.29).

Actually, field data suggest that under these conditions, the water spreads out into a maze of small anastomosing conduits with flow velocities that are too low to move significant amounts of sediment. If the adverse slope is steep enough, *frazel ice* (platelets of ice that are carried by the water) may form and further inhibit flow. Then, the products of glacial erosion will not be flushed from beneath the glacier, and a layer of till is likely to accumulate. Continuity considerations suggest that this layer will increase in thickness until the downglacier mass transfer by deformation within it equals sediment production by erosion. Such a sediment layer would protect the bed throughout the downglacier reaches of an overdeepening, thus concentrating erosion at its head. Hooke (1991) postulated that this is why overdeepenings exist, and why their longitudinal profiles are characteristically asymmetrical, with the deepest point at their upglacier ends. Because the subglacial drainage system is disrupted in overdeepenings, water is forced to follow englacial conduits through them (Hooke & Pohjola, 1994).

SUMMARY

In this chapter we investigated the glacier hydraulic system, starting with the vein network along intersections among three ice crystals and progressing through the englacial to the subglacial system. We found that, theoretically, water flow in the englacial system should be normal to equipotential surfaces that dip upglacier at an angle roughly 11 times the slope of the glacier surface. Along the bed, water flow is normal to the intersections of these equipotential surfaces with the bed.

Water moving through conduit systems in or beneath glaciers releases viscous energy. This energy melts ice on conduit walls. However, the pressure in the water is generally less than that in the surrounding ice, so tunnels close by creep. In the steady state, the water pressure is everywhere adjusted so that $u = \dot{m}$. Along the bed u is inhibited by drag. In addition, melting is concentrated low on the walls when tunnels are not full of water. Thus, the steady-state conduit shape is likely to be broad and low.

The water pressure in conduits increases upglacier, approximately in proportion to the increase in ice thickness over the conduit. In conduit systems consisting of tunnels, P_w *decreases* as Q increases, but in subglacial conduit systems consisting of cavities linked together by orifices on the lee sides of bedrock steps, P_w *increases* as Q increases. Thus, tunnel systems are believed to be arborescent, whereas linked-cavity systems are braided. In all probability, these are end members of a continuum of drainage system types.

Tunnel systems are likely to collapse when discharges are low, tunnels are shrinking, and water pressures are increasing. If the tunnel system cannot regenerate when discharges increase again, water pressures may rise high enough to initiate a surge.

In conduits between a glacier and a deformable bed, the steady-state condition is one in which, in addition to $u = \dot{m}$, erosion of sediment by the flowing water must balance the flux of sediment into the conduit. Under these conditions, conduit systems are likely to be braided. Analysis suggests that this condition is most likely to be met when glacier surface slopes are low and water pressures high, as is commonly the case beneath ice sheets. Beneath valley glaciers, water pressures are typically lower, so bed deformation is not as likely.

Finally, we investigated two problems of geomorphic interest: the formation and courses of eskers and the erosion of cirques and overdeepenings. Many esker characteristics can be understood in terms of gradients of hydraulic potential at the base of an ice sheet. Glacier erosion is greatly accelerated by fluctuations in water pressure at the bed.

9

Stress and Deformation

In this chapter we will derive general equations for calculating the force per unit area, or traction, on a plane that is not parallel to the coordinate axes, and then use these equations to determine the orientation of the plane on which tractions are a maximum. We will see how this leads to the concept of the *invariant* of a tensor and show that this provides the fundamental basis for Glen's flow law. We then derive the stress equilibrium equations.

In the second half of the chapter we derive expressions for strain rates in terms of velocity derivatives, and we develop some relations based on these expressions and some other basic equations. This will set the stage for calculating stresses and velocities in a very simple ice sheet, consisting of a slab of ice of uniform thickness on a uniform slope (Chapter 10), and for investigating some more realistic problems (Chapter 11).

STRESS

While we have been referring to stresses and strain rates throughout the last few chapters, we will now enter into a much more detailed discussion, involving the tensor properties of these quantities. The reader may find it helpful, therefore, to review the section on Stresses and Strain Rates in Chapter 2.

General Equations for Transformation of Stress in Two Dimensions

Consider a domain in a slab of material of unit thickness (measured normal to the page) as shown in Figure 9.1. Stresses are uniformly distributed over the domain; in terms of the x-y coordinate system shown, they are σ_{xx} and σ_{yx} in the x-direction, and σ_{yy} and σ_{xy} in the y-direction. Forces on any small element of the domain of unit size in the x- and y-directions must be in equilibrium if there is to be no tendency for this element to rotate. Thus, $\sigma_{xy} = \sigma_{yx}$, so we will use σ_{xy} to represent both. Now cut the domain along the plane $\overline{AC}$, which makes an angle θ with the y-axis. This plane has an area dA. As a consequence of the stress field in the slab, the edges of the cut will have a tendency to move with respect to one another. We will ignore the part of the domain to the right of the cut, and ask what forces must be applied on dA to balance this tendency. Specifically, we wish to find the stress vectors $\sigma_N(\theta)$ and $\sigma_S(\theta)$ on this plane, where the subscripts N and S refer to normal and shear, respectively.

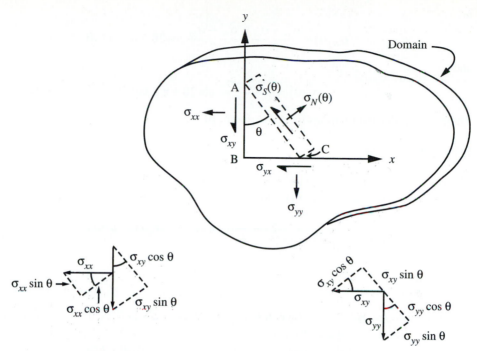

Figure 9.1   Stresses on a prism of material isolated from a domain.

To do this, we consider the prism ABC, and sum forces on it that act normal (F_N) and parallel (F_S) to dA, remembering that a force is a stress times an area, and set the sums equal to 0, the condition for static equilibrium. Note that surface $\overline{AB}$ has area $dA \cos \theta$ and surface $\overline{BC}$ has area $dA \sin \theta$. The force summation yields

$$\sum F_N = \sigma_N(\theta)\,dA - \sigma_{xx}\cos^2\theta\,dA - \sigma_{xy}\sin\theta\cos\theta\,dA - \sigma_{xy}\cos\theta\sin\theta\,dA - \sigma_{yy}\sin^2\theta\,dA = 0$$

and

$$\sum F_s = \sigma_S(\theta)\,dA + \sigma_{xx}\sin\theta\cos\theta\,dA - \sigma_{xy}\cos^2\theta\,dA + \sigma_{xy}\sin^2\theta\,dA - \sigma_{yy}\cos\theta\sin\theta\,dA = 0.$$

Simplifying results in

$$\sigma_N(\theta) = \sigma_{xx}\cos^2\theta + \sigma_{yy}\sin^2\theta + \sigma_{xy}(2\sin\theta\cos\theta)$$

and

$$\sigma_S(\theta) = -\frac{1}{2}\left(\sigma_{xx} - \sigma_{yy}\right)(2\sin\theta\cos\theta) + \sigma_{xy}\left(\cos^2\theta - \sin^2\theta\right).$$

These relations may be simplified with the use of the trigonometric identities

$$\sin 2\theta = 2\sin\theta\cos\theta$$

and

$$\cos 2\theta = \cos^2\theta - \sin^2\theta = 2\cos^2\theta - 1 = 1 - 2\sin^2\theta$$

to yield

$$\sigma_N(\theta) = \sigma_{xx}\frac{(1+\cos 2\theta)}{2} + \sigma_{yy}\frac{(1-\cos 2\theta)}{2} + \sigma_{xy}\sin 2\theta$$

or

$$\sigma_N(\theta) = \frac{\sigma_{xx} + \sigma_{yy}}{2} + \frac{\sigma_{xx} - \sigma_{yy}}{2}\cos 2\theta + \sigma_{xy}\sin 2\theta \tag{9.1}$$

and

$$\sigma_S(\theta) = -\frac{\sigma_{xx} - \sigma_{yy}}{2}\sin 2\theta + \sigma_{xy}\cos 2\theta. \tag{9.2}$$

These are the desired relations for $\sigma_N(\theta)$ and $\sigma_S(\theta)$.

Principal Stresses

We now wish to find the orientation, θ, of the plane on which $\sigma_N(\theta)$ is either a maximum or minimum. Take the derivative of equation (9.1) with respect to θ and set the result equal to 0; thus,

$$\frac{\partial \sigma_N(\theta)}{\partial \theta} = -2\frac{\sigma_{xx} - \sigma_{yy}}{2}\sin 2\theta + 2\sigma_{xy}\cos 2\theta = 0 \tag{9.3a}$$

or

$$\tan 2\theta = \frac{2\sigma_{xy}}{(\sigma_{xx} - \sigma_{yy})}. \tag{9.3b}$$

This equation may be satisfied by either of two values of 2θ, $180°$ apart. Thus, there are two solutions for θ that are $90°$ apart. One is the plane of maximum $\sigma_N(\theta)$ and the other is the plane of minimum $\sigma_N(\theta)$. We call the stresses acting in these directions the *principal stresses*. This is an important concept to understand, and we will return to it frequently.

The magnitude of the principal stresses is obtained by substituting for 2θ in equation (9.1). Equation (9.3b) and the diagram in Figure 9.2 are used to get expressions for $\cos 2\theta$ and $\sin 2\theta$. The result is

$$\sigma_{1,2} = \frac{\sigma_{xx} + \sigma_{yy}}{2} + \frac{\sigma_{xx} - \sigma_{yy}}{2}\frac{\sigma_{xx} - \sigma_{yy}}{\sqrt{(\sigma_{xx} - \sigma_{yy})^2 + 4\sigma_{xy}^2}} + \sigma_{xy}\frac{2\sigma_{xy}}{\sqrt{(\sigma_{xx} - \sigma_{yy})^2 + 4\sigma_{xy}^2}}$$

or

$$\sigma_{1,2} = \frac{\sigma_{xx} + \sigma_{yy}}{2} \pm \frac{1}{2}\sqrt{(\sigma_{xx} - \sigma_{yy})^2 + 4\sigma_{xy}^2} \tag{9.4}$$

where σ_1 is σ_{Nmax} and σ_2 is σ_{Nmin}. Thus, $(\sigma_1 + \sigma_2) = (\sigma_{xx} + \sigma_{yy})$.

Comparing equations (9.2) and (9.3a), it will be seen that $\dfrac{\partial \sigma_N(\theta)}{\partial \theta} = 2\sigma_S(\theta)$. Thus, when $\dfrac{\partial \sigma_N(\theta)}{\partial \theta} = 0$, $\sigma_S(\theta) = 0$. This is another important principle: *Shear stresses vanish on planes on which the normal stresses are a maximum or minimum.*

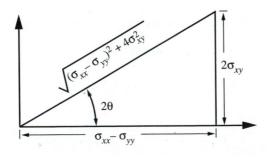

Figure 9.2   Illustration of relation among $(\sigma_{xx} - \sigma_{yy})$, σ_{xy}, and 2θ in equation (9.3b).

The orientations and magnitudes of the maximum shear stresses can be obtained in a similar manner. This is left as an exercise for the reader.

Mohr's Circle

A convenient way to illustrate the dependence of σ_{xx}, σ_{yy}, and σ_{xy} on 2θ is to use a graphical construction known as Mohr's circle (Fig. 9.3). To construct the figure:

1. Draw a rectangular coordinate system with normal stresses (σ_N) on the horizontal axis and shear stresses (σ_S) on the vertical axis, and plot points A and A′ at (σ_{xx}, σ_{xy}) and (σ_{yy}, $-\sigma_{xy}$), respectively.

2. Connect points A and A′ with a straight line, and draw a circle with B as the center and passing through A and A′.

In this figure, $\overline{BE} = (\sigma_{xx} - \sigma_{yy})/2$ so the radius of the circle is

$$\sqrt{\left(\frac{\sigma_{xx} - \sigma_{yy}}{2}\right)^2 + \sigma_{xy}^2} = \frac{1}{2}\sqrt{\left(\sigma_{xx} - \sigma_{yy}\right)^2 + 4\sigma_{xy}^2}.$$

Thus, from equation (9.4) the magnitudes of σ_1 and σ_2 are represented by the lengths of lines $\overline{OD}$ and $\overline{OC}$, respectively, and angle ABD is 2θ.

Invariants of a Tensor

Regardless of the orientation of the axes in Figure 9.1, the magnitudes and orientations of σ_1 and σ_2 cannot change as long as the overall stress field does not change. This is because σ_1 and σ_2 are functions of the state of stress in the domain and not of θ. We now use this fact and Mohr's circle to illustrate another fundamental principle.

Because the magnitudes of σ_1 and σ_2, as represented by $\overline{OD}$ an $\overline{OC}$, respectively, determine the size of the circle and its position on the σ_N-axis, the size and position will not change as θ varies. However, σ_{xx}, σ_{yy}, and σ_{xy} will vary. This is represented by movement of points A and A′ around the circle. On the other hand, $(\sigma_{xx} + \sigma_{yy})/2$ and $[(\sigma_{xx} - \sigma_{yy})^2 + 4\sigma_{xy}^2]^{1/2}$ (represented by $\overline{OB}$ and the radius of the circle, respectively) do not change. These two quantities are thus independent of the orientation of the axes, or θ; they are known as the *invariants* of the tensor.

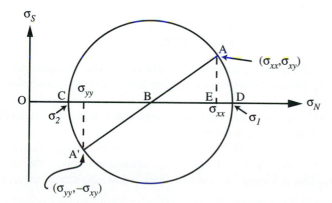

Figure 9.3   Mohr's circle.

Phrased in terms of equation (9.4), σ_1 and σ_2 will remain constant and independent of θ only if the quantities $(\sigma_{xx} + \sigma_{yy})/2$ and $[(\sigma_{xx} - \sigma_{yy})^2 + 4\sigma_{xy}^2]^{1/2}$ are independent of θ. Thus, these two quantities must be invariant.

Extension to Three Dimensions and Introduction of Deviatoric Stresses

To a first approximation it has been found empirically that deformation of ice subjected to a normal stress is independent of the hydrostatic pressure or mean stress, a fact that might be anticipated from the observation that ice is nearly incompressible. In three dimensions, the mean stress is given by

$$P = \frac{1}{3}\left(\sigma_{xx} + \sigma_{yy} + \sigma_{zz}\right). \tag{9.5}$$

This empirical result motivates us to define *deviatoric normal stresses*, denoted by primes, by $\sigma'_{xx} = \sigma_{xx} - P$, $\sigma'_{yy} = \sigma_{yy} - P$, and $\sigma'_{zz} = \sigma_{zz} - P$, or

$$\sigma'_{ij} = \sigma_{ij} - \frac{1}{3}\delta_{ij}\sigma_{kk} \qquad i, j = x, y, z. \tag{9.6}$$

In equation (9.6) we have introduced the Kronecker δ; δ_{ij} takes the values

$$\begin{aligned}\delta_{ij} &= 1 \qquad i = j \\ \delta_{ij} &= 0 \qquad i \neq j\end{aligned}$$

We have also introduced the summation convention. Whenever two subscripts are repeated in the same term, as in σ_{kk}, that term is summed over all possible combinations of the subscripts. Equation (9.6) therefore represents nine equations, of which three are identical due to the symmetry of the tensor. Two of the nine are

$$\sigma'_{xx} = \sigma_{xx} - \frac{1}{3}\left(\sigma_{xx} + \sigma_{yy} + \sigma_{zz}\right)$$

$$\sigma'_{xy} = \sigma_{xy}.$$

As you see, deviatoric shear stresses are identical to their nondeviatoric (or total) counterparts. Only the normal stresses are different.

If we were to go through a derivation similar to that on pages 152–155 in three dimensions (Johnson and Mellor, 1962, pp. 23–25), we would find there were three invariants, having the form

$$\begin{aligned}J_1 &= \sigma'_{xx} + \sigma'_{yy} + \sigma'_{zz} \\ J_2 &= \sigma'^2_{xy} + \sigma'^2_{yz} + \sigma'^2_{zx} - \sigma'_{xx}\sigma'_{yy} - \sigma'_{yy}\sigma'_{zz} - \sigma'_{zz}\sigma'_{xx} \\ J_3 &= \sigma'_{xx}\sigma'_{yy}\sigma'_{zz} + 2\sigma'_{xy}\sigma'_{yz}\sigma'_{zx} - \sigma'_{xx}\sigma'^2_{yz} - \sigma'_{yy}\sigma'^2_{xz} - \sigma'_{zz}\sigma'^2_{xy}\end{aligned} \tag{9.7}$$

[If total stresses were used instead of deviatoric stresses, the right-hand sides of equations (9.7) would look the same, except for the primes. However, on the left-hand sides, by convention, we would use I rather than J.]

The first invariant, J_1, can readily be shown to be 0. Just use equation (9.6) to express the deviatoric stresses in terms of their total counterparts, and simplify.

The second invariant has taken on considerable significance in glaciology. It is, in fact, a basic constituent of Glen's flow law, $\dot{\varepsilon}_e = (\sigma_e/B)^n$, as both $\dot{\varepsilon}_e$ and σ_e are defined in terms of the second invariants of the respective tensors. This is because, as we will show below, the second invariant is related to a yield criterion.

Before discussing the yield criterion, it is useful to develop a more convenient expression for σ_e (and by extension $\dot{\varepsilon}_e$) in the flow law. To do this, square the first of equations (9.7); thus,

$$J_1^2 = \sigma'^{\,2}_{xx} + \sigma'^{\,2}_{yy} + \sigma'^{\,2}_{zz} + 2\left(\sigma'_{xx}\sigma'_{yy} + \sigma'_{yy}\sigma'_{zz} + \sigma'_{zz}\sigma'_{xx}\right) = 0 \qquad (9.8a)$$

This expression equals zero because $J_1 = 0$, so we have:

$$\sigma'^{\,2}_{xx} + \sigma'^{\,2}_{yy} + \sigma'^{\,2}_{zz} = -2\left(\sigma'_{xx}\sigma'_{yy} + \sigma'_{yy}\sigma'_{zz} + \sigma'_{zz}\sigma'_{xx}\right). \qquad (9.8b)$$

Substituting this into the expression for J_2 yields

$$2J_2 = \sigma'^{\,2}_{xx} + \sigma'^{\,2}_{yy} + \sigma'^{\,2}_{zz} + 2\sigma'^{\,2}_{xy} + 2\sigma'^{\,2}_{yz} + 2\sigma'^{\,2}_{zx}$$

or, using the summation convention:

$$J_2 = \frac{1}{2}\left(\sigma'_{ij}\sigma'_{ij}\right). \qquad (9.9)$$

We define $\sigma_e = [\frac{1}{2}(\sigma'_{ij}\,\sigma'_{ij})]^{1/2}$ as the *effective shear stress*. The *effective shear strain rate* is defined identically, except that σ' is replaced with $\dot{\varepsilon}$. As noted in Chapter 2 [equations (2.10) and (2.11)], the σ_e and $\dot{\varepsilon}_e$ in Glen's flow law are the effective shear stress and shear strain rate. Thus, in the most general case, both of these symbols in the flow law represent a summation of nine terms.

A Yield Criterion

A yield criterion is a statement of the conditions under which deformation will occur. If the condition is not met, there is no deformation, and conversely. The simplest imaginable yield criterion is that of Tresca (1864):

$$|\sigma_k - \sigma_\ell| = \text{const} \qquad k, \ell = 1, 2, 3$$

or when the difference between any two principal stresses exceeds a material constant (determined experimentally for any given material) yielding will occur. An alternative, the von Mises (1913) yield criterion, is

$$(\sigma_1 - \sigma_2)^2 + (\sigma_2 - \sigma_3)^2 + (\sigma_3 - \sigma_1)^2 = \text{const.}$$

In this case, each one of the principal stresses contributes.

Let us investigate the relation between the von Mises criterion and J_2. After some manipulation we can obtain

$$(\sigma_1 - \sigma_2)^2 + (\sigma_2 - \sigma_3)^2 + (\sigma_3 - \sigma_1)^2 = 2\left(\sigma_1'^2 + \sigma_2'^2 + \sigma_3'^2\right) + 2J_2 \qquad (9.10)$$

where the primes denote deviatoric stresses as before. Note that we started with total stresses on the left side. Had we started with deviatoric stresses, we would have obtained the same result, as P drops out. Thus, *the yield criterion is unchanged if we use deviatoric stress instead of total stress* Expanding equation (9.9) and noting that the shear stresses vanish because we are here dealing with principal stresses, we find that the term in parentheses on the right-hand side of equation (9.10) is equal to $2J_2$. Thus, the von Mises yield criterion reduces to $6J_2 = \text{const}$, or since $J_2 = \sigma$ we have $\sigma_e = \text{const.}$

Yield criteria are often associated with perfect plasticity. A perfectly plastic material does not deform at stresses below its yield strength, k. However, once the applied stress reaches k, the material begins to deform, and it deforms at a rate such that the stress does not exceed k (Fig. 9.4). In terms of Glen's flow law, a perfectly plastic material would be represented by $n \to \infty$, so there would be no strain until σ_e equaled B where B would be related to the constant in the von Mises yield criterion.

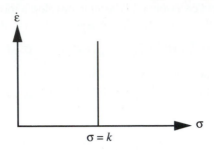

Figure 9.4 Variation of stain rate, $\dot{\varepsilon}$, with applied stress, σ, in a perfectly plastic material.

In fact, ice is not perfectly plastic, although there may, indeed, be a stress below which it does not deform. Thus, by using σ_e in the flow law, we are simply saying that we think the strain rate in any given direction is a function of all of the stresses acting on the material and not of just the stresses in that direction. For example, the flow law states that ice will shear faster under a stress σ_{xy} if there is also a deviatoric normal stress, σ'_{xx}, on it. Although some experimental work questions this conclusion (e.g., Baker, 1987), recent experiments by Li et al. (1996) firmly support it.

The Yield Criterion in Plane Strain

Let us now examine the yield criterion in plane strain. By *plane strain* we mean that there is no deformation in one of the coordinate directions, in this case the z-direction. As deformation is caused by deviatoric stresses, this implies that σ'_{zz}, σ'_{xz}, and σ'_{yz} are all 0. Thus, expanding J_2 yields

$$\sigma_e^2 = \frac{1}{2}\left(\sigma'^2_{xx} + 2\sigma'^2_{xy} + \sigma'^2_{yy} + 2\sigma'_{xx}\sigma'_{yy} - 2\sigma'_{xx}\sigma'_{yy}\right)$$

where the last two terms (that cancel each other) have been added to complete the square. Thus,

$$\sigma_e^2 = \frac{1}{2}\left[\left(\sigma'_{xx} - \sigma'_{yy}\right)^2 + 2\sigma'^2_{xy} + 2\sigma'_{xx}\sigma'_{yy}\right].$$

Changing to total stress by substituting $\sigma'_{ij} = \sigma_{ij} - P$, yields

$$\sigma_e^2 = \frac{1}{2}\left[\left(\sigma_{xx} - \sigma_{yy}\right)^2 + 2\sigma_{xy}^2 + 2\left(\sigma_{xx}\sigma_{yy} - \sigma_{xx}P - \sigma_{yy}P + P^2\right)\right].$$

Because $\sigma'_{zz} = 0 = \sigma_{zz} - P$, we find that P reduces to $\frac{1}{2}(\sigma_{xx} + \sigma_{yy})$. Therefore, after some manipulation we obtain

$$\sigma_e = \pm\frac{1}{2}\sqrt{\left(\sigma_{xx} - \sigma_{yy}\right)^2 + 4\sigma_{xy}^2}, \qquad (9.11)$$

which is, in fact, the maximum shear stress in plane strain. You can show this by setting the derivative of equation (9.2) equal to 0. If the axes were chosen to be parallel to the principal stresses, σ_{xy} would vanish, leaving

$$\sigma_e = \pm\frac{1}{2}(\sigma_1 - \sigma_2).$$

The directions of the maximum shear stresses then turn out to be 45° and 135° to the principal axes. [To show this, use a procedure similar to that which we used to obtain equations (9.3a and 9.3b) to determine the orientation of the planes on which $\sigma_S(\theta)$ is a maximum, and compare these with the orientations of the planes on which $\sigma_N(\theta)$ is a maximum].

The right-hand side of equation (9.11) will be recognized as the second term on the right-hand side of equation (9.4). Thus, we have shown that the invariants we discussed on page 155 are the two-dimensional equivalents of the first two of those in equations (9.7). (It can further be shown that the third invariant equals the second in two dimensions.)

STRESS EQUILIBRIUM

The stress-equilibrium equations are derived by balancing forces in the directions of the coordinate axes. Consider forces in the x-direction on a block of size $dx\, dy\, dz$ as shown in Figure (9.5):

$$\sum F_x = -\sigma_{xx} dy\, dz + \left(\sigma_{xx} + \frac{\partial \sigma_{xx}}{\partial x} dx\right) dy\, dz - \sigma_{yx} dx\, dz + \left(\sigma_{yx} + \frac{\partial \sigma_{yx}}{\partial y} dy\right) dx\, dz$$

$$-\sigma_{zx} dx\, dy + \left(\sigma_{zx} + \frac{\partial \sigma_{zx}}{\partial z} dz\right) dx\, dy + \rho g_x dx\, dy\, dz = 0.$$

The first two terms on the right are the normal forces on the faces of the block that are normal to the x-axis. Note that in each case, the stress (shown in Fig. 9.5) is multiplied by the area of the face, in this case $dy\, dz$, to obtain a force. The next four terms are the shear forces in the x-direction on faces normal to the y- and z-axes, respectively. The last term is the body force; g_x in this term represents the component of the gravitational acceleration that is parallel to the x-axis. Canceling like terms of opposite sign and dividing by $dx\, dy\, dz$ yields

$$\frac{\partial \sigma_{xx}}{\partial x} + \frac{\partial \sigma_{yx}}{\partial y} + \frac{\partial \sigma_{zx}}{\partial z} + \rho g_x = 0.$$

Similar expressions are readily obtained in the y- and z-directions. Using the summation convention, these become

$$\frac{\partial \sigma_{ij}}{\partial x_i} + \rho g_j = 0. \tag{9.12}$$

As i is repeated in the first term, this represents three terms. However, j is not repeated, so we can write separate equations for $j = x$, y, and z. Thus, equation (9.12) represents three equations.

DEFORMATION

Having applied a stress to a medium, we expect strain or deformation to occur. Suppose $\bar{x}$, $\bar{y}$ and $\bar{z}$ (Fig. 9.6) represent the displacement of a particle from P to P′, respectively, in the directions of the coordinate axes. We will consider infinitesimal displacements so that the time required for the deformation $\to 0$.

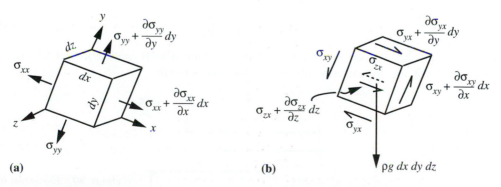

(a) (b)

Figure 9.5 Stresses on a block of size $dx\, dy\, dz$. (a) Normal stresses. (b) Shear stresses.

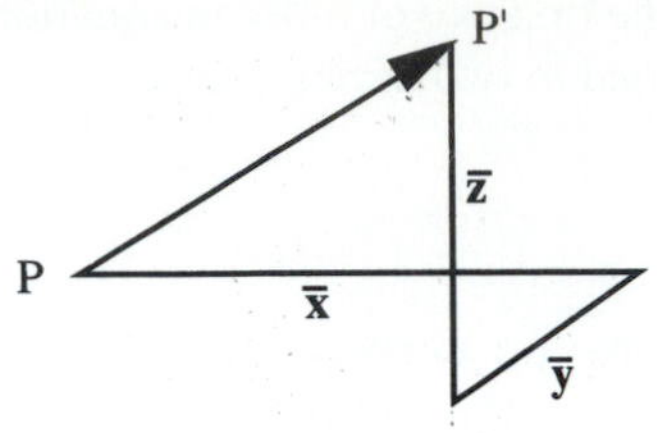

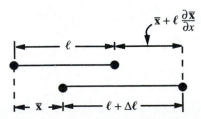

Figure 9.6 Components of a displacement from P to P′.

Normal strain in the x-direction at P is defined as

$$\varepsilon_{xx} = \mathrm{Lim}_{\ell \to 0} \frac{\Delta\ell}{\ell}, \tag{9.13}$$

where ℓ is the length of a line drawn in the x-direction, and $\Delta\ell$ is the elongation of that line, so $\Delta\ell/\ell$ is the elongation of the line per unit length. Referring to Figure 9.7, if a line, initially of length ℓ, is translated such that its left end moves a distance $\bar{x}$ in the x-direction, its right end moves a distance $\bar{x} + \ell(\partial\bar{x}/\partial x)$ in this direction, and the x-component of its new length is $\ell + \Delta\ell$, then substituting into equation (9.13) yields

$$\varepsilon_{xx} = \mathrm{Lim}_{\ell \to 0} \frac{\bar{x} + \ell\dfrac{\partial\bar{x}}{\partial x} - \bar{x}}{\ell} = \frac{\partial\bar{x}}{\partial x}. \tag{9.14}$$

By taking the limit as $\ell \to 0$ we eliminate the variation with x, thus obtaining ε_{xx} at point P. Similarly: $\varepsilon_{yy} = \partial\bar{y}/\partial y$ and $\varepsilon_{zz} = \partial\bar{z}/\partial z$.

Shear strain is defined as one half the decrease in an initially right angle. Referring to Figure 9.8, this can be expressed as

$$\varepsilon_{xy} = \frac{\Delta\theta}{2} = \frac{1}{2}\mathrm{Lim}_{\ell \to 0}\left[\tan^{-1}\frac{\ell(\partial\bar{x}/\partial y)}{\ell} + \tan^{-1}\frac{\ell(\partial\bar{y}/\partial x)}{\ell}\right].$$

For infinitesimal changes, $\theta \approx \tan\theta$ so in the limit:

$$\varepsilon_{xy} = \frac{1}{2}\left(\frac{\partial\bar{x}}{\partial y} + \frac{\partial\bar{y}}{\partial x}\right) \tag{9.15a}$$

and similarly

$$\varepsilon_{yz} = \frac{1}{2}\left(\frac{\partial\bar{z}}{\partial y} + \frac{\partial\bar{y}}{\partial z}\right) \tag{9.15b}$$

and

$$\varepsilon_{zx} = \frac{1}{2}\left(\frac{\partial\bar{x}}{\partial z} + \frac{\partial\bar{z}}{\partial x}\right). \tag{9.15c}$$

As before (Chapter 2), there are nine components of strain. This is another second rank tensor, the *strain tensor*. It, too, is symmetric because $\varepsilon_{xy} = \varepsilon_{yx}$ and so forth.

Figure 9.7 Elongation of a line during deformation.

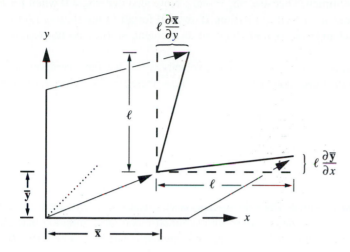

Figure 9.8 Change in a right angle during deformation.

In general, shear results in rotation as well as distortion. For example, if $\partial\bar{x}/\partial y \neq \partial\bar{y}/\partial x$ in Figure 9.8, the dotted line inclined at 45° to the x-axis will be rotated through an angle:

$$\omega_{xy} = \frac{1}{2}\left(\frac{\partial\bar{x}}{\partial y} - \frac{\partial\bar{y}}{\partial x}\right).$$

Similar expressions may be written for other rotations.

To obtain rates, which are normally of greater interest in a deforming ice mass, we differentiate with respect to time. Thus, the normal strain rate in the x-direction, $\dot{\varepsilon}_{xx}$, is

$$\dot{\varepsilon}_{xx} = \frac{d\varepsilon_{xx}}{dt} = \frac{d}{dt}\frac{\partial\bar{x}}{\partial x}.$$

Velocity is defined as a change in distance with time, or if u is the velocity in the x-direction, $u = d\bar{x}/dt$. Thus, we obtain

$$\dot{\varepsilon}_{xx} = \frac{\partial u}{\partial x}. \tag{9.16}$$

Similarly, shear strain rates become

$$\dot{\varepsilon}_{xy} = \frac{1}{2}\left(\frac{\partial u}{\partial y} + \frac{\partial v}{\partial x}\right) \tag{9.17}$$

and so forth.

The symmetry of equation (9.17) suggests the possibility of again using the summation convention to write expressions for the strain rates, thus;

$$\dot{\varepsilon}_{ij} = \frac{1}{2}\left(\frac{\partial u_i}{\partial x_j} + \frac{\partial u_j}{\partial x_i}\right). \tag{9.18}$$

If $i = j$ in this expression, it reduces to equation (9.16), so this formulation represents both shear and normal strain rates.

Similarly, the *rotation rate tensor* is

$$\dot{\omega}_{ij} = \frac{1}{2}\left(\frac{\partial u_i}{\partial x_j} - \frac{\partial u_j}{\partial x_i}\right). \tag{9.19}$$

The rotation rate tensor is antisymmetric because $\dot{\omega}_{ij} = -\dot{\omega}_{ji}$. Note also that $\dot{\omega}_{ij} = 0$ when $i = j$. In other words, pure stretching does not result in rotation. If $\dot{\omega}_{ij} = 0$ for all i,j the flow is said to be *irrotational*. Rotations do not change the size or shape of an element, so they do not require the application of a stress.

As implied by the notation $\partial u_i / \partial x_j$ $(i, j = x, y, z)$, the velocity vector has three components, and each of them can vary in each of the three coordinate directions. Thus, there are nine velocity derivatives:

$$\frac{\partial u}{\partial x} \quad \frac{\partial u}{\partial y} \quad \frac{\partial u}{\partial z}$$

$$\frac{\partial v}{\partial x} \quad \frac{\partial v}{\partial y} \quad \frac{\partial v}{\partial z}$$

$$\frac{\partial w}{\partial x} \quad \frac{\partial w}{\partial y} \quad \frac{\partial w}{\partial z}$$

This is called the *velocity derivative tensor*. The velocity derivative tensor is not symmetric, as $\partial u_i/\partial x_j \neq \partial u_j/\partial x_i$ in general. Therefore, it can be decomposed into symmetric and antisymmetric parts. The symmetric part is represented by equation (9.18), and the antisymmetric part by equation (9.19).

Logarithmic Strain

Earlier [equation (9.13)] we defined strain by

$$\varepsilon = \mathrm{Lim}_{\ell \to 0} \frac{\Delta \ell}{\ell}.$$

where ℓ is the initial length of a line and $\Delta \ell$ is the elongation of that line. This definition is suitable for small (infinitesimal) strains. However, when calculating strains or strain rates from measurements, the total strain is normally not infinitesimal. This is, in part, because deformations must be large enough to exceed the uncertainty in the measurement method.

If strains are infinitesimal, we can replace $\Delta \ell$ with $d\ell$. The total strain is then the sum of the infinitesimal strains, or

$$\varepsilon = \sum_{\ell=\ell_0}^{\ell=\ell_1} \frac{d\ell}{\ell} = \int_{\ell_0}^{\ell_1} \frac{d\ell}{\ell},$$

where ℓ_0 is the initial length of the line and ℓ_1 is its final length. Integrating yields

$$\varepsilon = \ln \frac{\ell_1}{\ell_0}$$

or in terms of rates

$$\dot{\varepsilon} = \frac{1}{\Delta t} \ln \frac{\ell_1}{\ell_0},$$

where Δt is the time interval between measurements. This is known as logarithmic strain.

General Equations for Transformation of Strain in Two Dimensions

Our next objective is to develop an expression for the strain rate in an arbitrary direction, θ. To simplify the equations, we will restrict the analysis to the case of plane strain. We will then take the derivative of this expression and set it equal to zero to find the directions in which $\dot{\varepsilon}(\theta)$ is maximum and minimum (the *principal strain rates*). Finally, we will look at the implications of this in terms of the flow law.

Let us examine the elongation of line $\overline{OP}$ in Figure 9.9a. The line has an initial length ℓ and makes an angle θ with the x-axis. The line is stretched to a final length $\overline{OP'}$ through a displacement with components $\overline{\mathbf{x}}$ and $\overline{\mathbf{y}}$ in the x- and y-directions, respectively. The elongation is given by (Fig. 9.9a)

$$\Delta\ell = \overline{\mathbf{x}}\cos\theta + \overline{\mathbf{y}}\sin\theta. \tag{9.20}$$

The displacement $\overline{\mathbf{x}}$ is a consequence of strain parallel to the x-axis and a shear strain which results in tilting of any line that is initially normal to the x-axis (Fig. 9.9b); thus,

$$\overline{\mathbf{x}} = \ell\cos\theta\frac{\partial\overline{\mathbf{x}}}{\partial x} + \ell\sin\theta\frac{\partial\overline{\mathbf{x}}}{\partial y}. \tag{9.21a}$$

The origin of the two terms on the right hand side may be clarified by reference to Figures 9.7 and 9.8, respectively. Similarly,

$$\overline{\mathbf{y}} = \ell\sin\theta\frac{\partial\overline{\mathbf{y}}}{\partial y} + \ell\cos\theta\frac{\partial\overline{\mathbf{y}}}{\partial x}. \tag{9.21b}$$

Substituting equations (9.21) into equation (9.20) yields

$$\Delta\ell = \ell\cos^2\theta\frac{\partial\overline{\mathbf{x}}}{\partial x} + \ell\sin^2\theta\frac{\partial\overline{\mathbf{y}}}{\partial y} + \ell\left(\frac{\partial\overline{\mathbf{x}}}{\partial y} + \frac{\partial\overline{\mathbf{y}}}{\partial x}\right)\cos\theta\sin\theta.$$

Finally, dividing by ℓ, noting that $\mathrm{Lim}_{\ell\to 0}\,\Delta\ell/\ell = \varepsilon\,(\theta)$, the strain along the length of line $\overline{OP}$, and using equations (9.14) and (9.15a) to express the derivatives in terms of strains yields

$$\varepsilon(\theta) = \varepsilon_{xx}\cos^2\theta + \varepsilon_{yy}\sin^2\theta + 2\varepsilon_{xy}\cos\theta\sin\theta.$$

With the trigonometric identities used earlier (page 153) this becomes

$$\varepsilon(\theta) = \frac{\varepsilon_{xx} + \varepsilon_{yy}}{2} + \frac{\varepsilon_{xx} - \varepsilon_{yy}}{2}\cos 2\theta + \varepsilon_{yx}\sin 2\theta. \tag{9.22}$$

Strain rates can be obtained by taking the derivative with respect to time. Equation (9.22) is a useful relation for obtaining normal strains or strain rates in a direction, θ, when values in the coordinate directions x, y are known.

To obtain the maximum and minimum values of $\dot{\varepsilon}(\theta)$, the principal strain rates, we proceed as before [equations (9.3)] to take the derivative with respect to θ, set it to zero, and solve for θ; thus,

$$\tan 2\theta_{sr} = \frac{2\dot{\varepsilon}_{yx}}{\dot{\varepsilon}_{xx} - \dot{\varepsilon}_{yy}}. \tag{9.23}$$

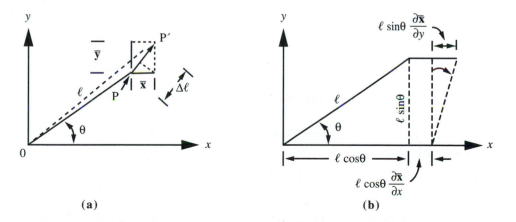

Figure 9.9 (a) Components of strain of a line of length ℓ, and (b) details of the shear component of the strain.

The two solutions for θ are the directions in which the strain rate is a maximum and minimum. The magnitudes of the principal strain rates can be obtained as we did for the principal stresses [equation (9.4)].

CONDITION THAT PRINCIPAL AXES OF STRESS AND STRAIN RATE COINCIDE

In calculating the stress and velocity fields in a glacier in Chapter 10, we will need to assume that the principal axes of stress and strain rate coincide. Let us explore the consequences of this condition.

We found earlier that the angle which the principal stresses make with the x-coordinate can be obtained from

$$\tan 2\theta_{stress} = \frac{2\sigma_{yx}}{\sigma_{xx} - \sigma_{yy}} = \frac{2\sigma'_{yx}}{\sigma'_{xx} - \sigma'_{yy}}.$$

Note that it does not matter whether we use deviatoric or total stresses here, as $\sigma'_{yx} = \sigma_{yx}$ and $\sigma'_{xx} - \sigma'_{yy} = \sigma_{xx} - P - (\sigma_{yy} - P) = \sigma_{xx} - \sigma_{yy}$.

The condition that the principal axes of stress and strain rate coincide is $\theta_{stress} = \theta_{sr}$, or at any given point in the medium:

$$\frac{2\sigma'_{yx}}{\sigma'_{xx} - \sigma'_{yy}} = \frac{2\dot{\varepsilon}_{yx}}{\dot{\varepsilon}_{xx} - \dot{\varepsilon}_{yy}}.$$

The only way to satisfy this condition in the general case is to let

$$\dot{\varepsilon}_{yx} = \lambda\sigma'_{yx}; \quad \dot{\varepsilon}_{xx} = \lambda\sigma'_{xx}; \quad \dot{\varepsilon}_{yy} = \lambda\sigma'_{yy},$$

where λ is a scalar; that is, its value at the particular point in the medium is independent of the direction in which the stress is acting. However, λ may vary from one point to another in the medium, so $\lambda = \lambda(x, y, z)$.

The fact that λ is a scalar implies that the deforming material is isotropic and incompressible. Thus, under a given stress, it will deform at the same rate regardless of the direction in which the stress is applied. Obviously, this is an approximation in a material such as ice that first has an hexagonal crystal structure, and secondly can develop a fabric with a preferred orientation. [If the material were compressible, a compressive deviatoric stress in, say, the x-direction, σ'_{xx}, would cause more deformation than the corresponding tensile deviatoric stress in the y-direction (which must equal σ'_{xx} in magnitude in plane strain). Thus, λ would differ between the two directions.]

We generalize this assumption to three dimensions and formalize it by writing

$$\dot{\varepsilon}_{ij} = \lambda\sigma'_{ij} \tag{9.24}$$

remembering that

$$\dot{\varepsilon}_{ij} = \frac{1}{2}\left(\frac{\partial u_i}{\partial x_j} + \frac{\partial u_j}{\partial x_i}\right) \tag{9.25}$$

and

$$\sigma'_{ij} = \sigma_{ij} - \frac{1}{3}\delta_{ij}\sigma_{kk}.$$

Because the stress and strain rate in the flow law are defined in terms of either the effective or the octahedral stress and strain rate, we can write out the first few terms of the effective strain rate and substitute equation (9.24) into the right-hand side, as follows:

$$\varepsilon_e^2 = \frac{1}{2}\left[\dot{\varepsilon}_{xx}^2 + \dot{\varepsilon}_{xy}^2 + \dot{\varepsilon}_{xz}^2 + \ldots\right]$$
$$= \frac{1}{2}\left[\lambda^2\sigma_{xx}'^2 + \lambda^2\sigma_{xy}'^2 + \ldots\right]$$
$$= \frac{\lambda^2}{2}\left[\sigma_{xx}'^2 + \sigma_{xy}'^2 + \ldots\right]$$
$$= \lambda^2\sigma_e^2$$

Dropping the subscript e for convenience (and also because the flow law can be written in terms of either the effective or the octahedral stress and strain rate), we obtain

$$\dot{\varepsilon} = \lambda\sigma = f(\sigma) \quad \text{so} \quad \lambda = \frac{f(\sigma)}{\sigma}. \qquad (9.26)$$

Here, $f(\sigma)$ is used to emphasize that, in the general case, λ is a function of the applied stress.

A few examples will serve to illustrate the meaning of λ.

- Newtonian fluid: $\lambda = 1/\mu$ where μ is the Newtonian viscosity, so in this case λ is not a function of σ.
- Power law fluid: $\lambda = \sigma^{n-1}/B^n$ so that $\dot{\varepsilon} = \lambda\sigma = (\sigma/B)^n$.
- Perfectly plastic material: As noted earlier (Fig. 9.4), in a perfectly plastic material there is no deformation below a critical stress, $\sigma = k$, so $\lambda = 0$ for $\sigma < k$. When $\sigma = k$, the material deforms at a rate such that the stress does not exceed k. In other words, λ depends on $\dot{\varepsilon}$: $\lambda = f_1(\dot{\varepsilon})$.

In the case of the power law fluid, λ varies with σ and B, and because B is a function of temperature, density, crystal size, and orientation, and perhaps other factors, λ also varies with these properties.

Combining equations (9.24), (9.25), and (9.26), writing $\dot{\varepsilon}$ in terms of velocity derivatives, and using the summation convention, we now have the following relation between individual components of the stress and strain-rate tensors:

$$\frac{1}{2}\left[\frac{\partial u_i}{\partial x_j} + \frac{\partial u_j}{\partial x_i}\right] = \frac{f(\sigma)}{\sigma}\sigma_{ij}'. \qquad (9.27a)$$

This represents nine equations, only six of which are independent. Together with equation (9.12),

$$\frac{\partial \sigma_{ij}}{\partial x_i} + \rho g_j = 0, \qquad (9.27b)$$

which represents an additional three equations, we have nine independent equations that can be solved for the three normal stresses, three shear stresses, and three velocities. Our objective in Chapter 10 will be to do this, but to simplify the problem we will confine our attention to plane strain.

SUMMARY

We reviewed in this chapter some elementary principles of continuum mechanics, with a particular focus on those principles needed for an understanding of much of both the classical and the modern literature on the flow of glaciers.

In the first part of the chapter we defined stress and showed that if we know the stresses in one coordinate system, we can calculate them in another system rotated with respect to the first. This allowed us to calculate the direction and magnitude of the maximum and minimum normal stresses, the *principal stresses*. We did the calculation in two dimensions, but the extension to three dimensions is straightforward. We found that shear stresses vanish in coordinate systems chosen with axes aligned parallel to the principal stresses.

The orientation and magnitude of the principal stresses is a property of the stress field and not of the orientation of the axes. Thus, there are certain combinations of the stresses that must be independent of the orientation of the axes: the *invariants* of the stress tensor. Glen's flow law for ice is based on the second of these invariants. This is logical because it is invariant, and also because the von Mises yield criterion can be expressed in terms of this invariant. Recent experimental data have validated this approach.

In the second part of the chapter we derived the stress equilibrium equations.

In the third part we defined strain and derived equations for calculating strains or strain rates in coordinate systems rotated with respect to one another. These equations are similar to those for transformation of stress. As with stresses, we introduced the concept of *principal strain rates*.

Finally, we showed that if a material is isotropic and incompressible, the principal axes of stress and strain rate coincide. Ice is clearly not isotropic and incompressible, but this approximation has proven to be a convenient starting point for calculations of glacier flow.

10

Stress and Velocity Distribution in an Idealized Glacier

Let us now use equations (9.27) to calculate stresses and velocities in an idealized glacier consisting of a slab of ice of infinite horizontal extent resting on a bed with a uniform slope. By appropriate choice of the coordinate system, the problem is thus reduced to two dimensions, or plane strain. The ice is assumed to be isotropic and incompressible. We will consider first the case of a perfectly plastic rheology. Then a more realistic nonlinear flow law is used. Our discussion is based on papers of Nye (1951, 1957), which are classics in glaciology.

Although glaciers consisting of such slabs are uncommon, to say the least, there are several reasons for undertaking this calculation. First, it provides an opportunity to apply some of the material discussed in the previous chapter. Second, the stress distributions are representative of those that we expect to find in glaciers, and are commonly used approximations when the required assumptions can be justified by the geometry of a problem. Third, the calculations demonstrate the limitations of analytical methods in situations in which boundary conditions are complex. For calculations involving glaciers with realistic shapes, numerical models are required for all but the simplest situations. Finally, the effect of longitudinal stresses on velocity profiles is elucidated.

SOLUTIONS FOR STRESSES AND VELOCITIES IN PLANE STRAIN

The coordinate system to be used for the calculation is shown in Figure 10.1: x is parallel to the glacier surface in the direction of flow, and z is directed downward normal to the surface. The origin is on the surface of the slab, which has a thickness h. The velocities are u, v, and w in the x, y, and z directions, respectively.

In plane strain, $\dot{\varepsilon}_{yy} = 0$ so $\sigma'_{yy} = 0$. Therefore,

$$\sigma_{yy} = \sigma'_{yy} + P = \frac{1}{3}(\sigma_{xx} + \sigma_{yy} + \sigma_{zz}).$$

Solving the equation represented by the first and last terms in this expression yields

$$\sigma_{yy} = \frac{1}{2}(\sigma_{xx} + \sigma_{zz}). \tag{10.1}$$

It is then easily shown that $P = \frac{1}{2}(\sigma_{xx} + \sigma_{zz})$ also. Again, because we have confined our attention to a situation in plane strain,

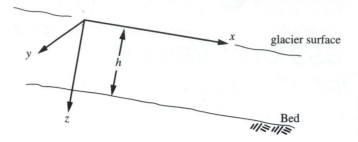

Figure 10.1 Coordinate system used in calculating stresses and velocities in plane strain.

$$\sigma_{xy} = 0$$
$$\sigma_{zy} = 0$$
$$v = 0. \tag{10.2}$$

Equations (10.1) and (10.2) are four of the equations needed to solve for the three normal stresses, three shear stresses, and three velocities, so we need five more from equations (9.27). These are:

$$\frac{\partial \sigma_{xx}}{\partial x} + \frac{\partial \sigma_{zx}}{\partial z} + \rho g_x = 0 \tag{10.3}$$

$$\frac{\partial \sigma_{zx}}{\partial x} + \frac{\partial \sigma_{zz}}{\partial z} + \rho g_z = 0 \tag{10.4}$$

$$\frac{\partial u}{\partial x} = \lambda \sigma'_{xx} = \lambda(\sigma_{xx} - P) = \lambda\left[\sigma_{xx} - \tfrac{1}{2}(\sigma_{xx} + \sigma_{zz})\right] = \frac{\lambda}{2}(\sigma_{xx} - \sigma_{zz}) \tag{10.5}$$

and similarly:

$$\frac{\partial w}{\partial z} = -\frac{\lambda}{2}(\sigma_{xx} - \sigma_{zz}) \tag{10.6}$$

and

$$\frac{1}{2}\left(\frac{\partial u}{\partial z} + \frac{\partial w}{\partial x}\right) = \lambda \sigma_{zx}. \tag{10.7}$$

Note that $\partial u/\partial x = -\partial w/\partial z$ as required by the incompressibility condition in plane strain [equation (2.5)].

Because $\lambda = f(\sigma)/\sigma$, we have to introduce an expression for σ, namely:

$$\sigma^2 = \frac{1}{4}(\sigma_{xx} - \sigma_{zz})^2 + \sigma_{zx}^2 \tag{10.8}$$

which, as we have seen, can be derived from the von Mises yield criterion in plane strain, and is, in fact, the effective stress in plane strain. Still needed is an expression for $f(\sigma)$; this will be based on the specific flow law chosen for any given solution.

Stress Solutions in a Perfectly Plastic Medium

Suppose that our slab is composed of perfectly plastic "ice." Suppose, further, that accumulation and ablation are occurring on the surface of the slab, but that it is in a steady state so the profile does not change. Thus, it must be deforming to accommodate the addition or loss of mass. In order to avoid a discontinuity at depth, the deformation must extend throughout the slab. Furthermore, in a deforming perfectly plastic material, the stress must reach but cannot exceed the yield stress, k (Chapter 9). Therefore, $\sigma = k$ everywhere.

Near the bed, although a small uniform longitudinal strain rate is present, deformation is principally by simple shear. In fact, in a perfectly plastic material any tendency toward an increase in stress above k is absorbed by more rapid deformation (Chapter 9), so $\dot{\varepsilon}_{zx} \to \infty$ at the bed. Thus, $\dot{\varepsilon}_{zx}$ is the dominant term in $\dot{\varepsilon}$, so σ_{zx} is the dominant term in σ [equation (10.8)], and σ_{zx} must approach $\sigma (= k)$ at the bed. Therefore, we adopt the following boundary conditions:

$$\sigma_{zx} = -k \quad \text{on} \quad z = h \quad \text{(bed)}$$
$$\sigma_{zx} = 0, \quad \sigma_{zz} = 0 \quad \text{on} \quad z = 0 \quad \text{(surface)}.$$

Note that σ_{zx} is set equal to $-k$ on the bed because the drag exerted by the bed on the ice is in the negative x direction. Because σ_{zx} is independent of x on the bed, we will seek a solution in which σ_{zx} is independent of x everywhere.

The following solution, previously unpublished, is contained in a manuscript by J.F. Nye, which he kindly provided to the author, and which is used with his permission. We take the derivative of equation (10.3) with respect to z and of equation (10.4) with respect to x; thus,

$$\frac{\partial^2 \sigma_{xx}}{\partial z \partial x} + \frac{\partial^2 \sigma_{zx}}{\partial z^2} = 0$$
$$\frac{\partial^2 \sigma_{zz}}{\partial x \partial z} + \frac{\partial^2 \sigma_{zx}}{\partial x^2} = 0. \tag{10.9}$$

If σ_{xx} and σ_{zz} are continuous functions, the order of differentiation can be reversed in the first of these equations. Subtracting then yields

$$\frac{\partial^2}{\partial x \partial z}(\sigma_{xx} - \sigma_{zz}) = \left(\frac{\partial^2 \sigma_{zx}}{\partial x^2} - \frac{\partial^2 \sigma_{zx}}{\partial z^2} \right). \tag{10.10}$$

Substituting for $\sigma_{xx} - \sigma_{zz}$ from equation (10.8) and setting $\sigma = k$ as noted above, we find

$$\pm \frac{\partial^2}{\partial x \partial z}\left(2\sqrt{k^2 - \sigma_{zx}^2} \right) = \left(\frac{\partial^2 \sigma_{zx}}{\partial x^2} - \frac{\partial^2 \sigma_{zx}}{\partial z^2} \right). \tag{10.11}$$

Making use of the condition that σ_{zx} must be independent of x allows us to simplify this to

$$\frac{\partial^2 \sigma_{zx}}{\partial z^2} = 0, \tag{10.12}$$

which has the solution

$$\sigma_{zx} = c_1 z + c_2. \tag{10.13}$$

To satisfy the boundary condition $\sigma_{zx} = 0$ on $z = 0$, we find that $c_2 = 0$. The boundary condition $\sigma_{zx} = -k$ on $z = h$ then yields $c_1 = -k/h$. Thus, the solution for σ_{zx} becomes

$$\sigma_{zx} = -\frac{kz}{h}. \tag{10.14}$$

In other words, the shear stress varies linearly with depth.

Using this solution for σ_{zx} in equations (10.3) and (10.4) yields

$$\frac{\partial \sigma_{xx}}{\partial x} = \frac{k}{h} - \rho g_x$$
$$\frac{\partial \sigma_{zz}}{\partial z} = -\rho g_z$$

which integrate to

$$\sigma_{xx} = \frac{kx}{h} - \rho g_x x + f_1(z)$$
$$\sigma_{zz} = -\rho g_z z + f_2(x) \tag{10.15}$$

where $f_1(z)$ and $f_2(x)$ are functions that are dependent only upon z and x, respectively, so that $\partial f_1(z)/\partial x = 0$ and $\partial f_2(x)/\partial z = 0$. They are analogous to the constants of integration in equation (10.13), and must be evaluated with the use of the boundary conditions. Substituting these solutions for σ_{xx}, σ_{zz}, and σ_{zx} back into equation (10.8), the yield criterion, we obtain

$$\frac{kx}{h} - \rho g_x x + f_1(z) + \rho g_z z - f_2(x) = \pm 2\sqrt{k^2 - \frac{k^2 z^2}{h^2}}$$

which is true for all x and z because, as noted, the yield criterion must be met throughout the slab. Thus, collecting the terms in x on the left-hand side and those in z on the right results in

$$\frac{kx}{h} - \rho g_x x - f_2(x) = -\rho g_z z \pm 2k\sqrt{1 - \left(\frac{z}{h}\right)^2} - f_1(z) = c. \qquad (10.16)$$

Because the left-hand side is a function of x alone and the right-hand side is a function of z alone, the two sides can be equal to each other in the general case only if each, individually, is equal to the same constant, c, as shown. We can thus solve equation (10.16) for $f_2(x)$ in terms of c, and for $f_1(z)$ in terms of c. These solutions are then inserted in equations (10.15) to yield

$$\sigma_{xx} = \frac{kx}{h} - \rho g_x x - \rho g_z z \pm 2k\sqrt{1 - \left(\frac{z}{h}\right)^2} - c$$

$$\sigma_{zz} = -\rho g_z z + \frac{kx}{h} - \rho g_x x - c. \qquad (10.17)$$

Making use of the boundary condition $\sigma_{zz} = 0$ on $z = 0$ in the second of equations (10.17) gives

$$c = \frac{kx}{h} - \rho g_x x = x\left(\frac{k}{h} - \rho g_x\right).$$

As this must hold for all x, it is clear that $k/h - \rho g_x$ must equal 0, and therefore c must also be 0. Thus,

$$\frac{k}{h} = \rho g_x.$$

This implies that the block must have some critical thickness, $h = k/\rho g_x$. Using these results in equations (10.17), and repeating equation (10.14) allows us to write the complete stress solutions thus:

$$\sigma_{xx} = -\rho g_z z \pm 2k\sqrt{1 - \left(\frac{z}{h}\right)^2}$$

$$\sigma_{zz} = -\rho g_z z \qquad (10.18)$$

$$\sigma_{zx} = -\frac{kz}{h}.$$

As a check on these solutions, they may be substituted back into the yield criterion, equation (10.8), to show that $\sigma = k$.

A plot (Fig. 10.2) will serve to illustrate the solutions. Scaling σ_{zz} by $k (= \rho g_x h)$ results in

$$\frac{\sigma_{zz}}{k} = -\frac{z \cos \alpha}{h \sin \alpha}.$$

σ_{zz} is 0 at the surface and it decreases linearly with depth. With $\tan \alpha = 1/5$, we find that $\sigma_{zz} = -5k$ on $z = h$ (Fig. 10.2). Similarly, $\sigma_{xx} = \pm 2k$ on $z = 0$ and $-5k$ on $z = h$, but in this case the distribution with depth is elliptic. Finally, σ_{zx} also decreases linearly with depth to $-k$ on the bed.

The deviatoric stresses can also be calculated. Noting that

$$P = \frac{1}{2}(\sigma_{xx} + \sigma_{zz}) = -\rho g_z z \pm k\sqrt{1 - \left(\frac{z}{h}\right)^2}$$

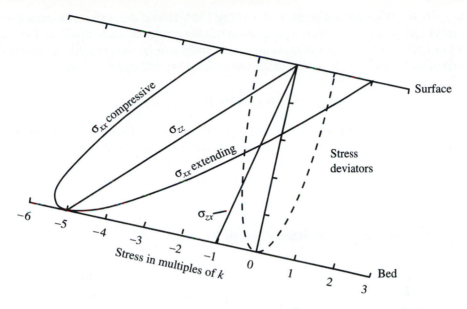

Figure 10.2   Depth variation of stress in a deforming slab of material with a perfectly plastic rheology. The slab is of uniform thickness and density, and is resting on a bed with a uniform slope (Modified from Nye, 1951, Fig. 2c. Reproduced with permission of the author and the Royal Society of London.)

we find that

$$\sigma'_{xx} = \sigma_{xx} - P = \pm k \sqrt{1 - \left(\frac{z}{h}\right)^2}$$

$$\sigma'_{zz} = \sigma_{zz} - P = \mp k \sqrt{1 - \left(\frac{z}{h}\right)^2}$$

(10.18a)

so σ'_{xx} is $\pm k$ at the surface and decreases to 0 at the bed, as shown by the dashed lines describing a semi-ellipse in Figure 10.2. As required by continuity, $\sigma'_{zz} = -\sigma'_{xx}$; that is, if our medium is homogeneous, isotropic, and incompressible, as assumed, a deformation in the x direction, $\dot{\epsilon}_{xx}$, caused by a stress σ'_{xx} must be accompanied by an equal deformation of opposite sign, $\dot{\epsilon}_{zz}$, in the z direction, and this requires a stress equal to σ'_{xx}, but in the opposite direction.

Negative or compressive deviatoric stresses in the x-direction result in longitudinal compression, and this flow regime is thus referred to as compressive flow. Similarly, positive deviatoric stresses in the x-direction result in what is called extending flow. The former is characteristic of ablation zones of glaciers, where melt must be replaced by upward flow of ice, and the latter is characteristic of accumulation zones.

It is also of interest to examine the orientation of the principal stresses and of the maximum shear stresses. In Chapter 9 [equation (9.3b)] we found that

$$\tan 2\theta = \frac{2\sigma_{zx}}{(\sigma_{xx} - \sigma_{zz})}$$

so:

$$\theta = \frac{1}{2}\tan^{-1}\frac{-2kz/h}{\pm 2k\sqrt{1 - \left(\frac{z}{h}\right)^2}} = \frac{1}{2}\tan^{-1}\left[\mp \frac{z}{\sqrt{h^2 - z^2}}\right].$$

On $z = 0$, $\theta = 0°$ or $90°$ and on $z = h$, $\theta = 45°$ or $135°$. Thus at the surface, the principal stresses are parallel and normal to the surface, and at the bed they make an angle of $45°$ or $135°$ with the bed. The orientation of the planes of maximum shear stress may be found by differentiating the equation derived earlier for $\sigma_s(\theta)$ [equation (9.2)] and setting the result equal to 0; thus,

$$\theta_{ss} = \frac{1}{2}\tan^{-1}\left[\pm\frac{\sqrt{h^2 - z^2}}{z}\right].$$

In this case, $\theta_{ss} = \pm45°$ at the surface and $0°$ or $90°$ at the bed. (In other words, the planes of maximum shear stress make an angle of $45°$ with respect to the principal stresses.) At any intermediate depth there are two solutions for θ_{ss} that are $90°$ apart. Thus, the loci of the zones of maximum shear stress are as shown in Figure 10.3. This is what is known as the *slip line field* for the particular stress configuration.

Velocity Solutions in a Perfectly Plastic Medium

We now use the stress solutions, equations (10.18), to obtain solutions for the velocities from equations (10.5) through (10.7). From equations (10.5) and (10.6), and making use of the continuity condition, $\partial u/\partial x = -\partial w/\partial z$, we obtain:

$$\frac{\partial u}{\partial x} = -\frac{\partial w}{\partial z} = \frac{\lambda}{2}(\sigma_{xx} - \sigma_{zz})$$

$$= \pm\lambda k\sqrt{1 - \left(\frac{z}{h}\right)^2}$$

(10.19)

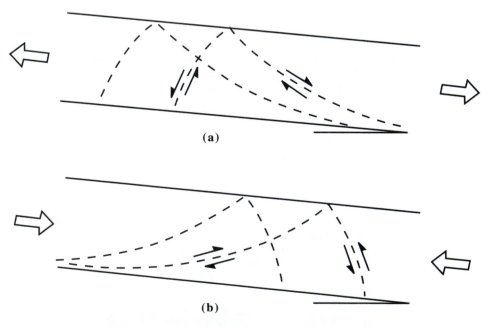

(a)

(b)

Figure 10.3 Slip-line fields in a deforming slab of material with a perfectly plastic rheology. (**a**) Extending flow, and (**b**) compressive flow. (Modified from Nye, 1951, Fig. 5.)

and from equation (10.7):

$$\frac{\partial u}{\partial z} + \frac{\partial w}{\partial x} = 2\lambda\sigma_{xz} = -2\lambda\frac{kz}{h}. \tag{10.20}$$

Let us first examine the applicable boundary conditions. The stress solutions are valid only for the thickness $h = k/\rho g_x$. Therefore, we seek a velocity solution that will maintain this thickness. Because there is accumulation, b_n (or ablation, $-b_n$) at the surface, we know that

$$w = b_n \quad \text{on} \quad z = 0$$
$$w = 0 \quad \text{on} \quad z = h$$

We will now show that $\partial w/\partial x = 0$. The stresses are independent of x, and the material is the limiting case of a purely viscous material in which stresses determine strain rates. Therefore, the strain rates must be independent of x. In particular, $\partial w/\partial z$ is independent of x, so

$$\frac{\partial}{\partial x}\left(\frac{\partial w}{\partial z}\right) = 0$$

or if w is continuous in x and z

$$\frac{\partial}{\partial z}\left(\frac{\partial w}{\partial x}\right) = 0.$$

Therefore, $\partial w/\partial x = $ const, independent of z. Because $\partial w/\partial x = 0$ on the upper and lower boundaries from the boundary conditions, $\partial w/\partial x = 0$ everywhere.

Equation (10.20) now becomes

$$\frac{\partial u}{\partial z} = -2\lambda\frac{kz}{h}. \tag{10.21}$$

Combining equations (10.19) and (10.21) to eliminate λ yields

$$\frac{\partial u}{\partial z} = \frac{2\dfrac{kz}{h}}{k\sqrt{1 - \left(\dfrac{z}{h}\right)^2}}\frac{\partial w}{\partial z} \tag{10.22}$$

$$= \frac{2z}{\sqrt{h^2 - z^2}}\frac{\partial w}{\partial z}.$$

Differentiating with respect to x and again making use of the fact that w is continuous in x and z gives

$$\frac{\partial^2 u}{\partial x \partial z} = \frac{2z}{\sqrt{h^2 - z^2}}\frac{\partial w}{\partial x \partial z}$$

$$= \frac{2z}{\sqrt{h^2 - z^2}}\frac{\partial}{\partial z}\left(\frac{\partial w}{\partial x}\right)$$

$$= 0$$

where the last equality results from the fact that $\partial w/\partial x = 0$. Thus from equation (10.19), differentiating with respect to z, we obtain

$$\frac{\partial^2 u}{\partial z \partial x} = -\frac{\partial^2 w}{\partial z^2} = 0 \tag{10.23}$$

which has the solution

$$w = c_1 z + c_2.$$

Using the first boundary condition, $w = b_n$ on $z = 0$, yields $c_2 = b_n$, whereupon the second boundary condition, $w = 0$ on $z = h$, yields $0 = c_1h + b_n$. The solution for the velocity in the z-direction thus becomes

$$w = b_n\left(1 - \frac{z}{h}\right). \tag{10.24}$$

Note that w varies linearly with depth. We will discuss this result in greater detail later.

Using this solution for w in equation (10.19) we obtain

$$\frac{\partial u}{\partial x} = -\frac{\partial w}{\partial z} = \frac{b_n}{h} \tag{10.25}$$

and from equation (10.22)

$$\frac{\partial u}{\partial z} = \mp \frac{2z}{\sqrt{h^2 - z^2}} \frac{b_n}{h}. \tag{10.26}$$

In the coordinate system we have chosen, σ_{zx} is negative for positive z. Therefore $\dot{\varepsilon}_{zx}$, and hence in equation (10.26), $\partial u/\partial z$ must be negative so that a negative stress produces a negative strain rate. (In other words, the horizontal velocity must decrease with depth.) Thus, when b_n is positive in equation (10.26) we use the upper sign and conversely. Equation (10.26) thus becomes

$$\frac{\partial u}{\partial z} = -2\frac{|b_n|}{h}\frac{z}{\sqrt{h^2 - z^2}}. \tag{10.27}$$

Integrating equation (10.25) yields

$$u = \frac{b_n x}{h} + f(z),$$

where, as in equation (10.15), $f(z)$ represents some function of z. Taking the derivative of this with respect to z, substituting the result into equation (10.27), and integrating gives

$$f(z) = 2\frac{|b_n|}{h}\sqrt{h^2 - z^2} + c.$$

Thus, the solution for the velocity in the x-direction is

$$u = \frac{b_n x}{h} + 2\frac{|b_n|}{h}\sqrt{h^2 - z^2} + c. \tag{10.28}$$

These velocity solutions are illustrated in Figure 10.4. On $z = 0$, $u = b_n x/h + 2\ |b_n| + c$ and $w = b_n$, while on $z = h$, $u = b_n x/h + c$ and $w = 0$. Thus, evidently $b_n x/h + c$ is the "sliding" speed. Note that equation (10.28) implies that the ice must be free to slide on the bed at a speed determined by b_n, and independent of σ_{zx} and bed roughness.

If, on the contrary, the sliding speed were presumed to be a function of σ_{zx} and bed roughness, the distribution of stress and hence of w could not be independent of x, and the ice mass would not remain a uniform slab. A reasonable presumption is that the sliding speed would not increase sufficiently rapidly with x, and that conservation of mass would then require that the ice thickness increase upglacier, leading to a convex surface profile. Depending on the degree of convexity and the consequent change in ice thickness, such a profile would offer the potential for increasing σ_{zx} downglacier. This, coupled with the increase in thickness, would provide the required increase in mass flux.

If the boundary conditions were selected such that $u = 0$ at $x = 0$, $z = h$, then c would be 0, but the original differential equations are not necessarily valid at the ends of the slab. Note also that

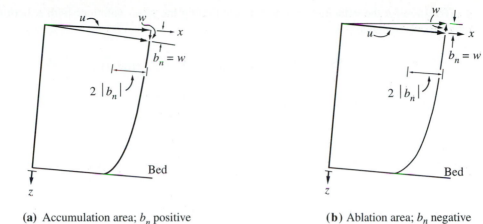

(a) Accumulation area; b_n positive   **(b)** Ablation area; b_n negative

Figure 10.4   Velocity solution for a deforming slab of material with a perfectly plastic rheology. Velocity is vector sum of components u and w.

between any two vertical sections, x_1 and $x_1 + \Delta x$, u increases by $b_n \Delta x / h$, which would be the increase in mean velocity required to transmit the additional flux, $b_n \Delta x$, downglacier.

If there is no accumulation or ablation, $b_n = 0$, and thus $u = c$ and $w = 0$ throughout the block. In this case there is no internal deformation. All movement is confined to sliding. The yield criterion is then satisfied only at the bed and σ_{xx} can take any value between the limits shown in Figure 10.2.

Stress and Velocity Solutions for a Nonlinear Material

Let us now relax the assumption that the material with which we are dealing is perfectly plastic, and instead allow it to have a nonlinear rheology, as is the case with real ice. We still consider a slab of infinite extent and uniform thickness resting on a bed with a uniform slope. The stress equations are not changed, so the solutions for the stresses [equations (10.18)] remain basically the same. However, now σ does not have a limiting value, k, the yield stress, so some changes must be made.

As before, we note that $\sigma_{xx} \approx \sigma_{zz}$ on the bed, as both are dominated by the hydrostatic pressure. As $\sigma \gg 0$ at the bed, it is clear from equation 10.8 that $\sigma \approx \sigma_{zx} \gg (\sigma_{xx} - \sigma_{zz})$ there. Again, this emphasizes that deformation at the bed is largely by simple shear. Thus $\sigma_{zx} \rightarrow \sigma$ on the bed, so we replace k with σ in equations (10.18). In other words, while we still require that σ and σ_{zx} be uniform (independent of x) on the bed, we do not require that they are necessarily equal to some specific value, as in a yield stress. In general, of course, σ_{zx} is likely to increase with the budget gradient (see Fig. 3.7), as flow rates must then be higher. In addition, we make use of the fact that $k/h = \rho g_x$ from the discussion following equation (10.17). With these changes, the stress solutions become

$$
\begin{aligned}
\sigma_{xx} &= -\rho g_z z \pm 2\sqrt{\sigma^2 - (\rho g_x z)^2} \\
\sigma_{zz} &= -\rho g_z z \\
\sigma_{zx} &= -\rho g_x z
\end{aligned}
\tag{10.29}
$$

As before, the upper sign is for extending, and the lower for compressive flow.

To evaluate $\sigma_{xx}(z)$, we need to know how σ varies with depth, z. This will emerge in the course of obtaining the velocity solutions.

To solve for the velocities, we start by combining the stress solutions with equations (10.5) through (10.7) to obtain

$$\frac{\partial u}{\partial x} = -\frac{\partial w}{\partial z} = \frac{\lambda}{2}(\sigma_{xx} - \sigma_{zz}) = \pm\lambda\sqrt{\sigma^2 - (\rho g_x z)^2}$$

(10.30)

$$\frac{1}{2}\left(\frac{\partial u}{\partial z} + \frac{\partial w}{\partial x}\right) = \lambda\sigma_{xz} = -\lambda\rho g_x z.$$

As before [discussion preceding equation (10.21)], because $w = 0$ on the lower boundary, and because w must thus be independent of x everywhere to avoid discontinuities, $\partial w/\partial x = 0$ everywhere. Thus, using the arguments outlined in equations (10.21–10.23) we obtain, as before:

$$w = c_1 z + c_2.$$

(10.31)

The boundary conditions now are not as clear as they were earlier because the thickness may change with time. Thus $w \neq b_n$ on $z = 0$.

From Equations (10.30 and 10.31) we have

$$\frac{\partial w}{\partial z} = c_1 = -\frac{\partial u}{\partial x}.$$

(10.32)

Therefore, from the first of equations (10.30),

$$c_1 = \mp\lambda\sqrt{\sigma^2 - (\rho g_x z)^2}.$$

(10.33)

Using this in the second of equations (10.30) yields

$$\frac{\partial u}{\partial z} = \pm 2\rho g_x z \frac{c_1}{\sqrt{\sigma^2 - (\rho g_x z)^2}}.$$

(10.34)

As λ is defined by $\dot{\varepsilon}_{ij} = \lambda\sigma_{ij}$, it must be positive if positive stresses are to produce positive strain rates. To ensure that this is the case, we give c_1 the values $\mp r$ where r is a positive constant. From equation (10.32) it is clear that r is the longitudinal strain rate. Equation (10.32) thus becomes

$$\frac{\partial u}{\partial x} = \pm r.$$

The solution for w now becomes

$$w = \mp rz + c_2.$$

Applying the boundary condition $w = 0$ on $z = h$, we get $c_2 = \pm rh$, so:

$$w = \mp r(z - h) = \pm rh\left(1 - \frac{z}{h}\right).$$

(10.35)

Note that w still varies linearly with depth as in the perfectly plastic case, despite the variation in σ with depth. This is implicit in the fact that the longitudinal strain rate is independent of depth, which in turn results from the fact that stresses must be independent of x in a slab on a uniform slope [see discussion following equation (10.20); $\partial u/\partial x = -\partial w/\partial z$ and by equation (10.23), $\partial w/\partial z = $ constant].

To obtain u, integrate equation (10.34) [with equation (10.33)]; thus,

$$u = -2g_x \int_0^z \lambda\rho z\, dz + f(x).$$

(10.36)

Setting the derivative of this with respect to x equal to $\pm r$, and integrating gives $f(x) = \pm rx + c$. Combining this with equation (10.36) and using the boundary condition $u = u_o$ at $x = z = 0$ yields:

$$u = \pm rx - 2g_x \int_0^z \lambda \rho z\, dz + u_o. \tag{10.37}$$

The velocity at the surface, u_s, is $\pm rx + u_o$; u_o is the velocity at the origin, $x = 0$, and rx is the increase (or decrease) in velocity between the origin and the point in question as a result of longitudinal straining. In a real glacier, $r = r(x)$, so one would have to integrate over x to obtain u_s in this way. In practice, we would be more likely to simply take u_s as known. Accordingly, we will replace $\pm rx + u_o$ with u_s in equation (10.37).

To proceed further, we must assume a flow law; as before, we use $\dot{\varepsilon} = \lambda\sigma = (\sigma/B)^n$ with B and n constant, independent of position. Hence,

$$\lambda = \frac{\sigma^{n-1}}{B^n} \tag{10.38}$$

and equation (10.37) becomes (assuming that ρ is independent of depth):

$$u = u_s - \frac{2\rho g_x}{B^n} \int_0^z \sigma^{n-1} z\, dz. \tag{10.39}$$

To integrate this, σ must be expressed in terms of z. From equation (10.33) with $c_1 = \mp r$:

$$\lambda = \frac{r}{\sqrt{\sigma^2 - (\rho g_x z)^2}} = \frac{\sigma^{n-1}}{B^n}.$$

Rearranging, this becomes

$$z = \frac{\sqrt{\sigma^{2n} - r^2 B^{2n}}}{\rho g_x \sigma^{n-1}} \tag{10.40}$$

whence:

$$\frac{dz}{d\sigma} = \frac{1}{\rho g_x \sigma^{n-1}} \left(\frac{n\sigma^{2n-1}}{\sqrt{\sigma^{2n} - r^2 B^{2n}}} \right) - \frac{\sqrt{\sigma^{2n} - r^2 B^{2n}}}{\rho g_x \sigma^{2(n-1)}} (n-1)\sigma^{n-2}$$

and noting that when $z = 0$, $\sigma^{2n} = r^2 B^{2n}$, or $\sigma = r^{1/n} B$, equation (10.39) becomes

$$u = u_s - \frac{2}{B^n} \int_{r^n B}^{\sigma} \sqrt{\sigma^{2n} - r^2 B^{2n}} \left[\frac{n\sigma^{2n-1}}{\rho g_x \sigma^{n-1} \sqrt{\sigma^{2n} - r^2 B^{2n}}} - (n-1)\frac{\sqrt{\sigma^{2n} - r^2 B^{2n}}}{\rho g_x \sigma^{2(n-1)}} \sigma^{n-2} \right] d\sigma.$$

Carrying out the integration, we obtain

$$u = u_s - \frac{2}{\rho g_x (n+1)} \left[\sigma\left(\frac{\sigma}{B}\right)^n - r^2 B^{2n}(n+1)\sigma^{1-n} + nr^{\frac{1}{n}+1} B \right] \tag{10.41}$$

which, together with equation (10.40), provides the desired solution for u in terms of z. Expressing u explicitly in terms of z is awkward. Rather, it is better to assume a value of σ, and to use it to calculate the depth z to that σ and the velocity at that depth.

When $r = 0$, these equations reduce to

$$u = u_s - \frac{2}{\rho g_x (n+1)} \left[\sigma\left(\frac{\sigma}{B}\right)^n \right]$$

and

$$z = \frac{\sigma^n}{\rho g_x \sigma^{n-1}} = \frac{\sigma}{\rho g_x}.$$

Thus:

$$u = u_s - \frac{2}{n+1}\left(\frac{\rho g_x}{B}\right)^n z^{n+1}$$

which is the same as equation 5.6. A fundamental assumption made in the derivation of equation (5.6) should now be more meaningful: namely that all strain rates other than shear strain parallel to the bed, $\dot{\varepsilon}_{zx}$, were negligible. In deriving equation (10.41) we added only one additional strain rate, $\dot{\varepsilon}_{xx}\, (= r)$, yet the complexity of the solution increased significantly.

Now that we have an expression relating σ and z, we can plot stress distributions from equations (10.29). This is done in Figure 10.5 for a glacier with a longitudinal strain rate of 0.1 a^{-1} resting on a bed with a slope of 0.1. As in the perfectly plastic case (Fig. 10.2), σ_{zz} and σ_{zx} vary linearly with depth, whereas σ_{xx} varies nonlinearly and is also double valued. Furthermore, for any given depth σ is a function of n [equation (10.40)]. Thus, σ_{xx} varies with n. As n becomes large, the solution for σ_{xx} converges on the elliptic distribution obtained earlier (Fig. 10.2). For $n = 1$, σ and hence σ_{xx} decrease linearly with depth. As $P\ [= \frac{1}{2}(\sigma_{xx} + \sigma_{zz})]$ also decreases linearly with depth with the same constant of proportionality, ρg_z, $\sigma_{xx}{}'$ becomes independent of depth.

Further insight may be gained by considering the case when $\sigma = 0$. If $r \neq 0$, z becomes indefinite. This is because although $\sigma_{zx} = 0$ at the surface, $\sigma = \frac{1}{2}\sigma_{ij}\sigma_{ij} > 0$ there. Thus, when there is longitudinal strain, there is no place in the slab where $\sigma = 0$.

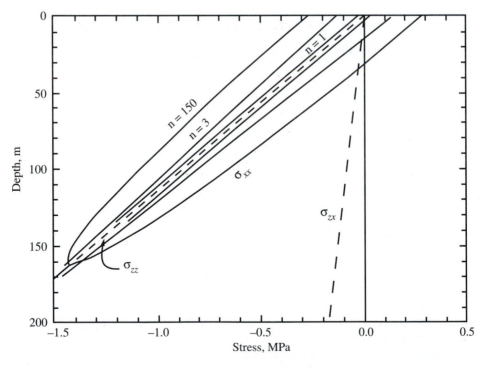

Figure 10.5   Depth variation of stress in an ice sheet with a power-law rheology. The ice sheet consists of a slab of uniform thickness and density resting on a bed with a uniform slope. The distribution of σ_{xx} is given by the curves labeled with values of n. There is one curve for extending and one for compressive flow. The distributions of σ_{zz} and σ_{zx} are shown by the dashed lines and are the same for all n. Curves are drawn for a glacier with a viscosity, B, of 0.141 MPa a$^{-1/3}$ and a longitudinal strain rate of 0.1 a^{-1} resting on a bed with a slope of 0.1. As $n \rightarrow \infty$, the thickness of the ice sheet is limited to $h = B/\rho g_x$, which in this case is ~160 m.

Two velocity profiles calculated from equations (10.40) and (10.41) are shown in Figure 10.6. One profile is calculated with $r = 0$, and the other with $r = 0.1$ a^{-1}. In both cases u_s is adjusted to yield $u_b = 20$ m a^{-1}. One might initially think that the higher speed represented by the dashed profile in Figure 10.6 was a consequence of longitudinal stretching. However, r is a positive constant and $\partial u/\partial x = \pm r$. Thus, the dashed profile is applicable to both compressive flow and extending flow. This is because the magnitude of the increase in σ, and hence in λ, resulting from the addition of a longitudinal stress, is independent of whether the longitudinal stress is compressive or extending. As λ increases so does $\dot\varepsilon_{ij}$ for any given σ_{ij} [equation (9.24)]. Specifically, $\dot\varepsilon_{zx}$ $(= \partial u/\partial z)$ increases, regardless of whether σ_{xx} is positive or negative.

COMPARISON WITH REAL GLACIERS

Real glaciers are not slabs of ice of uniform thickness, nor are they perfectly plastic. Thus, it is relevant to discuss what aspects of the solutions we have obtained are applicable in reality.

Consider first the result that σ_{zz} and σ_{zx} vary linearly with depth. Such a linear variation is often assumed in studies of real glaciers and is a reasonable approximation in many situations.

In addition we found that, in general, flow is extending in accumulation areas and compressive in ablation areas. This is because r varies directly with b_n (if the glacier is not too far from a steady state), and σ varies directly with r for any given depth [equation (10.40)]. Thus, σ_{xx} varies directly with b_n, being positive when b_n is positive and conversely. In an actual glacier, however, longitudinal stresses also depend on factors such as the curvature of the longitudinal surface profile and the rate of change in thickness. Thus, it is not productive to try to calculate longitudinal stresses in a real glacier with the use of the theory presented here.

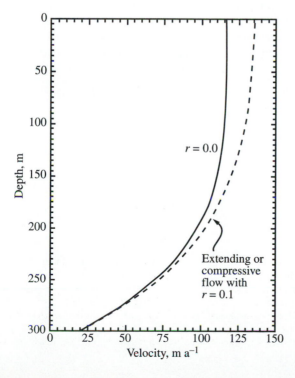

Figure 10.6 Velocity profiles, calculated from equations (10.40) and (10.41), in a glacier consisting of an infinite slab of ice, 300 m thick, resting on a bed with slope of 0.046 and sliding with a speed, u_b, of 20 m a^{-1}. Calculations use $n = 3$ and $B = 0.141$ MPa $a^{-1/3}$, and two different values of r as shown.

The physical processes by which the longitudinal strain rate is adjusted to balance b_n can be visualized qualitatively. If, in some location, r is too large so that thinning by extension exceeds thickening by accumulation, the profile will tend to become concave and this will have a tendency to decrease r.

Also relevant to real glaciers is the effect of longitudinal stress on velocities. Clearly, longitudinal compression should result in upward vertical velocities, and conversely. The linear decrease in w with depth, however, is an artifact of our slab model [see discussion following equation (10.35)], although it is a commonly used approximation in calculations. With respect to vertical profiles of u, longitudinal stresses, whether extending or compressive, increase the effective stress, so they increase $\partial u/\partial z$ throughout the profile. This may be particularly evident near a glacier surface where σ_{zx} is small so $\dot{\varepsilon}_{zx}$ would be negligible were it not for the contribution of σ'_{xx} to σ. We will find in the next chapter, however, that certain peculiarities of measured deformation profiles cannot be explained in this way.

SUMMARY

In this chapter we have shown that equations (9.27) can be solved for the three components of the velocity vector and nine components of the stress tensor in certain simple situations. Our solutions were for a slab of infinite horizontal extent resting on a bed with a uniform slope. We first obtained a solution for a perfectly plastic material, and found that the thickness of the slab was constrained by the yield strength of the material. We then obtained solutions for a nonlinear material, which are more relevant to real glaciers.

We found that σ_{zz} and σ_{zx} vary linearly with depth, which is probably a reasonable approximation to the situation in real glaciers. We also found that longitudinal stresses should be extending in accumulation areas and compressive in ablation areas, although the magnitude of the longitudinal stresses is not well constrained by our simple model.

Vertical velocities vary linearly with depth for the idealized situation that we studied, and this is commonly used as a first approximation in real glaciers (e.g., Chapter 6). Horizontal velocities decrease nonlinearly with depth, as we found in Chapter 5. In this chapter, however, we were able to investigate the effect of longitudinal stresses on the velocity profile, and we found that either longitudinal extension or longitudinal compression will increase $\partial u/\partial z$, leading to higher velocities.

11

Applications of Stress and Deformation Principles to Classical Problems

In this chapter, we will study some glaciologically significant problems for which an appreciation of the material presented in Chapters 9 and 10 is required. Our objective is not to provide a comprehensive overview of theoretical glaciology, but rather to solidify the gains made in the preceding two chapters by applying the principles developed therein. In the course of this discussion, the student will be introduced to some definitive studies frequently referenced in the glaciological literature.

Let us first consider the problem of closure of a cylindrical borehole, in part because this is relevant to our earlier discussion of glacier hydrology. We will then investigate efforts to calculate basal shear stresses using a force balance model, followed by study of the creep of ice shelves. Finally, the problem of using borehole deformation experiments to obtain estimates of the values of the parameters in the flow law will round out the chapter.

COLLAPSE OF A CYLINDRICAL HOLE

The first problem we address is that of the closure of a cylindrical hole in ice. This problem was studied by Nye (1953) in the context of using closure rates of tunnels in ice to estimate the constants in Glen's flow law, and our development is based on Nye's paper. More recently, the theory has been used to analyze two problems in water flow at the base of a glacier: (1) the closure of a water conduit, and (2) leakage of water into or away from a subglacial conduit. We used the first of these analyses in Chapter 8 [equation (8.3)].

Our approach is very similar to that used in Chapter 10 to obtain stresses and velocities in a "glacier" consisting of a slab of ice resting on a bed with a uniform slope. We first reduce the problem to one of plane strain by setting it up so that there is no deformation parallel to the axis of the hole. This reduces the number of unknowns from nine—the three components of the velocity vector and six components of the stress tensor—to five. Expressions for the stresses and strain rates are then obtained and inserted into the stress equilibrium and continuity equations. Finally, we use a constitutive relation between stress and stain rate, Glen's flow law, to obtain the desired solution.

Consider, first, the closure of a hole of radius a in an infinite weightless medium (Fig. 11.1). There is no strain parallel to its axis, the z-direction. Once the solution to this problem is obtained, we will modify it to apply to a real glacier. On the surface of the hole, at $r = a$, an internal tension, σ_a, is (somehow) applied. Eventually, this stress will be equated with that resulting from the difference between the pressure in the ice and that in the hole.

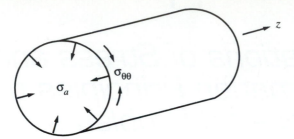

Figure 11.1 Stresses on the wall of a cylindrical hole in a weightless medium.

As there is no deformation in the z-direction, $\dot{\varepsilon}_{zz} = \lambda\sigma'_{zz} = 0$. Thus,

$$\sigma'_{zz} = 0 = \sigma_{zz} - \frac{1}{3}(\sigma_{rr} + \sigma_{\theta\theta} + \sigma_{zz})$$

or

$$\sigma_{zz} = \frac{1}{2}(\sigma_{rr} + \sigma_{\theta\theta}). \tag{11.1}$$

The mean stress thus becomes

$$P = \frac{1}{3}\sigma_{kk} = \frac{1}{3}\left[\frac{3}{2}(\sigma_{rr} + \sigma_{\theta\theta})\right],$$

so

$$\sigma'_{rr} = \frac{1}{2}(\sigma_{rr} - \sigma_{\theta\theta}) \tag{11.2}$$

and

$$\sigma'_{\theta\theta} = -\frac{1}{2}(\sigma_{rr} - \sigma_{\theta\theta}). \tag{11.3}$$

This gives us expressions for the three deviatoric stresses and the mean stress.

As stated, there is no deformation in the z-direction and, in addition, the hole wall cannot support shear tractions. Thus, in view of the radial symmetry, there can be no shear stresses in the θ- and z-directions in the medium away from the hole. Therefore, σ_{rr}, $\sigma_{\theta\theta}$, and σ_{zz} are principal stresses. The effective stress is then

$$\sigma^2 = \frac{1}{2}2\left[\frac{1}{4}(\sigma_{rr} - \sigma_{\theta\theta})^2\right]$$

or

$$\sigma = \frac{1}{2}|\sigma_{rr} - \sigma_{\theta\theta}|. \tag{11.4}$$

Let us now obtain another relation between σ_{rr} and σ by considering the condition for stress equilibrium. From Figure 11.2, we have

$$\sigma_{rr}(r\,d\theta) - \left(\sigma_{rr} + \frac{d\sigma_{rr}}{dr}\,dr\right)(r + dr)\,d\theta + 2\sigma_{\theta\theta}dr\frac{d\theta}{2} = 0.$$

Canceling like terms of opposite sign, dividing by $dr\,d\theta$, and ignoring the term still containing a differential yields

$$r\frac{d\sigma_{rr}}{dr} + \sigma_{rr} - \sigma_{\theta\theta} = 0$$

or, using equation (11.4)

$$\frac{d\sigma_{rr}}{dr} + \frac{2\sigma}{r} = 0. \tag{11.5}$$

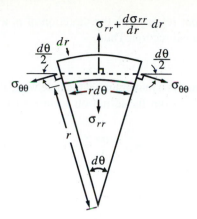

Figure 11.2 Stresses on a segment of the wall of a cylindrical hole.

At any radius $r \geq a$ the stress is σ_{rr}, while at infinity it is 0. Thus, equation (11.5) may be integrated:

$$\int_{\sigma_{rr}}^{0} d\sigma_{rr} = -\int_{r}^{\infty} \frac{2\sigma}{r} \, dr$$

to yield

$$\sigma_{rr} = \int_{r}^{\infty} \frac{2\sigma}{r} \, dr. \tag{11.6}$$

In order to integrate equation (11.6), we must express r in terms of σ. We will do this by determining the velocity field, and hence $\dot{\varepsilon}$, and inserting a flow law. Let u be the velocity in the radial direction. Other velocities vanish owing to the radial symmetry and the absence of deformation in the z-direction. Conservation of mass requires that the mass flux through two concentric cylindrical surfaces with radii r and $r + dr$ (Fig. 11.3) must be the same in an incompressibile medium. Thus, we have

$$2u\pi r = 2\left(u + \frac{du}{dr} dr\right)\pi (r + dr),$$

which may be simplified, thus:

$$\frac{du}{dr} = -\frac{u}{r},$$

and integrated to yield

$$ln\, u = -ln\, r + ln\, c.$$

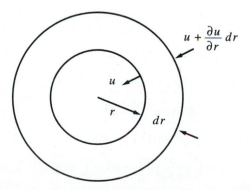

Figure 11.3 Radial velocity field around a cylindrical hole.

Because we obtained this without referring to the direction of u, a minus sign is now inserted to indicate that u is in the negative r-direction. Thus,

$$u = -\frac{c}{r}. \tag{11.7}$$

So u is directed toward the hole and is a maximum on the hole wall, at $r = a$. It decreases to 0 at $r = \infty$.

Now $\dot{\varepsilon}_{rr}$ may be calculated from its definition in terms of velocity derivatives [equation (9.18)]; thus,

$$\dot{\varepsilon}_{rr} = \frac{\partial u}{\partial r} = -\frac{u}{r},$$

and by continuity, since $\dot{\varepsilon}_{zz} = 0$,

$$\dot{\varepsilon}_{\theta\theta} = -\dot{\varepsilon}_{rr} = \frac{u}{r}.$$

The effective strain rate is thus:

$$\dot{\varepsilon}^2 = \frac{1}{2}\dot{\varepsilon}_{ij}\dot{\varepsilon}_{ij} = \frac{1}{2}\left(2\frac{u^2}{r^2}\right).$$

As $u = -c/r$, this becomes

$$\dot{\varepsilon} = \frac{c}{r^2}. \tag{11.8}$$

We can retain the generality of the solution a little longer before inserting a flow law. Because $\dot{\varepsilon} = f(\sigma)$, we have

$$f(\sigma) = \frac{c}{r^2}. \tag{11.9}$$

To obtain r in terms of σ, note that [using equation (11.9)]

$$\frac{df(\sigma)}{d\sigma} = \frac{df(\sigma)}{dr}\frac{dr}{d\sigma} = -\frac{2c}{r^3}\frac{dr}{d\sigma}.$$

Again using equation (11.9) to eliminate c:

$$\frac{df(\sigma)}{d\sigma} = -\frac{2}{r}f(\sigma)\frac{dr}{d\sigma}$$

so:

$$dr = -\frac{r}{2}\frac{df(\sigma)}{f(\sigma)}.$$

Substituting this into equation (11.6) yields

$$\sigma_{rr} = -\int \frac{2\sigma}{r}\frac{r}{2}\frac{df(\sigma)}{f(\sigma)} = -\int \frac{\sigma}{f(\sigma)}df(\sigma). \tag{11.10}$$

It is now necessary to make use of a flow law, specifically Glen's flow law, to obtain an analytical expression for the functional relation in equation (11.9); thus,

$$\dot{\varepsilon} = f(\sigma) = \left(\frac{\sigma}{B}\right)^n = \frac{c}{r^2} \tag{11.11}$$

whence, differentiating:

$$df(\sigma) = \frac{n\sigma^{n-1}}{B^n}d\sigma.$$

With the use of these last two relations, equation (11.10) becomes

$$\sigma_{rr} = \int_0^\sigma \frac{\sigma}{(\sigma/B)^n}\frac{n\sigma^{n-1}}{B^n}d\sigma = \int_0^\sigma n\,d\sigma$$

so,

$$\sigma_{rr} = n\sigma. \qquad (11.12)$$

In other words, the radial stress at any distance $r \geq a$ from the hole wall is simply n times the effective stress. One cannot help but be impressed by the simplicity and elegance of this result, considering the effort required to obtain it. Unfortunately, real life is rarely so conveniently uncomplicated. Furthermore, we still have some way to go before obtaining relations that can be applied to real glaciers.

In preparation for relaxing the assumption that the medium is weightless, let us now scale this solution to the normal stress, σ_a, on the hole wall. Using the last equality in equation (11.11) to obtain an expression for σ, σ_{rr} and σ_a become:

$$\sigma_{rr} = nB \left(\frac{c}{r^2} \right)^{1/n} \quad \text{and} \quad \sigma_a = nB \left(\frac{c}{a^2} \right)^{1/n},$$

whence:

$$\frac{\sigma_{rr}}{\sigma_a} = \left(\frac{a}{r} \right)^{2/n}. \qquad (11.13)$$

Solutions for the remaining stresses can now be written similarly, as follows: From equations (11.4) and (11.12):

$$\sigma = \frac{1}{2} (\sigma_{rr} - \sigma_{\theta\theta}) = \frac{1}{2} (n\sigma - \sigma_{\theta\theta})$$

whence, transposing and again using the last equality in equation (11.11):

$$\sigma_{\theta\theta} = (n-2)\,\sigma = (n-2)\,B \left(\frac{c}{r^2} \right)^{1/n}$$

so:

$$\frac{\sigma_{\theta\theta}}{\sigma_a} = \frac{n-2}{n} \left(\frac{a}{r} \right)^{2/n}. \qquad (11.14)$$

Then, from equations (11.1), (11.13), and (11.14):

$$\frac{\sigma_{zz}}{\sigma_a} = \frac{1}{2} \left(\frac{\sigma_{rr}}{\sigma_a} + \frac{\sigma_{\theta\theta}}{\sigma_a} \right) = \frac{1}{2} \left[\left(\frac{a}{r} \right)^{2/n} + \frac{n-2}{n} \left(\frac{a}{r} \right)^{2/n} \right] = \frac{1}{2} \left[\left(\frac{a}{r} \right)^{2/n} \left(1 + \frac{n-2}{n} \right) \right],$$

so

$$\frac{\sigma_{zz}}{\sigma_a} = \frac{n-1}{n} \left(\frac{a}{r} \right)^{2/n}. \qquad (11.15)$$

Finally, from equations (11.4), (11.13), and (11.14):

$$\frac{\sigma}{\sigma_a} = \frac{1}{2} \left(\frac{\sigma_{rr}}{\sigma_a} - \frac{\sigma_{\theta\theta}}{\sigma_a} \right) = \frac{1}{2} \left[\left(\frac{a}{r} \right)^{2/n} - \frac{n-2}{n} \left(\frac{a}{r} \right)^{2/n} \right].$$

Thus,

$$\frac{\sigma}{\sigma_a} = \frac{1}{n} \left(\frac{a}{r} \right)^{2/n}. \qquad (11.16)$$

We can now relax the assumption that the medium is weightless. Suppose we have a horizontal hole at atmospheric pressure at a depth h_o in a real glacier. The hydrostatic pressure in the glacier is $\mathcal{P} = \rho g h$, and around the hole it is $\mathcal{P} = \rho g h_o$. Note that $\mathcal{P}$ is not equal to the mean stress, $P \,(= \frac{1}{3}\sigma_{kk})$. If $h_o \gg a$, $\mathcal{P}$ will be nearly uniform around the hole. We have not previously specified the magnitude of σ_a, so let us now solve equations (11.13) through (11.16) for σ_{ii} ($i = r, \theta, z$) with $\sigma_a = \mathcal{P}$.

Furthermore, because $\mathcal{P}$ is hydrostatic, let us add a compressive stress, $-\mathcal{P}$, to the solutions. This is valid because a hydrostatic pressure influences all of the stresses equally, and therefore does not affect the local differences among the stresses given by equations (11.13) through (11.16). The resulting equations are

$$\sigma_{rr} = \mathcal{P}\left(\frac{a}{r}\right)^{2/n} - \mathcal{P} = \mathcal{P}\left[\left(\frac{a}{r}\right)^{2/n} - 1\right]$$

$$\sigma_{\theta\theta} = \mathcal{P}\left(\frac{n-2}{n}\right)\left(\frac{a}{r}\right)^{2/n} - \mathcal{P} = \mathcal{P}\left[\left(\frac{n-2}{n}\right)\left(\frac{a}{r}\right)^{2/n} - 1\right]$$

$$\sigma_{zz} = \mathcal{P}\left(\frac{n-1}{n}\right)\left(\frac{a}{r}\right)^{2/n} - \mathcal{P} = \mathcal{P}\left[\left(\frac{n-1}{n}\right)\left(\frac{a}{r}\right)^{2/n} - 1\right] \qquad (11.17)$$

$$\sigma = \frac{1}{2}(\sigma_{rr} - \sigma_{\theta\theta}) = \frac{\mathcal{P}}{n}\left(\frac{a}{r}\right)^{2/n},$$

and the mean stress is

$$\frac{1}{3}\sigma_{kk} = \frac{\mathcal{P}}{3}\left[\left(\frac{a}{r}\right)^{2/n} - 1 + \left(\frac{n-2}{n}\right)\left(\frac{a}{r}\right)^{2/n} - 1 + \left(\frac{n-1}{n}\right)\left(\frac{a}{r}\right)^{2/n} - 1\right] \qquad (11.18)$$

$$= \mathcal{P}\left[\left(\frac{n-1}{n}\right)\left(\frac{a}{r}\right)^{2/n} - 1\right].$$

Now, the stress causing closure is no longer a hypothetical traction on the inside of the hole, σ_a. Rather, it is the real hydrostatic stress in the medium. Note that all of the stresses decrease to $-\mathcal{P}$ (i.e., compressive) at large distances from the hole.

It is easy to show that the corresponding deviatoric stresses are

$$\sigma'_{rr} = -\sigma'_{\theta\theta} = \frac{\mathcal{P}}{n}\left(\frac{a}{r}\right)^{2/n} \qquad (11.19)$$

$$\sigma'_{zz} = 0$$

That $\sigma'_{rr} = -\sigma'_{\theta\theta}$ and $\sigma'_{zz} = 0$ are a consequence of our assumption of plane strain.

Setting $r = a$ in equations (11.17) we obtain the stresses on the hole wall:

$$\sigma_{rr} = 0$$

$$\sigma_{\theta\theta} = -\frac{2\mathcal{P}}{n}$$

$$\sigma_{zz} = -\frac{\mathcal{P}}{n} \qquad (11.20)$$

$$\sigma = \frac{\mathcal{P}}{n}$$

Tunnel and Borehole Closure

These relations have been used to determine values of the constants n and B in the flow law with the use of measurements of the rate of closure of a tunnel or borehole. To do this, it is necessary to incorporate the relations into the flow law. Deviatoric stresses are thus required. Because we are interested in closure, only $\sigma'_{rr(r=a)}$ is needed. Thus, using the first of equations (11.19) with $r = a$ and noting that $\sigma_{r=a} = \mathcal{P}/n$ [see equations (9.26) and (10.38)]:

$$\dot{\varepsilon}_{rr(r=a)} = \frac{\sigma^{n-1}}{B^n}\sigma'_{rr(r=a)} = \frac{(\mathcal{P}/n)^{n-1}}{B^n}\frac{\mathcal{P}}{n}. \qquad (11.21)$$

Because $\dot{\varepsilon}_{rr(r=a)} = -u_a/a$ where u_a is the closure rate, we obtain

$$-\frac{u_a}{a} = \left(\frac{\mathcal{P}}{nB}\right)^n. \tag{11.22}$$

To use equation (11.22) to estimate the constants in the flow law, one needs values of u_a, a, and $\mathcal{P}$ at two or more places. Inserting values for two such places in equation (11.22) would yield two equations with two unknowns (n and B). With three or more sets of data, it is useful to plot $\log \dot{\varepsilon}$ against $\log \mathcal{P}$ as in Figure 11.4.

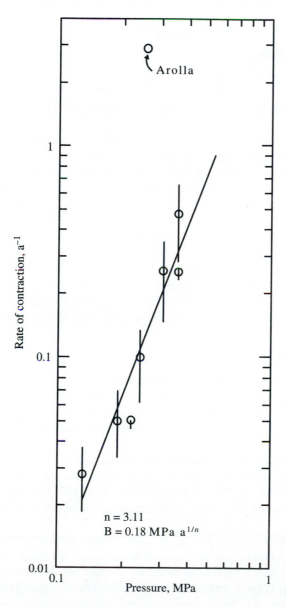

n = 3.11
B = 0.18 MPa a$^{1/n}$

Figure 11.4 Rate of contraction, u_a, of tunnel sections plotted against overburden pressure, $\mathcal{P}$. (Replotted from Nye, 1953, Fig. 1. Reproduced with permission of the author and The Royal Society of London.)

Some caution is required, however. In tunnel-closure studies, for example, pegs are normally inserted in the tunnel walls, and closure is measured by determining the change in distance between the heads of pegs on opposite sides of the tunnel. In this case, particularly in temperate glaciers, the point where the pegs are actually gripped by the ice may be some distance back in the wall. Furthermore, such tunnels are rarely if ever circular in cross section. Thus, the correct value of a must be guessed.

In borehole-closure studies, closure rates are measured with calipers, so determining the appropriate value of a is not a problem. However, the time interval between measurements is often fairly large, and a substantial amount of closure may occur between measurements. In this case, approximating u_a by $\Delta a/\Delta t$ is likely to yield a poor estimate (Paterson, 1977). The correct procedure is to use the temporal mean value of u_a/a, which is, by the definition of a mean,

$$\overline{\dot{\varepsilon}_a} = -\frac{1}{\Delta t}\int_{t_1}^{t_2}\frac{u_a}{a}\,dt.$$

Noting that $u_a = da/dt$, this becomes

$$\overline{\dot{\varepsilon}_a} = -\frac{1}{\Delta t}\int_{t_1}^{t_2}\frac{1}{a}\frac{da}{dt}\,dt = -\frac{1}{\Delta t}\ln\frac{a_2}{a_1}.$$

Some results from four borehole-closure studies are presented in Figure 11.5 (points labeled P, H, and G).

It is instructive to look at the results in Figures 11.4 and 11.5 in somewhat greater detail. In Figure 11.4, it will be noted that several of the sets of tunnel-closure data fall along a line with $n = 3.11$ and $B = 0.18$ MPa a$^{1/n}$, values that are quite consistent with other data. (This value of B is also plotted as point N in Figure 11.5.) However, the point representing data from the Arolla ice tunnel falls well above the line. The Arolla tunnel is at the base of an ice fall. Thus, owing to the contribution of longitudinal stresses, σ may be significantly higher than $\mathcal{P}/n$ here, and, as observed, one would expect the actual closure rate to be higher than that calculated using $(\mathcal{P}/n)^{n-1}$ to approximate σ^{n-1} in equation (11.21).

The problem with the borehole-closure rates in Figure 11.5, which seem to be too low and therefore yield values of B that appear to be too high, is different. Here, we speculate that the crystallographic fabric in the ice is adjusted to a stress regime in which the dominant deviatoric stress is simple shear normal to the axis of the hole. Such a fabric may have inhibited closure.

Subglacial Water Conduits

We have already seen equation (11.22) applied to closure of subglacial water conduits in Chapter 8. As noted there, problems arise when one attempts to estimate closure rates of semicircular conduits, owing to drag on the bed. Even more profound difficulties arise in attempting to estimate closure rates of broad low conduits, as stresses in the ice are no longer symmetrically distributed about the conduit.

Here, we look into another problem of interest: the normal stresses on the bed at the boundaries of a semicircular conduit, and in particular, the gradient in these stresses outward from the conduit (Fig. 11.6). This problem was first studied by Weertman (1972). If pressures are higher adjacent to the conduit, water in a film at the ice-bed interface will be forced away from the conduit, and conversely.

The significance of this problem lies in its application to water flow beneath polar ice sheets. Several authors have suggested that for conduits to exist beneath such ice masses in the absence of water inputs from the glacier surface, there must be a flux of water into the conduit from adjacent parts of the bed (Alley, 1989a; Walder, 1982; Weertman & Birchfield, 1983). The problem of the

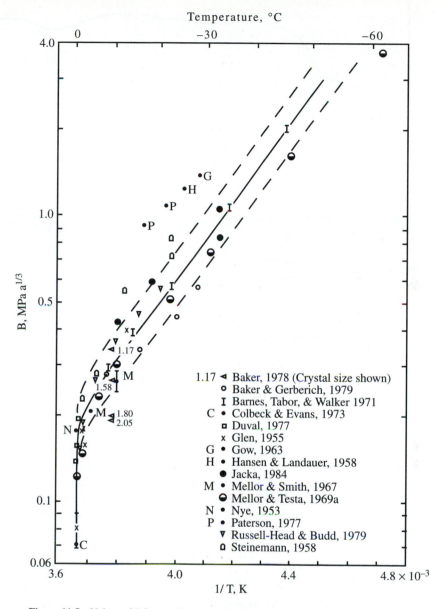

Figure 11.5 Values of B from various experiments in which the minimum strain rate was measured or estimated. Octahedral stresses and strain rates were used in calculating B. (After Hooke, 1981, Fig. 2.)

existence of such conduits is fundamental; where they are present, subglacial water pressures are probably appreciably lower than otherwise. Thus, any attempt to explain, for example, the fast flow of ice streams hinges upon an understanding of the nature of the water flow system.

The relevant stress in this problem is $\sigma_{\theta\theta}$. Thus, let us start with the expression for $\sigma_{\theta\theta}$ in equation (11.14). Note that in so doing, we tacitly assume that the bed is flat and slippery so that shear stresses do not impede movement of ice inward toward the tunnel. The appropriate value for σ_a is now the difference between the pressure in the ice and that in the water in the conduit, $\Delta\mathcal{P}$.

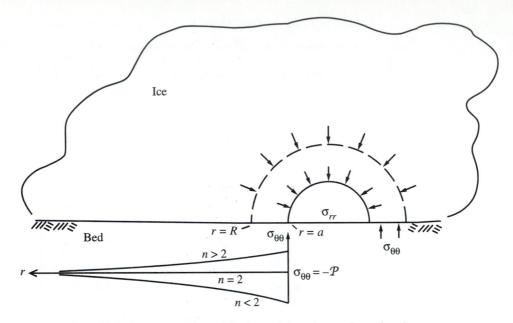

Figure 11.6   Stresses around a semicircular conduit. σ_{rr} is extending and $\sigma_{\theta\theta}$ is compressive (hence the minus sign on $\mathcal{P}$). The variation in $\sigma_{\theta\theta}$ away from the tunnel is shown schematically. For $n > 2$ $\sigma_{\theta\theta}$ is less compressive than $\mathcal{P}$, and conversely.

As before, we add a pressure, $-\mathcal{P}$, everywhere to account for the weight of the ice. With a little rearranging, equation (11.14) thus becomes

$$\sigma_{\theta\theta} = \frac{n-2}{n}\Delta\mathcal{P}\,a^{\frac{2}{n}}\,r^{-\frac{2}{n}} - \mathcal{P}. \tag{11.14a}$$

Note that $\sigma_{\theta\theta}$ is negative, or compressive, as $\mathcal{P}$ always exceeds the first term on the right.

It may appear from equation (11.14a) that $\sigma_{\theta\theta}$ will not support the weight of glacier when $n > 2$, as $\sigma_{\theta\theta} \to -\mathcal{P}$ as $r \to \infty$ but $\sigma_{\theta\theta} > -\mathcal{P}$ (ie, less compressive) near the conduit. In other words, $\sigma_{\theta\theta}$ is sufficiently compressive to support the glacier far from the conduit but not near or beneath it. However, $\sigma_{rr(r=R)}$ is more compressive than $\sigma_{rr(r=a)}$ [see first of equations (11.17)], and this provides the additional support. In other words, referring to Figure 11.6, the vertical force acting on the surface at radius R balances that on the bed, $2\int_0^a (\Delta\mathcal{P} - \mathcal{P})\, dr + 2\int_a^R (\sigma_{\theta\theta} - \mathcal{P})\, dr$.

Let us now consider a semicircular conduit at a depth h_o on a horizontal bed beneath an ice sheet of uniform thickness and infinite horizontal extent. Taking the derivative of $\sigma_{\theta\theta}$ with respect to r along the bed yields

$$\frac{d\sigma_{\theta\theta}}{dr} = -\frac{2}{n}\left(\frac{n-2}{n}\right)\Delta\mathcal{P}\,a^{\frac{2}{n}}\,r^{-\left(\frac{2}{n}+1\right)}. \tag{11.23}$$

If $n > 2$, as might be expected, $d\sigma_{\theta\theta}/dr$ is negative. Thus, $\sigma_{\theta\theta}$ decreases, or becomes more negative, or more compressive, away from the tunnel (Fig. 11.6). In this case, water in a film will be forced toward the conduit, enhancing discharge in it. However, when one considers coupling of stresses, particularly where there is a shear stress on the bed parallel to the conduit, the situation is not so simple. It appears that in this case, water flow may be away from the tunnel (Weertman, 1972, pp. 299–300).

The physical reason for the change in behavior of $d\sigma_{\theta\theta}/dr$ with n is not obvious. We might expect that if a cavity is introduced at the base of a glacier, compressive stresses adjacent to the cavity would increase in order to support that part of the weight of the glacier that is no longer supported by the bed under the cavity. However, toward the tunnel u, and hence $\dot{\varepsilon}_{rr}$, increase and this requires an increase in σ'_{rr}. The way in which the stress field is modified to satisfy this requirement, and hence the way in which the pressure on the bed is redistributed, depends upon n. A more intuitive explanation of this effect is elusive.

CALCULATING BASAL SHEAR STRESSES USING A FORCE BALANCE

To a first approximation, the basal drag can be estimated from $\tau_b = \rho g h \alpha$ (or $\tau_b = S_f \rho g h \alpha$ in a valley glacier). However, if longitudinal forces are unbalanced, τ_b may be either greater or less than $\rho g h \alpha$. For example, in Figure 11.7, the body force, $\rho g h$, has a downslope component, $\rho g h \alpha$. In addition, there are longitudinal forces F_u and F_d. If $F_u > F_d$, as suggested by the lengths of the arrows in Figure 11.7, τ_b will clearly have to be greater than $\rho g h \alpha$ in order to balance forces in the x-direction, and conversely. We now explore this effect in greater detail. The first part of the development is a three-dimensional generalization of an approach suggested by B. Hanson (Hooke & Hanson, 1986, p. 268).

Because our goal is to calculate the drag exerted on the glacier by the bed, the stress-equilibrium equations [equation (9.27b)] are the obvious starting point for the analysis. The coordinate system to be used is shown in Figure 11.8. The z-axis is vertical. Writing out the stress-equilibrium equations in the x- and z-directions, remembering that $\sigma'_{ij} = \sigma_{ij} - \delta_{ij}P$, leads to

$$\frac{\partial \sigma'_{xx}}{\partial x} + \frac{\partial \sigma_{yx}}{\partial y} + \frac{\partial \sigma_{zx}}{\partial z} + \frac{\partial P}{\partial x} = 0 \tag{11.24a}$$

and

$$\frac{\partial \sigma_{xz}}{\partial x} + \frac{\partial \sigma_{yz}}{\partial y} + \frac{\partial \sigma'_{zz}}{\partial z} + \frac{\partial P}{\partial z} = \rho g. \tag{11.24b}$$

τ_b is equal to σ_{zx} at $z = h$, so the procedure will be to solve equation (11.24b) for P, substitute the result into equation (11.24a), and integrate over the depth to obtain $\sigma_{zx(z=h)}$.

To solve equation (11.24b) for P, integrate from the surface to depth z:

$$\int_0^z \frac{\partial \sigma_{xz}}{\partial x} dz + \int_0^z \frac{\partial \sigma_{yz}}{\partial y} dz + \int_0^z d\sigma'_{zz} + \int_0^z dP = \int_0^z \rho g dz$$

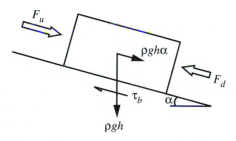

Figure 11.7 Longitudinal forces on a segment of a glacier on a sloping bed. If $F_u > F_d$, the basal drag will be greater than $\rho g h \alpha$, and conversely.

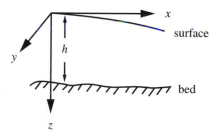

Figure 11.8 Coordinate system used in force-balance analysis.

or, noting that $P = 0$ at the surface:

$$\int_0^z \frac{\partial \sigma_{xz}}{\partial x} dz + \int_0^z \frac{\partial \sigma_{yz}}{\partial y} dz + \sigma'_{zz} - \sigma'_{zz}|_{z=0} + P = \rho g z.$$

Now take the horizontal derivative, assuming that $\partial \sigma'_{zz}/\partial x|_{z=0}$ is negligible, and substitute the result into equation (11.24a); thus,

$$\frac{\partial \sigma'_{xx}}{\partial x} + \frac{\partial \sigma_{yx}}{\partial y} + \frac{\partial \sigma_{zx}}{\partial z} - \int_0^z \frac{\partial^2 \sigma_{xz}}{\partial x^2} dz - \int_0^z \frac{\partial^2 \sigma_{yz}}{\partial x \partial y} dz - \frac{\partial \sigma'_{zz}}{\partial x} + \rho g \alpha = 0. \tag{11.25}$$

Equation (11.25) is in terms of stresses at any given level, z, in the glacier, whereas we are interested in summing the stresses over depth to obtain τ_b. Thus, as just noted, we integrate over depth:

$$\int_0^h \left(\frac{\partial \sigma'_{xx}}{\partial x} - \frac{\partial \sigma'_{zz}}{\partial x} \right) dz + \int_0^h \frac{\partial \sigma_{yx}}{\partial y} dz + \int_0^h d\sigma_{zx} - \int_0^h \int_0^z \left(\frac{\partial^2 \sigma_{xz}}{\partial x^2} + \frac{\partial^2 \sigma_{yz}}{\partial x \partial y} \right) dz dz + \rho g h \alpha = 0. \tag{11.26}$$

Clearly, $\sigma_{zx(z=h)}$ $(= \tau_b)$ will be obtained from integrating the third term, and the last term is the familiar $\rho g h \alpha$. Some simplification is obviously desirable, however.

The double integral term in equation (11.26), also sometimes referred to as the T-term, is difficult to interpret physically. Budd (1969, p. 116) has shown that, approximately:

$$\int_0^h \int_0^z \frac{\partial^2 \sigma_{xz}}{\partial x^2} dz dz \simeq \frac{1}{6} \left[\rho g \frac{\partial^2}{\partial x^2} \left(\alpha h^3 \right) \right]. \tag{11.27}$$

Van der Veen & Whillans (1989) argue that this term is related to "bridging" effects, in which the pressure on the bed varies spatially owing to the influence of bed irregularities and, particularly, cavity formation. Because ice is "soft," they suggest that these bridging effects should be small compared with the average normal pressure. Thus, they neglect the T-term in force-balance calculations, and we shall follow their lead in this respect. They acknowledge, however, that the "physical implications" of doing so are "not conceptually straightforward."

Turning to the first term in equation (11.26), σ'_{zz} can be eliminated by noting that, owing to the proportionality between deviatoric stress and strain rate, the incompressibility condition, $\dot{\varepsilon}_{xx} + \dot{\varepsilon}_{yy} + \dot{\varepsilon}_{zz} = 0$, leads to $\sigma'_{xx} + \sigma'_{yy} + \sigma'_{zz} = 0$. In addition, because σ_{zx} would be zero on a free horizontal surface, and is only slightly different from zero on the sloping glacier surface, the third term in equation (11.26) is the desired basal drag, τ_b, as just noted. With these modifications, equation (11.26) becomes

$$\int_0^h \frac{\partial}{\partial x} \left(2\sigma'_{xx} + \sigma'_{yy} \right) dz + \int_0^h \frac{\partial \sigma_{yx}}{\partial y} dz + \tau_b + \rho g h \alpha = 0. \tag{11.28}$$

[Adjusting for the fact the we have neglected bridging effects and have taken the z-direction to be positive downward, this is identical to van der Veen & Whillans' (1989) equation (25). The rest of our development follows theirs.]

Our objective now is to express equation (11.28) in a form that will allow evaluation of τ_b from strain-rate measurements at the glacier surface. To this end, we note that because

$$\sigma^{n-1} = \left(\dot{\varepsilon}^{\frac{1}{n}} B \right)^{n-1},$$

the flow law can be written

$$\dot{\varepsilon}_{ij} = \frac{\sigma^{n-1}}{B^n} \sigma'_{ij} = \frac{\dot{\varepsilon}^{\frac{(n-1)}{n}}}{B} \sigma'_{ij}.$$

Inserting this in equation (11.28), reversing the order of differentiation and integration, and rearranging terms yields:

$$\tau_b = -\rho g h \alpha - \frac{\partial}{\partial x} \int_0^h \frac{B}{\dot{\varepsilon}^{\frac{n-1}{n}}} \left(2\dot{\varepsilon}_{xx} + \dot{\varepsilon}_{yy}\right) dz - \frac{\partial}{\partial y} \int_0^h \frac{B}{\dot{\varepsilon}^{\frac{n-1}{n}}} \dot{\varepsilon}_{yx} dz. \tag{11.29}$$

The drag exerted by the bed on the glacier is in the negative x-direction in the coordinate system we are using; thus τ_b is negative. Van der Veen & Whillans (1989) developed a numerical procedure to carry out the integration over depth, z. However, for simple applications we assume that strain rates are independent of depth, and express equation (11.29) in finite difference form; thus,

$$-\tau_b = -\rho g h \alpha - B \dot{\varepsilon}^{\frac{1-n}{n}} \left[\frac{\left(2\dot{\varepsilon}_{xx} + \dot{\varepsilon}_{yy}\right)h\big|_{dwn} - \left(2\dot{\varepsilon}_{xx} + \dot{\varepsilon}_{yy}\right)h\big|_{up}}{\Delta x} \right] - B \dot{\varepsilon}^{\frac{1-n}{n}} \left[\frac{\dot{\varepsilon}_{yx} h\big|_{rgt} - \dot{\varepsilon}_{yx} h\big|_{lft}}{\Delta y} \right]. \tag{11.30}$$

Here, the symbols $\big|_{dwn}$, $\big|_{up}$, $\big|_{rgt}$, and $\big|_{lft}$ refer, respectively, to the downglacier and upglacier ends, and to the right and left sides (looking downglacier), of a "block" of the glacier of length Δx and width Δy. The first term on the right represents the contribution to τ_b of an imbalance in forces on the ends of the block; the second term represents the contribution of forces on the sides.

An example of an application of this procedure is provided by an experiment conducted on Storglaciären, Sweden (Hooke et al., 1989). Some stakes on the glacier surface (Fig. 11.9) were surveyed frequently between 1982 and 1985 to determine velocities (Fig. 11.10). The pattern of stakes was such that longitudinal and transverse strain rates could be calculated at the upglacier and downglacier ends of the "blocks" labeled **A** and **B** in Figure 11.9, and shear-strain rates could be calculated along the sides. Results of the calculations for six time periods are shown in Table 11.1. One time period represents mean winter conditions; τ_b was then –82 kPa beneath block **A** and –92 kPa beneath block **B**. The other five time periods were those during which high-velocity events occurred (Fig. 11.10). During these events, τ_b was reduced an average of nearly 30% beneath block **A**. Beneath block **B** the change in τ_b was more variable, but significant increases occurred during two events.

Study of the patterns of changes suggests that acceleration of block **A** was, in every case, accompanied by a decrease in Term 2 [the second term on the right in equation (11.30)]. From the strain-rate data, it can be seen that at the upglacier end of the block $\dot{\varepsilon}_{xx}$ became less compressive, and in two cases, even extending, whereas at the downglacier end it became more compressive in all but one case. Thus, the accelerations were not due to either push from upglacier or pull from downglacier. The clear implication is that they were a result of a reduction in resistive drag at the bed, presumably induced by increases in water pressure.

In the case of block **B**, the strain-rate data indicate that the marked increase in Term 2 reflects push from upglacier and, in the case of the June 1984 event, pull from downglacier. This combination of push and pull resulted in higher strain rates in the basal ice, and hence, owing to the proportionality between stress and strain rate, in higher basal drag.

Because we assumed that strain rates are uniform over the sides and ends of the blocks, and also owing to other uncertainties in the calculations, the values of τ_b obtained are only estimates. However, as the errors are probably of comparable magnitude and sign in all calculations, the direction and approximate magnitude of the *changes* in τ_b are probably reliable. These calculations thus help us understand the mechanisms by which the accelerations occurred in these instances. Through such analyses, we can gain insight into spatial and temporal variations in factors controlling the velocity of a glacier.

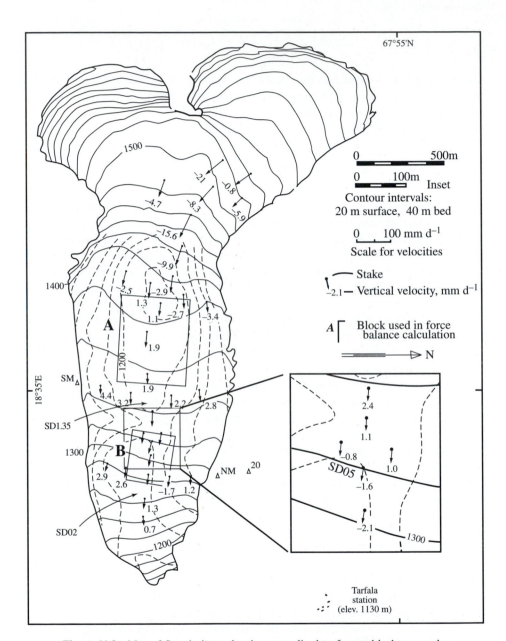

Figure 11.9 Map of Storglaciären showing generalized surface and bed topography (solid and dashed contours, respectively) and locations of stakes used for velocity measurements. Mean horizontal and vertical velocities of stakes are shown by arrows and numbers, respectively. (Modified from Hooke et al., 1989, Fig. 1a. Reproduced with permission of the International Glaciological Society.)

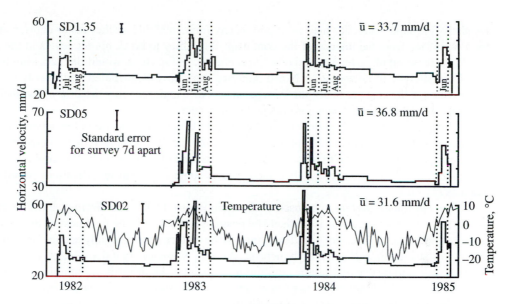

Figure 11.10 Horizontal velocities of three strain diamonds (SD) shown in Figure 11.9. Velocities are averages of those of the four (or five) stakes in each diamond. Mean daily temperature, smoothed using a five-day running mean, is shown in the bottom diagram. (Modified from Hooke et al., 1989, Fig. 3a. Reproduced with permission of the International Glaciological Society.)

TABLE 11.1 Force-balance calculations

Block A

Time	$\dot{\varepsilon}_{xx}$ up yr^{-1}	$\dot{\varepsilon}_{xx}$ dwn yr^{-1}	Term 2[a] kPa	Term 3[a] kPa	τ_b, kPa	$\Delta\tau_b$, %
Winter	−0.016	−0.004	1	−34	−82	—
July '83	0.001	−0.005	−25	−38	−52	−37
July '83	−0.006	−0.011	−17	−36	−62	−24
May '84	−0.008	−0.018	−19	−38	−58	−29
June '84	0.001	−0.002	−26	−40	−49	−40
June '85	−0.010	−0.007	−7	−38	−69	−16

Block B

Time	$\dot{\varepsilon}_{xx}$ up yr^{-1}	$\dot{\varepsilon}_{xx}$ dwn yr^{-1}	Term 2[a] kPa	Term 3[a] kPa	τ_b, kPa	$\Delta\tau_b$, %
Winter	0.006	−0.008	−28	−25	−92	—
July '83	Data incomplete					
July '83	Data incomplete					
May '84	0.000	−0.053	−23	−31	−91	−1
June '84	−0.002	0.037	40	−15	−170	+84
June '85	−0.002	−0.002	9	−25	−129	+40

$\rho g h \alpha$ was 115 kPa beneath block **A** and 145 kPa beneath block **B**.

[a]Terms 2 and 3 are the second and third terms on the right-hand side of equation (11.30), the longitudinal and transverse terms, respectively.

CREEP OF FLOATING ICE SHELVES

Ice shelves around Antarctica play an important environmental role, as they act as dams, restraining the flow of ice from the interior of the continent. Were they to break up, ice levels in the interior would decrease rapidly over a period of a few centuries, and sea level would rise accordingly. Break up of ice shelves in northeastern North America may have contributed to the collapse of the Laurentide Ice Sheet at the end of the Wisconsinan. Thus, understanding the flow of ice shelves is a problem of both academic and environmental significance.

The problem of ice-shelf flow is unique because τ_b is likely to be quite low where the shelf is grounded, and goes to zero in the limiting case when the shelf is afloat. Herein, we consider only the latter case. Weertman (1957b) was the first modern glaciologist to study this problem, and our approach follows his initially, but then incorporates some important modifications introduced by Thomas (1973a).

The coordinate system to be used is shown in Figure 11.11. The origin is at sea level, but is within the ice mass. The z-axis is vertical and positive upward. H is the thickness of the shelf, and h is the height of the surface above sea level. Inland, the surface rises gradually and the base drops further below sea level, so H and h both increase. As long as the ice shelf does not become grounded, however, we assume that hydrostatic equilibrium is maintained; therefore, assuming a constant density and thus ignoring the low-density snow and firn at the surface, $(H - h)\,\rho_w = H\rho_i$, where ρ_w and ρ_i are the densities of water and ice, respectively.

At the risk of being repetitive, it is convenient, once again, to write out the stress-equilibrium equation in the z-direction:

$$\frac{\partial \sigma_{xz}}{\partial x} + \frac{\partial \sigma_{yz}}{\partial y} + \frac{\partial \sigma_{zz}}{\partial z} = \rho_i g. \tag{11.31}$$

Our objective is to obtain an expression for σ'_{xx}, and then to use the flow law to solve for $\dot{\varepsilon}_{xx}$.

Because shear stresses are zero at the bed and surface, it is reasonable to assume that $\sigma_{xz} = \sigma_{zx} = \sigma_{yz} = \sigma_{zy} = 0$. This means that velocities and strain rates are independent of z. Equation (11.31) can thus be integrated:

$$\int_{\sigma_{zz}}^{0} d\sigma_{zz} = \rho_i g \int_{z}^{h} dz$$

to yield

$$\sigma_{zz} = \rho_i g (z - h). \tag{11.32}$$

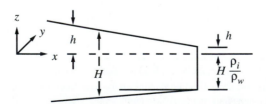

Figure 11.11   Coordinate system used in discussion of floating ice shelves.

Suppose that field measurements at some point give

$$\dot\varepsilon_{yy} = \eta\dot\varepsilon_{xx} \quad \text{and} \quad \dot\varepsilon_{xy} = \nu\dot\varepsilon_{xx}. \tag{11.33}$$

Then, from the incompressibility condition:

$$\dot\varepsilon_{zz} = -(1+\eta)\,\dot\varepsilon_{xx}. \tag{11.34}$$

Both η and ν are functions of position, but because strain rates are independent of z, η and ν are also independent of z. Then because we assume that the ice is isotropic, we have $\dot\varepsilon_{ij} = \lambda\sigma'_{ij}$, where, as before, $\lambda = \sigma^{n-1}/B^n$. Thus,

$$\sigma'_{yy} = \eta\sigma'_{xx}, \quad \sigma'_{xy} = \nu\sigma'_{xx}, \quad \text{and} \quad \sigma'_{zz} = -(1+\eta)\,\sigma'_{xx}.$$

From the last of these expressions, converting to total stresses, we obtain

$$\sigma'_{xx} - \sigma'_{zz} = \sigma'_{xx} + (1+\eta)\,\sigma'_{xx} = (\sigma_{xx} - P) - (\sigma_{zz} - P)$$

or

$$\sigma'_{xx} = \frac{\sigma_{xx} - \sigma_{zz}}{2+\eta}. \tag{11.35}$$

When $\eta = 0$, this expression reduces to one that often appears in analyses in plane strain. It can be derived, for example, from equations (10.18) and (10.18a).

It is interesting to consider the implications of this relation: σ_{zz} varies linearly with depth [equation (11.32)] but $\dot\varepsilon_{xx}$ is independent of depth. However, because the temperature of an ice shelf is normally well below 0°C at the surface and close to the pressure melting point at the base, B, and hence σ'_{xx}, also vary strongly with depth (Fig. 11.5). Thus, σ_{xx} varies with depth in a way that is not intuitively obvious. We will avoid this problem by seeking an expression for $\dot\varepsilon_{xx}$ in terms of the depth-integrated values of B and σ_{xx}.

Let us proceed by investigating the total longitudinal force per unit width. To do this, integrate equation (11.35) from the base, b, to the surface, s:

$$
\begin{aligned}
\int_b^s \sigma'_{xx}dz &= \frac{1}{2+\eta}\int_{-(H-h)}^h [\sigma_{xx} - \rho_i g\,(z-h)]\,dz \\
&= \frac{1}{2+\eta}\left[\int_b^s \sigma_{xx}dz - \rho_i g\left(\frac{z^2}{2} - hz\right)\Big|_{-(H-h)}^h\right] \\
&= \frac{1}{2+\eta}\left[\int_b^s \sigma_{xx}dz - \rho_i g\left(-\frac{h^2}{2} - \frac{(h-H)^2}{2} + h(h-H)\right)\right] \\
&= \frac{1}{2+\eta}\left[\int_b^s \sigma_{xx}dz + \rho_i g\frac{H^2}{2}\right],
\end{aligned}
$$

or defining

$$F = -\int_b^s \sigma_{xx}dz$$

we obtain

$$\int_b^s \sigma'_{xx}dz = \frac{1}{2+\eta}\left[\rho_i g\frac{H^2}{2} - F\right]. \tag{11.36}$$

F is the total force-opposing movement of a vertical section, of unit width, of the ice shelf.

We now need to use the flow law to express the left-hand side of equation (11.36) in terms of strain rates. First, the effective stress is

$$
\begin{aligned}
\sigma &= \left[\frac{1}{2}\sigma_{ij}\sigma_{ij}\right]^{\frac{1}{2}} \\
&= \left[\frac{1}{2}(\sigma_{xx}'^2 + \sigma_{yy}'^2 + \sigma_{zz}'^2 + 2\sigma_{xy}'^2)\right]^{\frac{1}{2}} \\
&= \left[\frac{1}{2}(1 + \eta^2 + 1 + 2\eta + \eta^2 + 2v^2)\sigma_{xx}'^2\right]^{\frac{1}{2}} \\
&= (1 + \eta + \eta^2 + v^2)^{\frac{1}{2}}|\sigma_{xx}'|.
\end{aligned}
$$

Thus, from the flow law:

$$
|\dot{\varepsilon}_{xx}| = \frac{(1 + \eta + \eta^2 + v^2)^{\frac{n-1}{2}}}{B^n}|\sigma_{xx}'|^n.
$$

If $n = 3$, we can drop the absolute value signs, which we now do. Thus, rearranging:

$$
\sigma_{xx}' = \frac{\dot{\varepsilon}_{xx}^{1/n}}{(1 + \eta + \eta^2 + v^2)^{(n-1)/2n}}B.
$$

As strain rates are assumed to be independent of z, equation (11.36) now becomes

$$
\int_b^s \sigma_{xx}'dz = \frac{\dot{\varepsilon}_{xx}^{1/n}}{(1 + \eta + \eta^2 + v^2)^{(n-1)/2n}}\int_b^s Bdz = \frac{1}{2+\eta}\left[\rho_i g\frac{H^2}{2} - F\right].
$$

Because B varies with depth, we define a depth averaged B by

$$
\overline{B} = \frac{1}{H}\int_b^s Bdz.
$$

We also define θ by

$$
\theta = \frac{(1 + \eta + \eta^2 + v^2)^{(n-1)/2}}{(2 + \eta)^n}.
$$

Solving for $\dot{\varepsilon}_{xx}$ now yields

$$
\dot{\varepsilon}_{xx} = \frac{\theta}{\overline{B}^n}\left[\rho_i g\frac{H}{2} - \frac{F}{H}\right]^n. \tag{11.37}
$$

To proceed further, we need to evaluate F, the force per unit width opposing motion. We do this for two special situations. In the first, the ice shelf is free to expand in both the x- and y-directions, and movement is restrained by seawater pressure only. Then, $\eta = 1$ and

$$
F_w = -\int_{-(H-h)}^0 \rho_w gz\, dz = \rho_w g\frac{(H - h)^2}{2}.
$$

Making use of the condition of hydrostatic equilibrium, $\rho_w(H - h) = \rho_i H$, yields

$$
F_w = \frac{1}{2}\rho_w g\left(\frac{\rho_i}{\rho_w}\right)^2 H^2 = \frac{1}{2}\rho_i g\left(\frac{\rho_i}{\rho_w}\right)H^2,
$$

and equation (11.37) becomes

$$
\dot{\varepsilon}_{xx} = \frac{\theta}{\overline{B}^n}\left[\frac{1}{2}\rho_i g\left(H - H\frac{\rho_i}{\rho_w}\right)\right]^n.
$$

The term in the inner brackets on the right-hand side is simply h, so this becomes

$$\dot{\varepsilon}_{xx} = \theta\left[\frac{\rho_i gh}{2\overline{B}}\right]^n. \tag{11.38}$$

As this expression is always positive, strain rates will always be extending. Note that the surface slope does not appear in this solution; thus, even an iceberg with a horizontal surface will deform. This solution does not apply very near a calving face where bending moments are present.

It is instructive to compare this expression with that developed in Chapter 5 [equation (5.3) with (5.2c)] for $\dot{\varepsilon}_{zx}$ at the bed of a land-based glacier in the absence of significant longitudinal strain:

$$\dot{\varepsilon}_{zx} = \left[\frac{\rho_i gH\alpha}{B}\right]^n.$$

Assuming that $n = 3$ and noting that $h = (1 - \rho_i/\rho_w)\,H \approx 0.1\,H$, and that $\theta = 1/9$ when $\eta = 1$, $v = 0$, equation (11.38) becomes

$$\dot{\varepsilon}_{xx} = \left[\frac{0.024\,\rho_i gH}{\overline{B}}\right]^3. \tag{11.39}$$

Thus, the driving stress ($\rho_i gh$) in an ice shelf is comparable to that in a land-based glacier of the same thickness with a surface slope of ≈ 0.024. However, because $\overline{B}$ increases with decreasing temperature, strain rates in ice shelves are normally less than those in land-based glaciers of comparable thickness.

The second example is that of an ice shelf between approximately parallel valley walls. In this case, $F = F_w + F_s$, where F_s is the shear force on the valley sides. Utilizing the expression for F_w just obtained, $\dot{\varepsilon}_{xx}$ becomes

$$\dot{\varepsilon}_{xx} = \theta\left[\frac{\rho_i gh}{2\overline{B}} - \frac{F_s}{H\overline{B}}\right]^n. \tag{11.40}$$

Now, $\dot{\varepsilon}_{xx}$ can be negative, or compressive, if F_s is sufficiently large.

F_s merits further comment. Suppose that $\mathbf{a}$ is the distance from the centerline of an ice shelf to the valley wall. Suppose further that the depth-averaged drag on a valley wall is $\overline{\tau}_s$. $\overline{\tau}_s H$ is then the force on the valley wall per unit length along the direction of flow. This force must balance forces acting in the direction of flow over the half-width of the ice shelf. In the absence of basal drag, it is reasonable to assume that any vertical slice of unit width extending through the ice shelf and parallel to the direction of flow will be restrained equally by this side drag. Thus, any such slice will experience a drag of $\overline{\tau}_s H/\mathbf{a}$ per unit length along the direction of flow. Noting that $\overline{\tau}_s$ is a negative quantity, as it is directed in the up-flow direction (Fig. 11.11), the resisting force per unit width is

$$F_s = -\int_x^L \overline{\tau}_s \frac{H}{\mathbf{a}}\,dx. \tag{11.41}$$

Here, x is the coordinate position where the calculation is being made, and L is the x-coordinate of the edge of the shelf. Note that, consistent with being a force per unit width, F_s has the dimensions N m^{-1}.

Equation (11.41) says that F_s increases as the distance to the edge of the shelf increases. Thus, from equation (11.40), $\dot{\varepsilon}_{xx}$ may change from extending nearer the shelf edge to compressive farther inland. This is the reverse of the normal situation in a grounded glacier, in which compressive flow is the rule in the ablation area and extending flow in the accumulation area. The implications of this are fascinating. With extending flow nearer the shelf edge, a positive emergence velocity would occur only if the product of the velocity times the surface slope were high enough to offset any downward vertical velocity resulting from the extension. In the absence of such conditions, a steady state can exist only if the mass balance is positive, as, in fact, is typically the case. This means that ice shelves with ablation (= melt) zones near the shelf edge should be uncommon. Furthermore, if the mass balance near the shelf edge is positive, it must also be positive at higher elevations farther inland. Thus, if F_s ever became large enough to make $\dot{\varepsilon}_{xx}$ compressive, the ice shelf would increase in thickness unstably until it became grounded.

ANALYSIS OF BOREHOLE-DEFORMATION DATA

Our next example is drawn from the work of Shreve & Sharp (1970) and deals with the analysis of inclinometry data collected in boreholes that are undergoing deformation. In the simplest case, we might assume that at depth d, $\sigma_{zx} = S_f \rho g d\alpha$, and that successive measurements of the inclination of a borehole would give $\partial u / \partial z$. Then $\dot{\varepsilon}_{zx} = \frac{1}{2}(\partial u / \partial z + \partial w / \partial x)$ and, if the deformation is entirely simple shear, $\partial w / \partial x = 0$. Thus, measurements of the change in inclination at several depths would permit a (double log) plot of σ_{zx} versus $\dot{\varepsilon}_{zx}$ and, if other stresses and strain rates were negligible, the slope and intercept of the resulting line could be used to obtain n and B, respectively. Such an approach would be valid if the borehole were in a slab of ice of uniform thickness and infinite horizontal extent. In other cases, nonzero vertical velocities and (or) longitudinal strain rates could result in errors.

Figure 11.12 illustrates the effect of the longitudinal strain rate on a borehole. In a zone of longitudinal extension, the inclination of a hole that is inclined with respect to the direction of extension will increase, even if there is no shear strain. Nye (1957) realized this and made a correction for this effect in his reanalysis of the Jungfraufirn borehole experiment. However, it was Shreve (Shreve & Sharp, 1970) who undertook the first complete study of the problem.

We start by looking at the difference in velocity between two points in a borehole from the point of view of motion of the ice. This is what we want to determine from the inclinometry measurements. The axes are as shown in Figure 11.13. The ℓ_i are direction cosines describing the orientation of the borehole, and $d\lambda$ is an increment of length along the hole. Two points in the hole a distance $d\lambda$ apart will be separated from one another by distances $\ell_i d\lambda$ in the i-direction (Fig. 11.14). The difference in the u_i velocity at depth $(z + \ell_z d\lambda)$ and that at depth z is du_i^λ. This is given by

$$du_i^\lambda = \ell_x \frac{\partial u_i}{\partial x} d\lambda + \ell_y \frac{\partial u_i}{\partial y} d\lambda + \ell_z \frac{\partial u_i}{\partial z} d\lambda.$$

Figure 11.12   Effect of longitudinal strain on an inclined borehole.

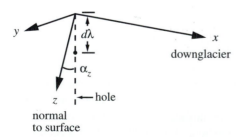

Figure 11.13   Coordinate system for analysis of borehole deformation.

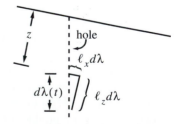

Figure 11.14   Distance between two points in a borehole expressed in terms of the direction cosines of the hole.

The first term on the right is the change in u_i as a result of moving a distance $\ell_x d\lambda$ in the x-direction, and so forth. Using the summation convention, this can be written:

$$du_i^\lambda = \ell_j \frac{\partial u_i}{\partial x_j} d\lambda. \tag{11.42}$$

In terms of motion of the borehole casing (holes are often cased to provide a smoother and more reliable path for the inclinometer), we again consider the difference in velocity between points a distance $d\lambda$ apart (Fig. 11.15). The point at depth z moves a distance ΔX, and a point at depth $(z + \ell_z d\lambda)$ moves a distance Δx, both in time Δt. The inclinometry measurements, when combined with an accurate survey of the motion of the hole top, provide us with these distances. They are related by

$$\Delta u^\lambda \Delta t = \Delta X - \Delta x = u_{(z)}\Delta t - [u_{(z)}\Delta t + \ell_x d\lambda_{t=t_1} - \ell_x d\lambda_{t=0}]$$

where $u = u_x$, the x-component of the velocity vector, and $u_{(z)}$ is the value of u at depth z. The quantity in brackets represents the length Δx; that is, it is the length ΔX plus the displacement of the upper point with respect to the lower one at the end of the time interval Δt, minus this displacement at the beginning. Including the changes in $\ell d\lambda$ in the y and z directions, allowing for a change in $\ell d\lambda$ with time (unsteady flow), and expressing the result in differential form yields

$$du^\lambda \Delta t = \frac{\partial \ell_x d\lambda}{\partial x} u\Delta t + \frac{\partial \ell_x d\lambda}{\partial y} v\Delta t + \frac{\partial \ell_x d\lambda}{\partial z} w\Delta t + \frac{\partial \ell_x d\lambda}{\partial t} \Delta t.$$

Here, the derivative with respect to x in the first term on the right-hand side gives the rate of change of $\ell_x d\lambda$ in the x-direction, and $u\Delta t$ gives the distance moved in the x-direction, and so forth. Dividing by Δt and using the summation convention, we obtain

$$du_i^\lambda = u_j \frac{\partial \ell_i d\lambda}{\partial x_j} + \frac{\partial \ell_i d\lambda}{\partial t} = \frac{D}{Dt}(\ell_i\, d\lambda), \tag{11.43}$$

where D/Dt is the substantial or Lagrangian derivative [see equation (6.12b)].

Equations (11.42) and (11.43) are both expressions for du_i^λ, the difference in velocity between two points a distance $d\lambda$ apart along the hole, so equating them yields

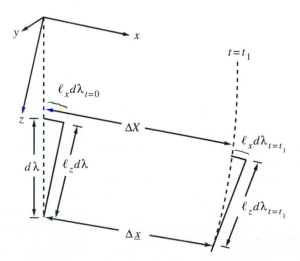

Figure 11.15 Deformation of a borehole casing through time.

$$\ell_j \frac{\partial u_i}{\partial x_j} d\lambda = \frac{D}{Dt}(\ell_i d\lambda) = \ell_i \frac{D}{Dt} d\lambda + \frac{D\ell_i}{Dt} d\lambda. \tag{11.44}$$

We would like to divide by $d\lambda$ to eliminate it from the first and last terms, but first we need an expression for $D(d\lambda)/Dt$. To obtain this, multiply both sides by ℓ_i:

$$\ell_i \ell_j \frac{\partial u_i}{\partial x_j} d\lambda = \ell_i \ell_i \frac{D}{Dt} d\lambda + \ell_i \frac{D\ell_i}{Dt} d\lambda.$$

Because the sum of the squares of the direction cosines is unity, $\ell_i \ell_i = 1$. Similarly,

$$2\ell_i \frac{D\ell_i}{Dt} = \frac{D}{Dt}(\ell_i \ell_i) = \frac{D}{Dt}(1) = 0.$$

Thus,

$$\frac{D}{Dt} d\lambda = \ell_i \ell_j \frac{\partial u_i}{\partial x_j} d\lambda. \tag{11.45}$$

Equations (11.44) and (11.45) can be combined to yield the desired expression. However, we need to be careful of the subscripts when doing this. Expanding equation (11.44) for $i = x$ and dividing by $d\lambda$ yields

$$\ell_x \frac{\partial u}{\partial x} + \ell_y \frac{\partial u}{\partial y} + \ell_z \frac{\partial u}{\partial z} = \frac{1}{d\lambda}\ell_x \frac{D}{Dt} d\lambda + u\frac{\partial \ell_x}{\partial x} + v\frac{\partial \ell_x}{\partial y} + w\frac{\partial \ell_x}{\partial z} + \frac{\partial \ell_x}{\partial t}. \tag{11.46}$$

Because the inclination of the casing is a function of z alone, $\partial \ell_x/\partial x = 0$ and $\partial \ell_x/\partial y = 0$. Expanding the right-hand side of equation (11.45), using the result to replace the term involving $D(d\lambda)/Dt$ in equation (11.46), and rearranging terms, we obtain:

$$
\begin{aligned}
\ell_x \frac{\partial u}{\partial x} + \ell_y \frac{\partial u}{\partial y} + \ell_z \frac{\partial u}{\partial z} - \ell_x \ell_x \ell_x \frac{\partial u}{\partial x} - \ell_x \ell_x \ell_y \frac{\partial u}{\partial y} - \ell_x \ell_x \ell_z \frac{\partial u}{\partial z} \\
- \ell_x \ell_y \ell_x \frac{\partial v}{\partial x} - \ell_x \ell_y \ell_y \frac{\partial v}{\partial y} - \ell_x \ell_y \ell_z \frac{\partial v}{\partial z} \\
- \ell_x \ell_z \ell_x \frac{\partial w}{\partial x} - \ell_x \ell_z \ell_y \frac{\partial w}{\partial y} - \ell_x \ell_z \ell_z \frac{\partial w}{\partial z} = w \frac{\partial \ell_x}{\partial z} + \frac{\partial \ell_x}{\partial t}.
\end{aligned}
\tag{11.47}
$$

Using the summation convention, this can be written as

$$\ell_k \frac{\partial u_i}{\partial x_k} - \ell_i \ell_j \ell_k \frac{\partial u_j}{\partial x_k} = w\frac{\partial \ell_i}{\partial z} + \frac{\partial \ell_i}{\partial t}$$

or even more compactly as

$$(\delta_{ij} - \ell_i \ell_j)\ell_k \frac{\partial u_j}{\partial x_k} = w\frac{\partial \ell_i}{\partial z} + \frac{\partial \ell_i}{\partial t}. \tag{11.48}$$

Because i is not repeated in any of the terms in equation (11.48), this equation represents three separate equations (for $i = x,y,z$). However, only two of these equations are independent because only two of the direction cosines are independent.

If the inclination of a borehole is known at two separate times, and if seven of the nine velocity derivatives in equation (11.47) can be measured or estimated, equations (11.48) can be solved for the two remaining velocity derivatives. Equations (11.48) are exact, but approximations have to be made in calculating the ℓ_i, $\partial \ell_i/\partial t$, and $\partial \ell_i/\partial z$ from observational data obtained at discrete points in time and space.

TABLE 11.2 Calculation of velocity derivatives in borehole-deformation studies of Hooke (1973b) and Hooke and Hanson (1986)

Derivative	1973	1986
$\partial u/\partial x$	Two boreholes	We assumed that this decreased with depth in proportion to the decrease in u with depth.
$\partial u/\partial y$	0 (assumed)	This was obtained from the measured u at the surface and the radius of curvature of the flow line.
$\partial v/\partial x$	0 (assumed)	$\dot{\varepsilon}_{xy}$ was measured at the surface and assumed to decrease with depth in proportion to the decrease in u with depth. Then, from equation (9.18): $\partial v/\partial x = 2\dot{\varepsilon}_{xy} - \partial u/\partial y$
$\partial v/\partial y$	This was measured at the surface, and we assumed that it decreased with depth in proportion to the decrease in $\partial u/\partial x$	
$\partial w/\partial z$	$= -\partial u/\partial x - \partial v/\partial y$ by continuity	Same as 1973.
	$\partial w/\partial z$ was then integrated over depth to obtain w as a function of depth, using either a no-slip boundary condition at the bed where temperatures are well below the melting point, or the measured w at the surface.	
$\partial w/\partial x$	Two boreholes	This was measured at the surface; we assumed that it decreased with depth in proportion to the decrease in w with depth.
$\partial w/\partial y$	We set $\dot{\varepsilon}_{yz} = \frac{1}{2}(\partial v/\partial z + \partial w/\partial y) = 0$ at the surface, and let $\partial w/\partial y$ decrease linearly with depth. $\partial v/\partial z$ was calculated (see below), so an iterative procedure was required.	Same as $\partial w/\partial x$.
$\partial u/\partial z, \partial v/\partial z$	These derivatives were then calculated from equations (11.48).	Same as 1973.

Two alternative approaches taken to this problem in two separate field experiments on Barnes Ice Cap (Hooke, 1973b; Hooke & Hanson, 1986) are outlined in Table 11.2. Strain nets were placed around the tops of the boreholes, so that some of the velocity derivatives could be measured directly at the surface. Assumptions were then made about how they varied with depth. In the first experiment, the boreholes were closely spaced so $\partial u/\partial x$ could be determined, as a function of depth, from the successive borehole profiles. As can be seen from Table 11.2, the two velocity derivatives that were calculated were $\partial u/\partial z$ and $\partial v/\partial z$. One might expect that measurements of the rate of tilting of the borehole would give these velocity derivatives directly, but this is not the case. Yet, as implied by our opening discussion, $\partial u/\partial z$ is, in fact, one of the most important velocity derivatives.

Sensitivity studies suggest that the solutions obtained in these two Barnes Ice Cap studies do not depend strongly on the assumptions. The most important term is $\partial \ell_i/\partial t$. In instances where the casing bends abruptly, as at joints, $w \, \partial \ell_i/\partial z$ also becomes important. In experiments on other glaciers, the results might be more sensitive to some of the other velocity derivatives, and hence to any assumptions made in obtaining them.

In plane strain, assuming incompressible flow and a uniform longitudinal strain rate, r, we can write $\partial u/\partial x = -\partial w/\partial z = r$, $\partial w/\partial x = 0$, $\ell_x = \sin \theta$, $\ell_z = \cos \theta$, and $\ell_y = 0$, where θ is the inclination of

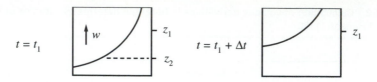

Figure 11.16   Effect of vertical advection on borehole inclination.

the borehole from the vertical. Equation (11.48) then reduces to

$$\frac{\partial u}{\partial z} = \frac{\partial}{\partial t} \tan \theta - 2r \tan \theta + w \frac{\partial}{\partial z} \tan \theta. \tag{11.49}$$

The first term on the right is the obvious one, involving a change in inclination of the borehole with time. The second is the one illustrated in Figure 11.12 and discussed earlier. The third is an advection effect. In an area of nonzero vertical velocity, a section of a borehole at depth z_2, measured with respect to some constant datum, and with inclination $\ell(z_2)$ will, at the end of a time interval Δt, be at, say, depth z_1 (Fig. 11.16). If the initial inclination of the borehole at depth z_1 was different from $\ell(z_2)$, our measurements would show that the inclination at depth z_1 had changed, and this would be true even if $\partial u/\partial z$ were 0.

The results of the borehole-deformation experiment reported by Hooke & Hanson (1986) will be used to illustrate an application of this analysis. Four boreholes, located approximately along a flowline on Barnes Ice Cap (Fig. 11.17), were drilled and cased, and inclinometry data were obtained from them over a period of up to four years. Figure 11.18a shows the deformation profiles, and Figure 11.18b shows values of $\partial u/\partial z$ calculated from equations (11.48).

The deformation profiles in most of the holes end at the top of a zone of white ice (Fig. 11.18a). Oxygen isotope data demonstrate that this ice is of Pleistocene age (Hooke & Clausen, 1982). The ice is white because it contains a lot of air bubbles. As a result of these bubbles, the density of this ice is only 870 kg m^{-3}, compared with a density of 920 kg m^{-3} in the overlying blue ice. We presume that the high concentration of air bubbles is a result of two processes:

1. When the climate warmed at the end of the Pleistocene, meltwater percolation increased, and

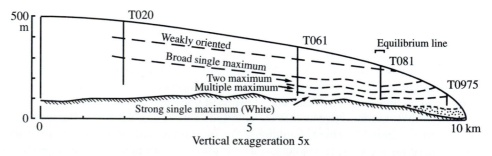

Figure 11.17   Longitudinal section along a flowline on Barnes Ice Cap showing types of ice encountered in boreholes. Stippled zone near margin is inferred to be deformed superimposed ice formed at the margin and overridden during an advance of the glacier (see Fig. 5.14). (After Hooke & Hanson, 1986, Fig. 2. Reproduced with the kind permission of Elsevier Science.)

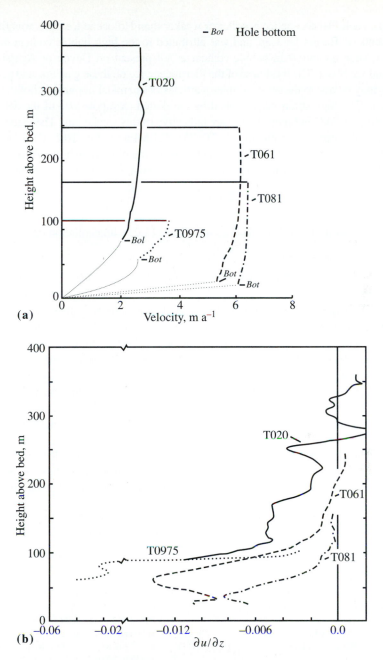

Figure 11.18 (a) Velocity profiles in boreholes, and (b) $\partial u/\partial z$ as a function of depth. (After Hooke & Hanson, 1986, Fig. 3. Reproduced with the kind permission of Elsevier Science.)

ice lenses formed. These lenses trapped air in the underlying porous firn.

2. As basal meltwater escaped into the underlying permeable bedrock, air may have been left behind in a sort of physical fractionation process.

It is commonly found (Paterson, 1977; Fisher & Koerner, 1986; Dahl-Jensen & Gundestrup,

1987) that such Pleistocene ice is softer or weaker than Holocene ice. This was first noted in the 1973 experiments on Barnes Ice Cap, and was attributed to the high bubble content of the Pleistocene ice. However, such a contrast in bubble content is not present on Devon or Agassiz ice caps or on the Greenland Ice Sheet. The weakness of the Pleistocene ice on these glaciers, and perhaps also on Barnes Ice Cap, may be due to the effect of microparticles or chemical impurities, both of which are known to be present. The high strain rates implied by the dotted extrapolations of the deformation profiles for holes T061 and T081 in Figure 11.18a are indicative of this weakening. The value of B obtained for this ice from the deformation profile in hole T0975 is 0.1 MPa $a^{1/3}$ (at -10.1 °C), which is much lower than those ranging from 0.23 to 0.30 MPa $a^{1/3}$ in the overlying blue ice in holes T061 and T020 (Table 11.3).

Also of interest are the values of the parameter Λ, defined by [see equation (9.24)]

$$2\Lambda = \frac{1}{\lambda} = \frac{\sigma_{zx}}{\dot{\varepsilon}_{zx}} = \frac{B}{\dot{\varepsilon}^{\frac{(n-1)}{n}}}. \tag{11.50}$$

$\dot{\varepsilon}_{zx}$ is obtained from the velocity derivatives (Table 11.2) using equation (9.18), while σ_{zx} is estimated with the use of

$$\sigma_{zx} = -\rho g z \alpha - \frac{\partial}{\partial x} \int_0^z \frac{B}{\dot{\varepsilon}^{\frac{n-1}{n}}} (2\dot{\varepsilon}_{xx} + \dot{\varepsilon}_{yy})\, dz - T. \tag{11.51}$$

which is derived from equation (11.26) in much the same way that we derived equation (11.29) except that we now retain the T term and also assume that changes in the transverse direction are negligible in an ice cap. Equation (11.27) was used to evaluate the T term. If B and n are constant, as might be expected, Λ should decrease as $\dot{\varepsilon}$ increases. The awkward fact is that near the surface where $\dot{\varepsilon}_{zx}$ is low, this does not appear to be the case. Figure 11.19a shows that Λ is effectively independent of $\dot{\varepsilon}$. Even the direction of change of Λ with depth is not consistent from one hole to the next, as indicated by the arrows on the curves in Figure 11.19a. This problem is not unique to Barnes Ice Cap; Raymond (1973) also found that Λ was independent of $\dot{\varepsilon}$ near the surface of Athabasca Glacier.

Somewhat deeper in the glacier the situation improves, and Λ decreases steadily with increasing $\dot{\varepsilon}$ (Fig. 11.19b). Here, the slope and intercept of the log Λ – log $\dot{\varepsilon}$ line can be used to determine B and n [equation (11.50)]. In the present case, however, Hooke and Hanson (1986) chose, instead, to

TABLE 11.3   Values of B in MPa $a^{1/3}$ for different ice types[a]

Borehole	T0975	T081	T061	T020
Ice type				
Weakly oriented	—	—	0.23 (–10.2)	—
Broad single maximum[b]	—	0.46 (–9.0)[c]	0.26 (–8.4)	0.30 (–7.4)
Two maxima	—	0.44 (–8.6)	0.24 (–7.5)	0.26 (–6.5)
Three or four maxima	—	0.50 (–8.3)	0.30 (–6.8)	—
White ice (clean)	0.10 (–10.1)	0.18[d] (–7.8)	0.10[d] (–6.4)	—
		0.18[e] (–7.8)		
White ice (dirty)	0.13 (–9.8)			

[a]Values given are for zones in which fabric is well developed, and thus exclude transition regions.

[b]Equivalent fabric in T020 is small circle.

[c]Numbers in brackets are mean temperatures in °C.

[d]Velocity profile calculated by assuming no slip on the bed.

[e]Measured over two weeks, starting three weeks after completion of hole in 1977. No smoothing used in calculation.

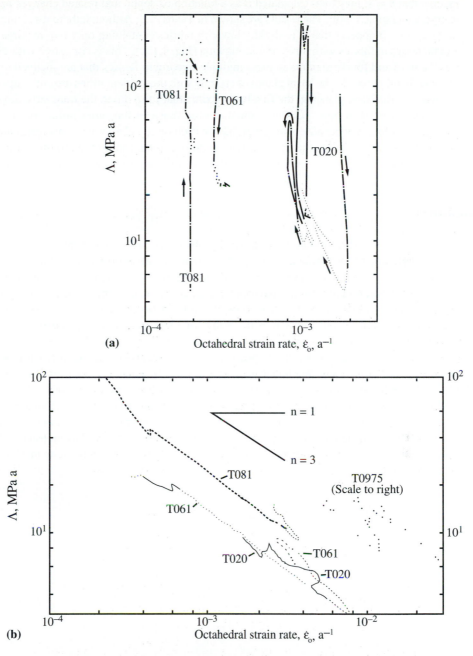

Figure 11.19 Octahedral shear strain rate, $\dot{\varepsilon}_o$, plotted against Λ for: (**a**) The upper 50 m of holes T061 and T081, and the upper 150 m of hole T020. Arrows show direction of increasing depth. (**b**) The lower parts of the holes. Depth increases from upper left, following lines of points. Reversals in trends reflect hardening of ice in zones where fabric is changing. (After Hooke & Hanson, 1986, Fig. 4. Reproduced with the kind permission of Elsevier Science.)

assume that $n = 3$; they then calculated B as a function of depth, and related changes in B to changes in crystallographic fabric. The results are shown in Table 11.3. Although there was quite a lot of noise in the record, it appears that B is slightly lower in fabrics containing only two maxima (Fig. 4.13b), and increases in fabrics with three or four maxima (Fig. 4.13c). This is consistent with expectation, as it is the third and fourth maxima in these multiple-maximum fabrics that are inclined to the direction of shear. In other words, the basal planes of crystals with these orientations dip either up- or downglacier, whereas the basal planes in the first two maxima that form dip in the transverse direction.

Values of B in hole T081 are about double those in the other holes. Hooke and Hanson assumed that this was because stresses at this location on the glacier were overestimated. However, they were unable to isolate the apparently erroneous assumption that led to this error, even though they undertook calculations with a numerical model utilizing the finite element method.

SUMMARY

In this chapter we have studied four classical problems in glacier mechanics: closure of cylindrical holes, calculation of force balances, creep of ice shelves, and deformation of boreholes. As examples of applications of the theory presented, we discussed problems such as the flow of water to, from, and in subglacial conduits, the mechanics of glacier accelerations, the stability of ice shelves, and the extraction of flow-law parameters from borehole-deformation data. From these examples, we gained insights into the dynamic and kinematic behavior of glaciers. These, however, were secondary objectives.

The primary objective of the chapter was to help students develop facility with the mathematics of stress and deformation as applied to problems in glacier mechanics. Such analyses are complicated because multiple stresses, strains, and strain rates are involved, and even more so because the strains in which we are interested are a consequence of deviatoric, not total, stresses. In many cases, once the physics of a problem have been formulated, prescribed mathematical procedures must be followed before a result with clear physical significance reappears. Students who have mastered the material in this chapter will be able to understand many papers in the glaciological literature that would otherwise be impenetrable.

Response of Glaciers to Changes in Mass Balance

Climatic changes, as we discussed in Chapter 3, may take the form of changes in precipitation, in temperature, in radiation balance, or in some combination of these three parameters. From the point of view of a glacier, however, the effect in each case is to increase or decrease the amount and the spatial distribution of accumulation and melt. Such changes lead to discrepancies between the specific net balance and the local emergence or submergence velocity, and hence to changes in glacier geometry, as analyzed in Chapter 5.

Were the climate of a region to remain constant for a long time, several decades or even centuries, the geometry of nonsurging glaciers in that region would adjust so that the specific net balance was everywhere equal to the local emergence or submergence velocity, and in addition (or as a consequence) the integral of the specific net balance over the glacier, $B_n = \int b_n dA$, would be zero. The glacier would then be said to be in a steady state, an ideal that may occasionally be approached but rarely, if ever, reached.

The principal adjustment that takes place is, of course, a change in length or size. A positive mass balance, maintained over a period of years, will result in an advance. As the glacier expands into lower elevations or more southerly latitudes, the summer balance becomes more negative until it becomes equal (in magnitude) to the winter balance, and the net balance returns to zero, and conversely.

The goal of this chapter is to study the details of the adjustment process. In particular, we will see that changes in mass balance lead to changes in thickness which influence the speed of the glacier, and hence the rate at which ice is transferred from the accumulation area to the ablation area. The changes in thickness propagate and diffuse down the glacier. Thus, the propagation and diffusion processes control the way in which the profile adjusts to the new mass-balance conditions. As these processes take time, years can elapse before the terminus gets the message that something has happened higher on the glacier, and decades may pass before it adjusts to the changes.

POSITIVE FEEDBACK PROCESSES

Before proceeding, it is appropriate to mention some feedback processes that can influence the way in which a glacier adjusts to climatic change, but which we will not consider in detail. The first is the *Bodvarsson instability* (Bodvarsson, 1955). In alpine regions, b_n may become increasingly positive with increased elevation. Thus if a positive net balance leads to an increase in thickness, there may be a positive feedback in which the increase in thickness reinforces the positive net balance.

Lliboutry (1970) has discussed a different positive-feedback process that may affect polar glaciers. This process results from the fact that a change in temperature has not only an immediate effect on the mass balance but also a delayed effect on the flow, owing to the temperature dependence of the flow law. For example, an increase in temperature may increase ablation and thus decrease the annual net balance. The decrease in net balance causes the glacier to thin. Then, as the temperature change gradually penetrates into the glacier, the flow rate increases. As this increases the mass flux to the terminus relative to the input of ice upglacier, this results in further thinning. The decrease in thickness, a combined effect of the changes in mass balance and temperature, leads to further warming of the glacier surface, the boundary condition, owing to the increase in temperature with decreasing elevation.

It is well to keep in mind, however, that if the climate is sufficiently cold, increases in temperature may actually increase the winter balance, as the atmosphere is then able to hold more moisture in the vapor state. For example, studies of the volume of air in bubbles in a core from Byrd Station on the West Antarctic Ice Sheet suggest that as the Pleistocene gave way to the Holocene, the ice sheet there became ~250 m thicker. This change is inferred to have been a result of an increase in precipitation as the climate warmed (Raynaud & Whillans, 1982). Eventually, however, as the warm wave penetrated deeper and the flow rate increased, the ice sheet began to thin (Alley & Whillans, 1991). Thus in this case, the processes did not reinforce one another. Measurements of strain rate and mass balance along a 160-km strain network upglacier from Byrd Station suggest that the thinning is continuing today (Whillans, 1977).

Superimposed on these positive-feedback loops in large ice masses is yet another delayed response, that of the Earth's crust to the additional ice load. As the crust is depressed isostatically, the surface elevation of the ice sheet is lowered, and in some areas the bed may become depressed below sea level. Where the bed is thus depressed, any subsequent climatic warming that results in thinning of the ice or a rise in sea level, or both, is likely to lead to buoyant forces at the bed that greatly increase flow speeds, potentially leading to collapse of the ice sheet. Numerical models of ice sheets are used to study the possibility that such collapses occurred in the past, and that the West Antarctic Ice Sheet might collapse in the future in response to greenhouse warming. MacAyeal (1993a, 1993b), among others, has also considered the possibility that several layers of ice-rafted sediment that have been detected in cores across the North Atlantic Ocean, the *Heinrich* layers, reflect repeated collapses of the Laurentide Ice Sheet.

RESPONSE OF A TEMPERATE GLACIER

Let us now consider, qualitatively, how a temperate glacier should respond to a change in mass balance. Suppose b increases, becoming more positive in the accumulation area and less negative in the ablation area. Initially, this leads to an increase in thickness. Suppose further that the longitudinal strain rate is extending, as is commonly the case in accumulation areas. Over the course of a year, this extension, operating on a block of ice of thickness h, results in thinning by an amount Δh (Fig. 12.1a). For a constant rate of extension, Δh is proportional to h. [For example, in Figure 12.1a, conservation of mass requires that $\Delta h\, x \cong (h - \Delta h)\, \Delta x$ or ignoring second order terms, $\Delta h \cong h\, (\Delta x/x)$.] Thus, as the glacier grows thicker, the amount of thinning, Δh, resulting from the stretching increases each year. After many years, Δh becomes large enough to absorb most of the increased accumulation and the glacier gradually approaches a new steady state. As we shall see, however, the time required for full adjustment is theoretically infinite.

The situation is quite different if the longitudinal strain rate is compressive, as is normally the case in ablation areas. Δh is still proportional to h, but now, because both the longitudinal strain rate

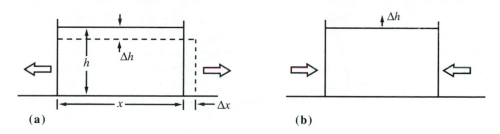

Figure 12.1 Sketch illustrating why adjustment toward a new steady state is **(a)** stable where flow is extending, and **(b)** unstable where flow is compressive.

and the change in mass balance cause the glacier to become thicker (Fig. 12.1b), Δh increases each year, unstably. Thus, in the absence of some mitigating effect, a new steady state would never be reached. This contrast in behavior between the accumulation and ablation areas, however, leads to kinematic waves and diffusional processes that restore stability.

As Nye (1960) emphasizes, kinematic waves are not dynamic waves like waves on a water body; indeed kinematic waves need not have a wave form. Dynamic waves are a consequence of inertial forces. Because velocities are low in glaciers, inertial forces are negligible in comparison with gravitational and viscous forces. Kinematic waves, on the other hand, are a consequence of a conservation law. On a glacier, it is mass (or volume at constant density) that is conserved, and the type of kinematic wave in which we are interested is a wave of constant ice flux. Kinematic waves move through a medium at a speed that is different from the speed of the medium itself.

Kinematic waves on glaciers arise from the fact that if the ice flux into an element of a glacier of length dx is greater than the flux out of it, the glacier becomes thicker there. Because both the ice velocity and the ice flux in the thicker ice in the resulting wave are greater than in the thinner ice on either side of it, the wave moves faster than the ice.

Numerical modeling experiments suggest that kinematic waves on glaciers are likely to be long and low, and that the increases in velocity and thickness associated with them should rarely exceed about 10% of the unperturbed values (van de Wal & Oerlemans, 1995). Thus, they will be difficult to detect in the field. Larger waves have been observed in the field, but these are probably a consequence of changes in other factors, such sliding speed. Of course, changes in sliding speed can be induced by perturbations in mass balance.

(For comparison, waves of denser traffic on a highway are also a form of kinematic wave. In this case, cars catching up to a wave from behind are forced to slow down, while those finally making their way through the wave can accelerate again. Thus, in this case, the wave speed is less than the speed of the individual cars.)

ELEMENTARY KINEMATIC WAVE THEORY

Let us now develop these ideas analytically. In this development, following an analysis by Nye (1960), we will consider a slab of ice on a slope, $\beta(x)$, with thickness, $h(x,t)$, and surface slope, $\alpha(x,t)$ (Fig. 12.2). We will assume that $\partial h/\partial x$ is small and that the slab is of infinite extent in the horizontal direction normal to the x-axis. Note that the surface slope is related to the bed slope by:

$$\alpha = \beta - \frac{\partial h}{\partial x}. \tag{12.1}$$

Here, if h decreases downglacier, $\partial h/\partial x$ is negative so $\alpha > \beta$.

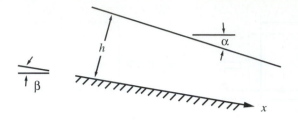

Figure 12.2 Relation among surface slope, α, bed slope, β, and thickness, h.

Consider conservation of mass in an element of a glacier of length dx (Fig. 12.3). For convenience, we will express mass fluxes in terms of the equivalent volumes of ice, based on a standard density. Ice flows into the element at a rate, $q(x,h,\alpha,t)$, and out of it at a rate, $q + (\partial q/\partial x)\, dx$. Here, q is the flux per unit of glacier width, and thus has the dimensions $m^3\, m^{-1}\, a^{-1}$. In addition, there is accumulation at a rate $b\, dx$. If more ice flows into, or accumulates in, the element than leaves it, the glacier increases in thickness at a rate $\partial h/\partial t$, so the increase in volume of ice in the element is $(\partial h/\partial t)\, dx$. Thus,

$$q - \left(q + \frac{\partial q}{\partial x}\, dx\right) + b\, dx = \frac{\partial h}{\partial t}\, dx$$

or, simplifying:

$$\frac{\partial q}{\partial x} + \frac{\partial h}{\partial t} = b. \tag{12.2}$$

Because q is a function of h and x, the functional dependence expressed by equation (12.2) leads to a general class of motions in flow systems known as kinematic waves (Lighthill & Whitham, 1955). Our objective next is to gain some appreciation for the nature of such waves on glaciers.

Let us begin by considering the wave speed. Suppose we multiply both sides of equation (12.2) by $(\partial q/\partial h)_x = c$, where c is the change in flux resulting from a change in thickness at point x; thus,

$$c\frac{\partial q}{\partial x} + \frac{\partial q}{\partial h}\frac{\partial h}{\partial t} = bc \qquad \text{or} \qquad c\frac{\partial q}{\partial x} + \frac{\partial q}{\partial t} = bc. \tag{12.3}$$

Equation (12.3) is known as the kinematic wave equation; c has the dimensions $m^3a^{-1}m^{-1}/m = m\ a^{-1}$. Thus, it is a speed. In fact, it is the celerity (or speed) of the wave. [Because $q = \bar{u}h$, where $\bar{u}$ is the mean (depth-averaged) speed, $\partial q/\partial h = \bar{u} + h\,(\partial\bar{u}/h)$.]

To gain some appreciation for the implications of equation (12.3), consider the situation in Figure 12.4. The ice flux into the element of the glacier is $1050\ m^3\, m^{-1}\, a^{-1}$, whereas that out is $1000\ m^3\, m^{-1}\, a^{-1}$.

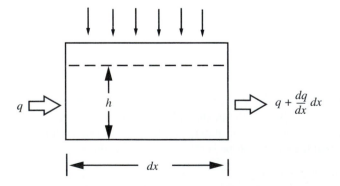

Figure 12.3 Contributions to change in mass in an element of a glacier of length dx.

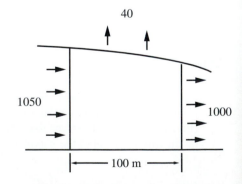

Figure 12.4 Numerical interpretation of the terms in equation (12.3).

The element is in the ablation area, and the ablation rate is -0.4 m a^{-1}, or -40 m^3 m^{-1} a^{-1} over the length of the element. As a result of this positive balance, with more ice entering the element than leaving it, the glacier is increasing in thickness. With the use of equation (12.3) we can calculate the rate of increase in mass flux, $\partial q/\partial t$, resulting from this increase in thickness. For example, the flux gradient, $\partial q/\partial x$, over the 100-m-long element is -50 m^3 m^{-1} a^{-1}/100 m or -0.5 m a^{-1}. Suppose $c = 200$ m a^{-1}. Then, $\partial q/\partial t$ is 20 m^3 m^{-1} a^{-1}. In other words, owing to the increase in thickness and the resulting increase in speed, the mass flux increases by 20 m^3 m^{-1} a^{-1}.

The relationship among $q, h, \bar{u},$ and c is illustrated in Figure 12.5, in which q is plotted against h. Because of the nonlinearity of the flow law, we expect q to increase nonlinearly with h as shown. The mean speed, $\bar{u}$, of a glacier with a thickness and ice flux given by the values of q and h at point P in the figure is $\bar{u} = q/h$. This is represented by the slope of the line connecting P with the origin. However, the speed, c, of a kinematic wave is $(\partial q/\partial h)_P$, which is the slope of a line drawn tangent to the q-h curve at point P. In other words, as mentioned earlier, the speed of the kinematic wave is appreciably larger than the mean speed of the glacier.

To get a sense of how much faster the kinematic wave moves, consider the case of a glacier moving entirely by internal deformation such that [see equation (5.19)]:

$$\bar{u} = \frac{2}{n+2}\left(\frac{S_f \rho g \alpha}{B}\right)^n h^{n+1}.$$
(12.4)

Noting that $q = \bar{u}h$, we have

$$q = \frac{2}{n+2}\left(\frac{S_f \rho g \alpha}{B}\right)^n h^{n+2}$$

so:

$$c = \left(\frac{\partial q}{\partial h}\right) = 2\left(\frac{S_f \rho g \alpha}{B}\right)^n h^{n+1} = (n+2)\bar{u}$$
(12.5)

or with $n \cong 3$:

$$c \cong 5\bar{u}.$$
(12.6)

In other words, the kinematic wave moves with a speed that is roughly five times the depth-averaged velocity of the glacier. If there is basal sliding and the sliding speed varies as τ^2 [Equation (7.10)], the ratio is likely to be slightly less than 5. This relation applies, rigorously, only to infinitesimal waves. Waves of finite amplitude may have higher speeds.

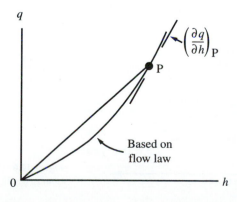

Figure 12.5 Relation between mean speed of a glacier and speed of a kinematic wave.

ANALYSIS OF THE EFFECT OF A SMALL CHANGE IN MASS BALANCE USING A PERTURBATION APPROACH

Let us now, following Nye (1960, pp. 561–562), use perturbation techniques to study the change in thickness with time after a small change in mass balance. Consider the situation in which the specific mass balance is shown by the solid line in Figure 12.6. We will refer to the situation represented by this solid line as the "0" or *datum* or *equilibrium* state, and analyze the effect of small perturbations from this state such as those represented by the dashed lines in the figure. For example, during a cold or unusually snowy year the mass balance may be increased everywhere by an amount $b_1(x,t)$, so we have

$$b(x, t) = b_0(x) + b_1(x, t)$$

Note that b_0 is a function only of x; it does not vary with time because the datum state is a steady state. Other properties of the datum state are $q_0(x, h, \alpha)$, $h_0(x)$, and $\alpha_0(x)$. In the perturbed state these become $q = q_0 + q_1$, $h = h_0 + h_1$, and $\alpha = \alpha_0 + \alpha_1$. Substituting these into the continuity equation [equation (12.2)] yields

$$\frac{\partial}{\partial x}(q_0 + q_1) + \frac{\partial}{\partial t}(h_0 + h_1) = b_0 + b_1. \tag{12.7}$$

We now write equation (12.2) in terms of the "0" state; thus,

$$\frac{\partial q_0}{\partial x} + \frac{\partial h_0}{\partial t} = b_0$$

and subtract this from equation (12.7) to obtain

$$\frac{\partial q_1}{\partial x} + \frac{\partial h_1}{\partial t} = b_1. \tag{12.8}$$

In passing, it is again worth noting that $\partial h_0/\partial t = 0$ because h_0 is a property of the steady state.

At any position, x, q varies with h and α so we can write.

$$dq = \frac{\partial q}{\partial h} dh + \frac{\partial q}{\partial \alpha} d\alpha$$

or for small perturbations, $dq = q_1$, $dh = h_1$, and $d\alpha = \alpha_1$ so:

$$q_1 = \frac{\partial q}{\partial h} h_1 + \frac{\partial q}{\partial \alpha} \alpha_1. \tag{12.9}$$

Previously we identified $\partial q/\partial h$ with the celerity of a kinematic wave, c, or in the "0" state: $(\partial q/\partial h)_0 = c_0$. Now we similarly define $D_0 = (\partial q/\partial \alpha)_0$, so in the "0" state, equation (12.9) becomes

$$q_1 = c_0 h_1 + D_0 \alpha_1. \tag{12.10}$$

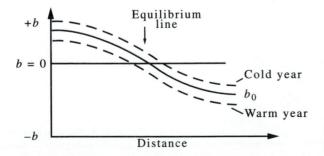

Figure 12.6 Perturbations in mass balance from an equilibrium state, b_0.

This relation is valid only for small perturbations. Were we interested in larger perturbations, terms involving h_1^2, h_1^3, ... α_1^2, α_1^3, ... would have to be included. Thus, our approach is referred to as a linearized theory.

Equations (12.8) and (12.10) are a pair of simultaneous differential equations that can be solved for the change in ice flux and thickness resulting from a perturbation in net balance. Let us first eliminate q_1 from the two equations; thus,

$$\frac{\partial c_0}{\partial x} h_1 + c_0 \frac{\partial h_1}{\partial x} + \frac{\partial D_0}{\partial x} \alpha_1 + D_0 \frac{\partial \alpha_1}{\partial x} + \frac{\partial h_1}{\partial t} = b_1. \tag{12.11}$$

Returning to equation (12.1) we see that in the "o" and perturbed states, respectively:

$$\alpha_0 = \beta - \frac{\partial h_0}{\partial x} \qquad \text{and} \qquad \alpha_0 + \alpha_1 = \beta - \frac{\partial h_0}{\partial x} - \frac{\partial h_1}{\partial x}. \tag{12.12}$$

Thus, subtracting the first of these expressions from the second: $\alpha_1 = -\partial h_1/\partial x$. This result can be substituted into equation (12.11) to yield, after some rearranging:

$$\frac{\partial h_1}{\partial t} = b_1 - \frac{\partial c_0}{\partial x} h_1 - \left(c_0 - \frac{\partial D_0}{\partial x} \right) \frac{\partial h_1}{\partial x} + D_0 \frac{\partial^2 h_1}{\partial x^2}. \tag{12.13}$$
$$\quad \text{(i)} \quad \text{(ii)} \qquad \text{(iii)} \qquad \text{(iv)}$$

Equation (12.13) was first derived by Nye (1960, p. 562). As he noted, the terms in it have the following meanings:

(i) h_1 increases at a rate given by the perturbation in accumulation.

(ii) This term results in an exponential decrease or increase in the rate of change of h_1, as we shall show below.

(iii) This represents a kinematic wave of constant h_1. The speed of propagation of the wave is $c_0 - \partial D_0/\partial x$ in the $+x$ direction. Note that both c_0 and $\partial D_0/\partial x$ have dimensions $\ell \cdot t^{-1}$.

(iv) This represents diffusive damping of the perturbation h_1, in accord with the diffusion equation, with diffusivity D_0.

Our objective now is to solve equation (12.13) for a simple case, neglecting diffusion. Then we will investigate the role of diffusion.

Solution for a Small Part of a Glacier with Uniform Longitudinal Strain Rate and without Diffusion

Consider a situation in which a glacier is initially in a steady state with an accumulation rate b_0 (Nye, 1960, p. 563). Then the accumulation rate increases abruptly by an amount b_1 to $b = b_0 + b_1$ and remains at this increased level indefinitely. Suppose $\partial c_0/\partial x$ is independent of x on this glacier. From equations (12.5) we see that

$$\frac{\partial c_0}{\partial x} \cong (n+2) \frac{\partial \bar{u}_0}{\partial x} \tag{12.14}$$

where $\partial u_0/\partial x$ is the longitudinal strain rate, r_0. This, thus, corresponds to a situation in which the longitudinal strain rate is uniform in the x-direction. We seek a solution to equation (12.13) such that h_1 is independent of x so $\partial h_1/\partial x = 0$. Thus, the increase in thickness is uniform over the glacier. We will let $\gamma_0 = \partial c_0/\partial x$.

With these simplifications, equation (12.13) becomes

$$\frac{dh_1}{dt} = b_1 - \gamma_0 h_1. \tag{12.15}$$

Separating variables we obtain

$$\int_o^{h_1} \frac{dh_1}{b_1 - \gamma_0 h_1} = \int_o^t dt.$$

Integration yields:

$$h_1 = \frac{b_1}{\gamma_0}(1 - e^{-\gamma_0 t}). \tag{12.16}$$

If γ_0 is positive, corresponding to a positive longitudinal strain rate such as we expect in the accumulation area of a glacier, the perturbation, h_1, asymptotically approaches the value b_1/γ_0 (Fig. 12.7). In other words, after a very long time, the glacier will have increased in thickness by this amount. This is the situation described earlier and illustrated in Figure 12.1a.

The quantity $1/\gamma_0$, which has the dimensions of time, is known as the *time constant* for this change. This is sometimes associated with the "response time" of a glacier, or the length of time required for the glacier to respond to a change in climate. When $t = 1/\gamma_0$, h_1 is $(1 - 1/e)$ or ~2/3 of the way to the new equilibrium state. Mathematically [equation (12.16)], it is clear that the new equilibrium state is never reached. Thus, it would be meaningless to try, instead, to define the response time as the total time required to attain a new steady state.

From equation (12.14) we see that $1/\gamma_0 \cong 1/5r_0$. In other words, in this simple model the response time is inversely proportional to the longitudinal strain rate. For example, typical longitudinal strain rates for Storglaciären, Barnes Ice Cap, and the Antarctic Ice Sheet are 0.015 a^{-1}, 0.005 a^{-1}, and 0.00005 a^{-1}, respectively. Thus, the response time of Barnes Ice Cap might be expected to be 3 times as long as that of Storglaciären, and that of the Antarctic Ice Sheet, 100 times as long as Barnes Ice Cap. While these multiples are not unrealistic, it turns out that $1/5r_0$ seriously underestimates the actual response time. As we will see, below, this is because diffusion has been neglected.

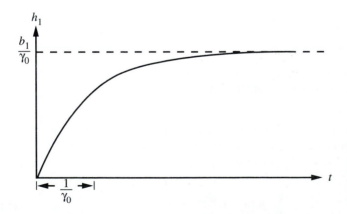

Figure 12.7 Asymptotic adjustment of a glacier toward a new steady state, b_1/γ_0, following a perturbation in accumulation rate, b_1, in an area of extending flow. $1/\gamma_0$ is the response time.

If γ_0 is negative, corresponding to longitudinal compression as would be typical in an ablation zone, there is an obvious problem. Equation (12.16) then predicts that h_1 will increase exponentially with time. Thus, a new steady state is never even approached. This is the situation we discussed in connection with Figure 12.1b.

Clearly, it is not possible to have the upper part of a glacier increasing in thickness slowly and stably while the lower part is increasing rapidly and unstably. In the absence of diffusion, Nye (1960) suggests that the initial response in the ablation area would, indeed, be unstable. At any location, however, stability would be restored when a kinematic wave, initiated in the vicinity of the equilibrium line and propagating down glacier, reached that location. With diffusion, however, such an unstable response may never develop.

EFFECT OF DIFFUSION

Diffusion occurs whenever fluxes are proportional to gradients. In the present case, the flux, q, is proportional to the slope (or gradient), α. Where α is largest, on the downslope side of a wave, q is highest. Conversely, q is lowest on the upslope side of the wave. Thus, the flux into the wave is diminished and that out of it is enhanced. This tends to decrease the amplitude and increase the wavelength of a wave.

As in the case of c (or c_0) [equation (12.5)], an analytical expression for D_0 can be obtained by differentiating q with respect to α, thus:

$$D_0 = \left(\frac{\partial q}{\partial \alpha}\right)_0 = n\left(\frac{2}{n+2}\right)\left(\frac{S_f \rho g}{B}\right)^n h^{n+2}\alpha^{n-1} = \frac{n\,q}{\alpha}$$

or with $n \cong 3$:

$$D_0 \cong \frac{3\bar{u}\,h}{\alpha}. \tag{12.17}$$

In other words, diffusion will be most significant where the glacier is thick, the speed high, and the slope low.

Unfortunately, it is difficult to probe this dependence more thoroughly at the level of the treatment herein. However, Nye (1963a, pp. 442–445) has shown that diffusion decreases the rate of thickening, a result that is intuitively logical. As a result, the response time increases quite markedly. In one example, the response time increases by more than an order of magnitude (Nye, 1963a, Fig. 4a). In addition, the increase in thickness of the glacier near the equilibrium line is substantially greater when diffusion is taken into consideration.

THE PROBLEM AT THE TERMINUS

It is difficult to use Nye's kinematic wave theory to study the details of the advance and retreat of real glaciers. This is, in part, because the mass flux, q, cannot go to zero at the terminus if the glacier is to respond to an increase in accumulation by advancing. However, equation (12.4) suggests that $u \to 0$ as $h \to 0$. To avoid this, Nye (1963b, p. 92) assumes that the glacier is sliding at the terminus so $q = u_{b0}\,(\ell_0)h$ where $u_{b0}\,(\ell_0)$ is the sliding speed at the terminus, ℓ_0 being the length of the glacier in the datum state. Then $c_0 = \partial q/\partial h = u_{b0}\,(\ell_0)$.

In addition, the amount of advance, $\Delta \ell$, is sensitive to the assumed geometry of the terminus. As shown in Figure 12.8:

$$\Delta \ell = \frac{h_1(\ell_0)}{\tan \theta_0} \qquad (12.18)$$

where $h_1(\ell_0)$ is the perturbation in ice thickness at the terminus. Thus, $\Delta \ell$ depends on θ_0.

FURTHER STUDY OF THE RESPONSE TIME

Jóhannesson and others (1989) have studied the question of response times and of conditions at the terminus in greater detail. They identify three possible natural time scales that might be used in the analysis of glacier responses:

$$t_C = \frac{\ell_0}{\overline{c}_0} \qquad \left[\frac{m}{m/a} \right] \qquad (12.19a)$$

$$t_D = \frac{\ell_0^2}{\pi^2 \overline{D}_0} \qquad \left[\frac{m^2}{m^2/a} \right] \qquad (12.19b)$$

$$t_V = \frac{V_1}{\overline{b}_1 \ell_0} \qquad \left[\frac{m^2}{m\,a^{-1} \cdot m} \right] \qquad (12.19c)$$

Here, t_C and t_D are time constants for propagation or diffusion[1] of a disturbance over the length of a glacier. In effect, they are measures of the time required to establish the general shape of the new thickness profile, $h_1(x,t)$. As the size of such a disturbance decreases with time, the rate of propagation or diffusion also decreases. Therefore, as with $1/\gamma_0$, t_C and t_D are measures of the time required for the processes to proceed to about 2/3 of the way to completion. Similarly, as we shall show below, t_V is the time required for accumulation (or loss) of about 2/3 of the volume (per unit width), V_1, required to re-establish an equilibrium geometry after a perturbation in mass balance, b_1.

Jóhannesson and others found that t_V is usually appreciably longer than t_C or t_D. This means that perturbations in ice thickness are spread out over the glacier by propagation and diffusion rather quickly in comparison with the time needed for accumulation of the additional mass.

In an extension of the Nye theory, t_V would be viewed approximately as follows: At any given time after a perturbation in mass balance, the mean perturbation in thickness, averaged over the length of the glacier, would equal the perturbation at the terminus, $h_1(\ell_0,t)$, multiplied by some function of the conditions in the datum state and the magnitude of the perturbation; thus,

$$\overline{h_1(x,t)} = f(c_0, D_0, b_1, t) \cdot h_1(\ell_0, t). \qquad (12.20)$$

Remember that ℓ_0 is the position of the terminus in the unperturbed state. Once a new equilibrium geometry has been attained, at $t = \infty$, the increase in volume of the glacier would be obtained by multiplying equation (12.20) by the length of the glacier; thus,

$$V_1 = \overline{h_1} \ell_0 = f(t = \infty) \cdot h_1(\ell_0, \infty) \ell_0. \qquad (12.21)$$

However, once a new steady state has been attained, the annual mass gain resulting from the perturbation, $b_1 \ell_0$, must equal the flux past the old terminus position, $u_{b0}(\ell_0) \cdot h_1$; thus,

$$\overline{b_1} \ell_0 = u_{b0}(\ell_0) \, h_1(\ell_0, \infty) \qquad (12.22)$$

---

1. The π^2 term in equation (12.19b) comes from the Fourier solution of the diffusion equation (Jóhannesson, 1996).

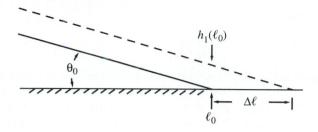

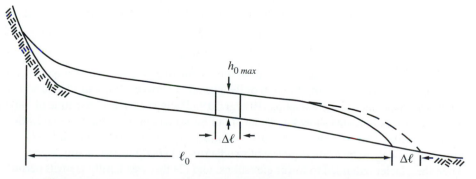

Figure 12.8   Geometry of the terminus region.

Eliminating h_1 from these two equations yields

$$V_1 = \frac{f(t = \infty)\,\overline{b}_1\,\ell_0^2}{u_{b0}(\ell_0)}.$$

(12.23)

Whence, from equation (12.19c):

$$t_V = \frac{f(t = \infty)\,\ell_0}{u_{b0}(\ell_0)}.$$

(12.24)

Thus, by this approach, t_V turns out to be sensitive to the unknown sliding speed at the terminus. In addition, the function, f, is highly sensitive to the details of the variations in c_0 and D_0, especially near the terminus (Jóhannesson et al., 1989, p. 364 and appendices).

Jóhannesson and others have developed a much simpler geometrical argument to estimate t_V. Consider the situation in Figure 12.9 in which the advance of a glacier by an amount $\Delta\ell$ is illustrated graphically by cutting the glacier at its point of maximum thickness and sliding the lower part forward by $\Delta\ell$. Then, the increase in volume of the glacier is approximately $\Delta\ell \cdot h_{0\,max}$. Detailed numerical modelling suggests that this is a good approximation to the response of a real glacier when the dynamical properties of the glacier are the same in the initial and final state, and thus influence the initial and final profiles in the same way. Now, rather than equate the annual mass gain resulting from the perturbation with the flux past the old terminus position, as in equation (12.22), we equate it with the mass loss over the new part of the glacier, $\Delta\ell$, thus:

$$\overline{b}_1\,\ell_0 = |b(\ell_0)|\Delta\ell$$

(12.25)

where $b(\ell_0)$ is the net balance rate at the terminus (a negative quantity). Therefore,

$$V_1 = \Delta\ell\,h_{0\,max} = \frac{h_{0\,max}\,\overline{b}_1\,\ell_0}{|b(\ell_0)|}$$

(12.26)

Figure 12.9   Geometrical argument for evaluating t_V. During an advance, $\Delta\ell$, the mass that must be added to a glacier is approximately $\Delta\ell \cdot h_{0\,max}$.

Whence, from equation (12.19c):

$$t_V = \frac{h_{0\,max}}{|b(\ell_0)|}. \qquad (12.27)$$

Thus, t_V can be estimated, quite easily, from knowledge of the thickness of the glacier and the net balance rate at the terminus.

We noted above that t_V is a time constant in the same sense that $1/\gamma_0$ is. Let us now demonstrate this. Immediately after a permanent change in balance rate, b_1, the rate at which additional mass, V_1, is acquired by the glacier, dV_1/dt, is $B_1 = \overline{b}_1 \cdot \ell_0$. However, as the glacier becomes longer [by an amount $\Delta \ell(t)$], some of the additional annual input is lost through ablation in the new part of the terminus region. Thus,

$$\frac{dV_1}{dt} = B_1 - |b(\ell_0)|\Delta \ell$$

Now from Figure 12.9, $\Delta \ell \cong V_1/h_{0\,max}$, so:

$$\frac{dV_1}{dt} = B_1 - \frac{|b(\ell_0)|}{h_{0\,max}}V_1.$$

Comparing this with equation (12.15), it is clear that $h_{0\,max}/|b(\ell_0)|$ is analogous to $1/\gamma_0$.

When calculating t_V in practice, the three-dimensional geometry of the glacier must be taken into consideration. For example, in the case of a glacier like Storglaciären that has a number of overdeepened basins in its longitudinal profile, $h_{0\,max}$ must be replaced by an appropriate longitudinally-averaged thickness. In addition, the terminus of Storglaciären is constrained between bedrock and morainal highs (Fig. 11.9) so that its width is about half the average width of the glacier. Thus, equations (12.25) to (12.27) need to be generalized to three dimensions. For example, if we write equation (12.25) as

$$\overline{b}_1 A_0 = |b(\ell_0)|\, W(\ell_0)\, \Delta \ell \qquad (12.28)$$

where A_0 is the initial area of the glacier and $W(\ell_0)$ is the width of the terminus, equation (12.19c) becomes

$$t_V = \frac{V_1}{\overline{b}_1 A_0}. \qquad (12.29)$$

Now V_1 (in m³) must be estimated based on the glacier geometry. For example, by analogy with Figure 12.9, one might consider that the new geometry could be approximated by (mentally) sliding forward the central part of the glacier of width $W(\ell_0)$. Then V_1 is $\overline{h}_{0\,max}\, W(\ell_0)\, \Delta \ell$ where $\overline{h}_{0\,max}$ is a mean thickness over this central part. Inserting this in equation (12.29) and using equation (12.28) then yields

$$t_V = \frac{\overline{h}_{0\,max}}{|b(\ell_0)|}. \qquad (12.30)$$

The ablation rate on the lower part of the tongue of Storglaciären averages ~1.3 m a⁻¹, and the mean thickness over the central region of the glacier is between 100 and 150 m, so t_V is ~100 years. For comparison, numerical modeling (Brugger, 1992) suggests a response time of ~80 years. In contrast, field measurements show that about 2/3 of Storglaciären's retreat from its Little Ice Age maximum position, which it reached in 1910, took place in ~45 years (Holmlund, 1987). This more rapid response is likely to be a consequence of two factors: (1) $b(\ell_0)$ was probably higher (more negative) when the glacier extended to lower elevations, and (2) this was hardly a small perturbation. In any case, all of these times are substantially longer than the $1/5r_0$ ($\approx$13 years) time scale mentioned above. As noted, this is because diffusion is neglected in the latter.

NUMERICAL MODELING OF GLACIER RESPONSES

In the absence of analytical solutions to equation (12.13), glaciologists have resorted to numerical modeling. In such models one can, in addition, retain nonlinear effects that are neglected in linearized theories. Thus, the models are not restricted to infinitesimal perturbations. Furthermore, one can use glacier shapes and mass-balance patterns specific to a particular glacier.

For example, van de Wal and Oerlemans (1995) modeled Hintereisferner in the Austrian Alps. First they calculated a profile for the glacier that would be in equilibrium with a certain mass-balance rate, $b_0(x)$ (Fig. 12.10a). Then, for one year, they increased the mass-balance rate by 0.5 m a^{-1}. This could represent the situation during an unusually positive balance year (see the curve labeled "+" in Figure 3.5a or that labeled "Cold year" in Figure 12.6). The following year, the net budget was returned to its normal value. Figure 12.10b shows the increase in thickness as a function of distance from the head of the glacier at various times after the perturbation. After two years, a wave has formed with its crest ~3 km from the head, or about a kilometer upglacier from the equilibrium line. By the sixth year, the crest is a little more than 4 km from the head, representing a wave speed of about 300 m a^{-1}. For comparison, the depth-averaged velocity over this part of the glacier is a little under 50 m a^{-1}. In addition, the wave has been dampened and lengthened by diffusion. With time, diffusion continues to smooth the wave, the surface in the accumulation area sinks back toward its original level, and the surface in the ablation area, particularly at the terminus, rises sharply. In Figure 12.10c it will be seen that the terminus begins to collapse back to its original form after about 30 years, but that a significant thickening remains after 100 years.

The Hintereisferner modeling experiment serves to emphasize that kinematic waves on glaciers are likely to be long and low, as mentioned earlier in this chapter. Thus, sophisticated survey techniques are likely to be required to detect them in the field. In addition, the modeling suggests that the wave speed is ≥ 6 times the depth-averaged velocity, $\bar{u}$, rather than $\leq 5\bar{u}$ as suggested by equation (12.6) and the following discussion. Van de Wal and Oerlemans argue that this is due to changes in the longitudinal strain rate which appear in the numerical model but which are not taken into consideration in the linear model. Such changes are likely to affect the $q - h$ relation. Finally, the response at the terminus is stable, as shown in Figure 12.10c.

COMPARISON WITH OBSERVATION

Let us now discuss some actual examples of how glaciers have responded to climatic perturbations. We have already mentioned Storglaciären briefly, and noted that estimates of the response time based on equation (12.30), on a numerical model, and on observation are reasonably consistent with each other, and suggest a time of decades to a century, whereas $1/\gamma_0$ is only ~13 years. Nisqually and South Cascade glaciers in the state of Washington are two others that have been studied extensively.

Nisqually Glacier

Nisqually Glacier on Mt. Rainier retreated several hundred meters during the first part of the twentieth century. A trimline on the valley side above and downglacier from the present terminus (Fig. 12.11) shows the shape of the glacier at its nineteenth-century maximum position, a position that it occupied, more or less, from about 1840 to 1910 (Meier, 1965, p. 803). Thus, the difference in elevation between the present (debris-covered) glacier surface and the trimline represents the amount of thickening that would need to occur in order for the glacier to readvance to that maximum position. The amount of thickening increases rapidly toward and down-valley from the terminus.

The response of Nisqually Glacier to perturbations in mass balance is illustrated in Figures 12.12 and 12.13. The upper part of Figure 12.12 shows that the net budget was generally positive

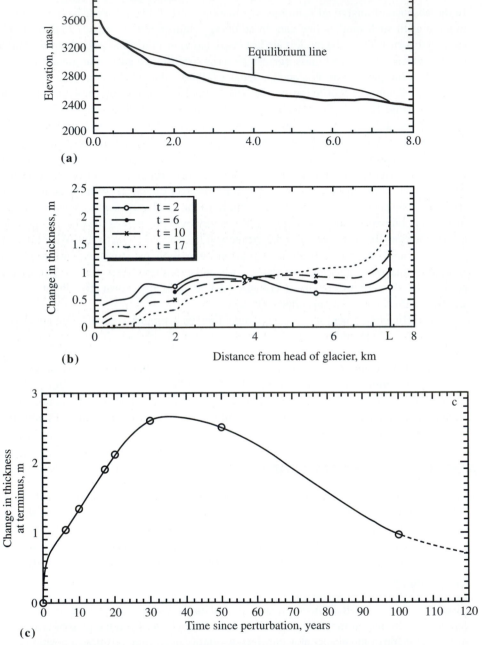

(a)

(b)

(c)

Figure 12.10 (a) Longitudinal profile of Hintereisferner in a stable state. (b) Changes in thickness of Hintereisferner resulting from a 0.5-m perturbation in mass balance, b_1, at time $t = 0$. Times are in years. Thickness changes are larger than 0.5 m because they include a contribution from the unperturbed mass balance, b_0. (c) Change in thickness at the terminus of Hintereisferner as a function of time after the perturbation. (Both **a** and **b** are reproduced from van de Wal & Oerlemans, 1995; Figs. 7a and 9b, with permission of the authors and the *Journal of Glaciology*; **c** is calculated from data in Figs. 9b and 9c of van de Wal and Oerlemans, 1995.)

Figure 12.11   Nisqually Glacier, Mt. Rainier, Washington, in September 1964, showing distinct trimline. Glacier tongue, covered with volcanic debris, lies between isolated snow patch in valley bottom and snowfields low on the mountain.

between 1942 and 1951. In fact, the retreat rate of many temperate alpine glaciers in the Northern Hemisphere decreased during this time period, and some actually advanced. Thus, this represents a major climatic event (Meier, 1965, p. 803). However, in the middle part of Figure 12.12 it can be seen that the terminus was still retreating during this time; it was responding to negative mass budgets of the early 20th century. The total retreat between 1918 and 1960 was slightly over 1000 m.

In the mid-1940s, a wave of thickening was detected ~1500 m upglacier from the terminus, and this wave was tracked downglacier until it reached the terminus in about 1960 (lower part of Fig. 12.12). This wave was presumably the response to the positve mass budgets of the 1940s. The progress of the wave is documented in Figure 12.13a, which is based on surveys, conducted almost every year, of the elevation of the glacier surface along three profiles across the glacier. The average elevation of the ice surface on each profile is shown as a function of time. At Profile 3, which is 2.7 km from the mid-20th-century terminus, thickening began in about 1945. Profile 2 is 1.6 km from the terminus; thickening began here in 1949. The wave reached Profile 1, 0.8 km from the terminus, in 1955. In Figure 12.13b the ice surface slope, surface elevation, and velocity are shown as functions of time at Profile 2. As noted earlier, this is probably not a pure kinematic wave as the changes in thickness and velocity are rather large. Thus, some other mechanisms, such as an increase in sliding speed, were probably involved.

The reader may find it of interest to compare the change in velocity in Figure 12.13b with that predicted by equation (5.7) with $u_b = 0$:

$$u_s = \frac{2}{n+1}\left(\frac{\rho g \alpha}{B}\right)^n h^{n+1}. \tag{12.31}$$

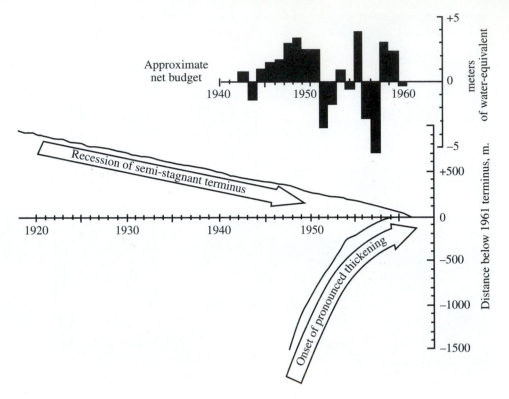

Figure 12.12   Recession of Nisqually Glacier between 1918 and 1961, together with the advance of a wave of pronounced thickening and the approximate net budget. (Reproduced from Meier, 1965, Fig. 8, with permission of the author.)

To do this, take the differential of equation (12.31) and divide the result by equation (12.31) to yield

$$\frac{du_s}{u_s} = n\frac{d\alpha}{\alpha} + (n+1)\frac{dh}{h}. \tag{12.32}$$

To make this calculation you need the ice thickness, which is about 80 m at Profile 2. Despite the approximations inherent in equations (12.31) and (12.32) and in estimating the values of the parameters in equation (12.32) from the field data, the calculated du_s is surprisingly close to that observed. (The numerical computations are left as an exercise for the reader; Problem 12.2).

South Cascade Glacier

Owing to the availability of an impressive database, South Cascade Glacier is another that has been analyzed in some detail. For example, Nye (1963b) used these data to test the kinematic wave theory. To do this, he had to take into consideration the three-dimensional character of the glacier. Thus, our equations (12.8) and (12.10) become

$$\frac{\partial Q_1}{\partial x} + w_0\frac{\partial h_1}{\partial t} = w_0 b_1$$
$$Q_1 = c_0 h_1 + D_0 \alpha_1 \tag{12.33}$$

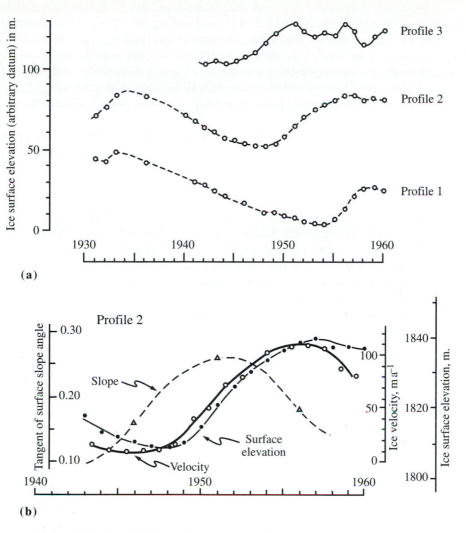

Figure 12.13 (a) Variation in ice-surface elevation on three transverse profiles on Nisqually Glacier, 1931–1960. (b) Variations in ice-surface elevation, velocity, and surface slope at Profile 2, 1943–1960. (*a* is from Johnson, 1960, Fig. 2; *b* from Meier, 1965, Fig. 4. Reproduced with permission of the author.)

where $Q(x)$ is the ice flux through a cross section of the glacier at position x, and $w_0(x)$ is the width of the glacier as a function of x. In addition, c_0 and D_0 have to be redefined as

$$c_0 = \frac{1}{w_0}\left(\frac{\partial Q}{\partial h}\right)_0 \qquad \text{and} \qquad D_0 = \frac{1}{w_0}\left(\frac{\partial Q}{\partial \alpha}\right)_0.$$

As was the case with equations (12.8) and (12.10), equations (12.33) are a pair of simultaneous differential equations that can be solved for the change in ice flux, $Q_1(x, t)$, and in thickness, $h_1(x, t)$, resulting from a perturbation in mass balance, $b_1(x, t)$.

Previously [equations (12.6) and (12.17)] we found that, in the absence of sliding, c_0 and D_0 could be related to certain measures of the speed and ice flux. Thus, if the geometry and velocity field of a glacier are known, reasonable estimates of $c_0(x)$ and $D_0(x)$ can be made. Nye calculated these parameters for South Cascade Glacier (Fig. 12.14) and used the results to solve equations (12.33) for the situation in which perturbations in b varied sinusoidally with period, T, in years, or frequency, $\omega = 2\pi/T$. The solution is expressed in terms of series approximations, and detailed study of it is beyond the scope of this book. Numerical results for South Cascade Glacier are shown in Figure 12.15.

The curve of ø in Figure 12.15 is the phase lag between the variation in budget and the response of the terminus. For example, for an oscillation in mass balance that has a period of 100 years, the phase lag is approximately 110°. This means that the maximum thickness of the glacier at the terminus (and hence the maximum extent of the glacier) would occur $(110/360)\cdot100 \cong 31$ years after the maximum in the mass balance. This latter number can be read from the curve of ø/ω, using the inner scale on the left side of the figure. Thus, ø/ω is the time lag between the maximum accumulation rate and the maximum thickness. For variations in budget with very long periods, the phase lag decreases, but the time lag does not change appreciably. For example, for an oscillation with a period of 1000 years, the time lag is ~43 years. Because of the way in which the mass balance is changing throughout this oscillation, this time lag turns out to be essentially equivalent to t_V. Conversely, for oscillations with a period of only a year, which would represent the seasonal cycle from winter accumulation to summer melt, ø = 90° so the time lag is ¼ year. In other words, the maximum thickness does not occur when the rate of snow fall is a maximum, but rather at the end of the accumulation season when accumulation gives way to melt.

The curve of $|\mathrm{H}|/b_1$ shows the change in thickness of the glacier at the terminus, expressed in terms of the perturbation in accumulation. For a perturbation with an amplitude of 0.1 m and a period of 100 years, the increase in thickness here would be about 100 times the amplitude of the perturbation, or ~10 m.

The ultimate objective of an analysis such as this might well be to solve the inverse problem—namely, given a history of advance and retreat of a glacier, to deduce the mass-balance history and thus to learn something about the climatic changes that produced the fluctuations. Nye (1965b) did this for South Cascade Glacier and for Storglaciären with mixed results. He concluded that the records of terminus position of the two glaciers were not sufficiently well known to accurately

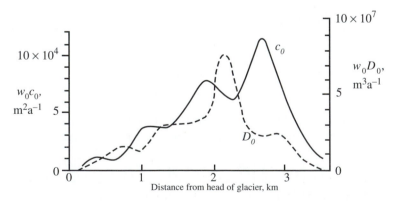

Figure 12.14 w_0c_0 and w_0D_0 as functions of x for South Cascade Glacier. (After Nye, 1963b, p. 104, Fig. 7. Reproduced with permission of author and the Royal Society of London.)

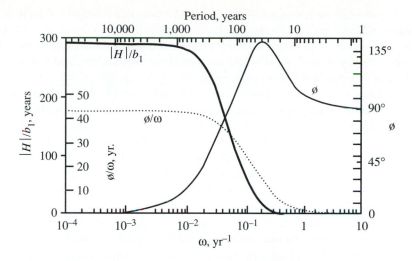

Figure 12.15 Theoretical response of the terminus of South Cascade Glacier to a perturbation of amplitude b_1, period T, and frequency ω. Curves shown are the phase lag, ø, the time lag ø/ω , and the amplitude of the response $|H|/b_1$. (After Nye, 1963b, p. 107, Fig. 9. Reproduced with permission of author and the Royal Society of London.)

deduce annual changes in net budget, but that coarser features of the records yield net-balance figures that are in agreement with decadal means of recent observations.

It is of interest to use the data in Figure 12.14 to estimate the respective times scales from equations (12.19). As South Cascade Glacier averages about 800 m in width, $t_C \cong 47$ years and $t_D \cong 33$ years. Jóhannesson (1996) suggests that time scales calculated in this way, however, are likely to be maximum estimates because many perturbations do not cover the entire glacier and thus are advected and diffused over the glacier more rapidly. Nevertheless, the relative magnitudes should be correct. Because $t_D < t_C$, disturbances should be damped by diffusion before a significant unstable response is generated. Jóhannesson (1995) finds that this is generally the case, and thus argues that diffusion cannot be neglected.

As with Storglaciären, it is difficult to estimate t_V for South Cascade Glacier because, again, there is a riegel beneath the middle of the ablation area. However, it appears that $100 < \overline{h}_{0\,max} < 200$ m and $b(\ell_0) \cong 5$ m a^{-1} so $20 < t_V < 40$ a. These values are reasonably consistent with those for t_C and t_D obtained above, particularly considering that the latter are likely to be maximum values. This is somewhat unusual, however, as Jóhannesson and others (1989) find that t_V is usually significantly longer than t_C or t_D, as noted earlier. The value of t_V is also consistent with Nye's estimate of 43 years. For comparison, the $1/r_0$ time scale for South Cascade Glacier is about 15 years.

SUMMARY

In this chapter, we have reviewed Nye's kinematic wave theory for predicting the response of a glacier to changes in mass balance, and have solved the resulting linearized equation [equation (12.13)] for a simplified situation neglecting diffusion. Largely because it neglects diffusion, this solution predicts response times that are, in general, too short. The more complete approaches that Nye (1963a,1963b) used in his later papers, however, are sensitive to conditions at the terminus of the glacier, so while the linearized theory can yield reasonable estimates of the response time if these terminus conditions are well known, attempts to generalize from it have often led to times that are too long. Nevertheless, evidence from real glaciers is consistent with at least two of the conclusions from Nye's theory: that the most visible response is at the terminus, and that this response lags the perturbation by years, decades, or even centuries.

Jóhannesson and others (1989) have suggested three alternative time scales for adjustment. Their time scales for propagation and diffusion of a disturbance over a glacier, t_C and t_D, provide measures of the time required for the glacier to adjust its shape (but not size) to changed conditions. Their volume time scale, t_V, on the other hand, utilizes a conservation of mass argument. That is, after a change in climate a glacier will be either too large or too small, and thus will not be in equilibrium with the changed conditions. It takes time for the surplus or deficit in mass balance to bring about the necessary change in volume. Thus, the volume time scale is more consistent with "response times" based on observation, and indeed with those based on numerical modeling.

Numerical modeling suggests that kinematic waves such as those that Nye envisioned should form on glaciers, but they are likely to be long and low, and the increase in speed within them, small. Thus, they will be difficult to detect. Additional factors, such as major changes in conditions at the bed, are probably responsible for the impressive waves that have been documented by field observations. Because diffusive processes dampen kinematic waves relatively rapidly, unstable responses (Fig. 12.1b) in areas of compressive flow are unlikely.

Appendix: Problems

CHAPTER 3

3.1 Determine the changes in b_w, R, and T_a that would, if they occurred alone, result in a 100-m increase in equilibrium line altitude. Assume a 120-d melt season and a lapse rate of -0.007 °C·m^{-1}.

CHAPTER 4

4.1 Determine the activation energy for creep for the following two sets of data:

Data set 1		Data set 2	
$\dot{\varepsilon}$, a^{-1}	T, °C	$\dot{\varepsilon}$, a^{-1}	T, °C
18.65	−5.5	1.33	−30.4
9.06	−9.9	0.0047	−61.0

All experiments were run at the same stress. Express the activation energy in kJ·mol^{-1}.

4.2 a. The temperature dependence of ice creep can be represented by an Arrhenius-type relation:

$$\dot{\varepsilon} = \left(\frac{\sigma}{B_o}\right)^n \exp\left(-\frac{Q}{R\theta_K}\right)$$

By differentiating this with respect to θ_K and expressing the result in terms of differentials, determine the fractional change in $\dot{\varepsilon}$, $d\dot{\varepsilon}/\dot{\varepsilon}$, due to a change, $d\theta_K$, in θ_K.

b. In a laboratory experiment run at a temperature of −15°C, what would be the approximate percentage variation in $\dot{\varepsilon}$ if the temperature were allowed to vary by 0.5°C? Use $Q = 79$ kJ·mol^{-1}.

4.3 Demonstrate analytically that the temperature dependence of ice creep can be reasonably approximated by

$$\dot{\varepsilon} = \dot{\varepsilon}_o e^{k\theta},$$

where θ is the temperature in Celsius degrees and $\dot{\varepsilon}_o$ is the strain rate at 0°C. What are reasonable values of k for different temperatures?

CHAPTER 5

5.1 Calculate the difference between the surface velocity and the bed velocity in a glacier 300 m thick with a surface slope of 0.046. Use $n = 3$ and $B = 0.14$ MPa a$^{1/3}$. Use:

- The infinitely wide approximation,
- An approximation based on Raymond's estimate of the appropriate shape factor for Athabasca Glacier (0.58), and
- The semicircular approximation.

Which result comes closest to the values measured by Raymond on Athabasca Glacier? Why?

5.2 At the equilibrium line on the Barnes Ice Cap Trilateration Net, the surface velocity is 6.7 m a^{-1}, the ice is 185 m thick, and the surface slope is 0.07. Using $B = 0.317$ MPa a$^{1/3}$ (appropriate for ice at about $-5°C$) and $n = 3$, calculate and plot a velocity profile through the ice cap. What is the basal velocity? Is your basal velocity consistent with the above ice temperature?

5.3 An infinitely wide glacier has a velocity of 1 m a^{-1} at the surface and 0.7 m a^{-1} at a depth of 16 m. Determine the thickness of the glacier. Assume $u_b = 0$ and $n = 3$.

5.4 An ice sheet has a surface profile given by $h = \sqrt{cx}$ where h is the height in meters and x is the distance from the margin, also in meters. Differentiate this to obtain an expression for the surface slope, S. By inserting this in the expression for the basal drag, $\tau = \rho ghS$, show that τ is independent of x. Obtain a numerical value for τ if $c = 16$ m.

5.5 On the ice sheet of problem 5.4, the ablation rate is A_b m a^{-1}. By equating the discharge through any cross section to the volume of ice lost by melting downglacier from that cross section, show that the average horizontal velocity in the ablation zone is $u = -\alpha\sqrt{x}$ where $\alpha = A_b/\sqrt{c}$. (The minus sign indicates that u is in the $-x$ direction.)

5.6 At $x = 1500$ m the glacier in Problem 5.4 flows over a bump in the bed, 0.5 m high, and quarries a cobble from the lee slope of the bump. The ice closes under the cobble, so at the start of its journey to the margin it is 0.5 m above the bed. Determine the x and z coordinates (z vertical) of the point where the cobble will melt out, and its time en route. Plot the path of the particle. Assume plug flow (ie, u is constant, independent of the depth) and incompressibility. Use $A_b = 0.6$ m a^{-1} and $c = 16$ m, and assume the ablation zone is 2 km wide. [*Hint*: Use the incompressibility condition, $du/dx = -dw/dz$, and the result from problem 5.5 to get $w(x)$. Then use the definition of velocity, $u = dx/dt$, and the initial condition, $x(t = 0) = x_o$, to integrate the expression for u to get $x(t) = (x_o^{1/2} - \frac{1}{2}\alpha t)^2$. Then use $w = dz/dt$, and the initial condition, $z(0) = z_o$ to obtain $z(t) = z_o (x_o/x)^{1/2}$.]

5.7 The accumulation zone in the above problem is 10 km wide (see figure below). At the end of the Pleistocene a mammoth dies 500 m from the divide. Determine the x and z coordinates of the point where he melts out, and his time en route. Assume the glacier has had a balanced budget for the last 10,000 y, that the accumulation rate is uniform over the accumulation zone, and that the ablation rate is 0.6 m a^{-1} over the 2-km-wide ablation zone, as before. Plot the path. [*Hint*: Do as in problems 5.5 and 5.6, remembering that the horizontal velocity is now $u = -A_c(L - x)/\sqrt{cx}$, where L is the distance from the margin to the divide. You will encounter an integral $\int \sqrt{x}/(L-x)\,dx$, which may be transformed using $r = L - x$ and then evaluated using the tabulated integral:

$$\int \frac{\sqrt{L-r}}{r}\,dr = 2\sqrt{L-r} + L\left(\frac{1}{\sqrt{L}}\ln\frac{\sqrt{L-r}-\sqrt{L}}{\sqrt{L-r}+\sqrt{L}}\right).$$

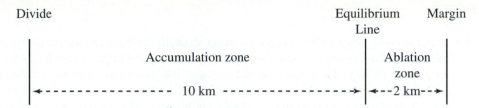

Divide   Accumulation zone   Equilibrium Line   Margin   Ablation zone   ← 10 km → ← 2 km →

5.8 If the mammoth was 3 m long and, when he died, he was lying down with his tail 3 m closer to the divide than his head, determine the time required for his body to pass completely beneath the equilibrium line, and his approximate length when he is at this point in his journey.

CHAPTER 6

6.1 Calculate and plot a temperature profile for an ice sheet that is 1368 m thick, assuming that $\theta_s = -24°C$, $b_n = 0.35$ m a^{-1}, $\kappa = 37.2$ m^2 a^{-1}, $\bar{u} = 0$, and $\beta_G = -0.0228$ °C m^{-1}. Obtain temperatures at least at 0, 200, 500, and 900 m above the bed.

6.2 a. Determine the influence of strain heating on the temperature profile by integrating the energy-balance equation, simplified with the use of the following assumptions:

- Horizontal temperature gradients are negligible
- Steady state
- Zero accumulation (or $w_s = 0$)
- $\dot{\varepsilon} = (\sigma/B)^n$.

Note: This is easier if the z-axis points downward. If you retain a z-axis pointing upward, the signs of β_G below must be changed.

b. Plot the profile for a glacier that is 1000 m thick with:

$$\theta_s = -35 °C \qquad \rho = 900 \text{ kg m}^{-3} \qquad \alpha = 0.01$$
$$\beta_G = 0.0228 °C \text{ m}^{-1} \qquad K = 6.6 \times 10^7 \text{ J m}^{-1} \text{ a}^{-1} °C^{-1} \qquad n = 3$$
$$B = 0.397 \text{ MPa a}^{1/3}$$

To see the effect, you will need to make calculations at about 25-m intervals between 900 and 1000 m depth, and you will need to use a long temperature axis.

6.3 Solve Problem 6.2 but with the additional assumption that strain heating is negligible, and calculate the temperature at the base of this same 1000-m-thick glacier. How much does strain heating increase the basal temperature?

6.4 a. To get a sense of the influence of longitudinal advection, calculate and plot a temperature profile for the glacier in Problem 6.1, assuming $\bar{u} = 15$ m a^{-1}. Use the Column model with the values of θ_s, b_n, κ, H, and β_G given in Problem 6.1, the value of K in Problem 6.2, $\alpha = -0.01$, $\lambda = -0.01$ °C m^{-1}, and $w_b = 0$ m a^{-1}.

b. Compare the result with that from Problem 6.1 in detail.

6.5 a. Obtain an expression for the temperature gradient and the temperature distribution in a stagnant sheet of ice of infinite horizontal extent, and thickness H. Assume that the climate has been warming at a rate of $\dot{\theta}$, and that the interior of the glacier is warming at the same rate.

b. By examining the original differential equation after simplification, explain how the uniform warming rate is accomplished.

CHAPTER 8

8.1 Water flowing along a glacier bed must warm up as the ice thins and the pressure melting point increases. Water flowing up an adverse bed slope must warm up more rapidly, as the ice is thinning more rapidly. The energy needed to warm the water comes from viscous dissipation. Determine how steep the bed slope can get, relative to the surface slope, without exceeding the amount of viscous energy available. Obtain a numerical value for the constant of proportionality between the two slopes.

8.2 The discharge in a horizontal subglacial conduit with a circular cross section is 0.025 m^3 s^{-1}. The water pressure in the conduit is 1.5 MPa and the hydrostatic pressure in the adjacent ice is 2 MPa. The Manning roughness of the conduit is 0.1 $m^{-1/3}$s and the viscosity parameter, B, is 0.16 MPa $a^{1/3}$. Determine the pressure gradient in the conduit, the radius of the conduit, the water velocity in the conduit, and the melt rate on the conduit walls (or closure rate).

8.3 An esker splits as shown on the map below. Stratigraphic relations suggest that the branch around the end of the ridge is younger. Explain why the esker changed course, and estimate the

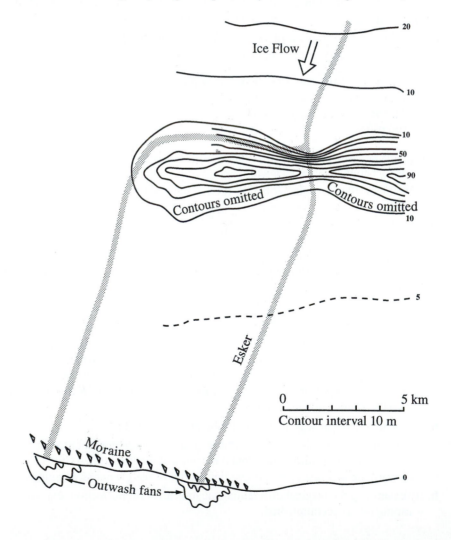

basal shear stress at the time of the change in course. Assume that the glacier had a parabolic profile, $h = \sqrt{cx}$. Assume further that water flow down the potential gradient could be maintained even though some water might be forced to refreeze to keep the temperature at the pressure melting point.

8.4 Consider a glacier with a parabolic profile, $h = \sqrt{16x}$ where x is the horizontal coordinate in meters and h is the surface elevation. Assume that the glacier is 2 km long and is on a horizontal bed. It is drained by a circular conduit at the bed. Calculate and plot the height of the energy grade line (the height to which water would rise in piezometer tubes penetrating to the conduit) as a function of distance from the terminus for discharges of 0.015 m³ s⁻¹, a winter discharge, and 1.0 m³ s⁻¹, a summer discharge. Use a channel Manning roughness of 0.1 m⁻¹ᐟ³s and ice viscosity parameter, $B = 0.06$ MPa a¹ᐟ³ . Assume that the conduit is at atmospheric pressure within 50 m of the margin. (*Hint*: As the integration has to be carried out numerically, you might want to write a short program to do the calculations.)

CHAPTER 9

9.1 Using equation (9.2) for σ_S in terms of σ_{xx}, σ_{yy}, and θ:

   a. Determine the angle θ of the planes on which σ_S is a maximum.

   b. What is the orientation of these planes relative to those on which σ_N is a maximum?

   c. Determine the normal stress, σ_N, on the plane on which σ_S is a maximum.

   d. Determine the magnitude of σ_{Smax}.

   Express all answers in terms of σ_{xx}, σ_{yy}, and σ_{xy}.

9.2 We have shown (Chapter 6) that $\frac{1}{2}\dot{\varepsilon}_{xz}\sigma_{xz} + \frac{1}{2}\dot{\varepsilon}_{zx}\sigma_{zx}$ is the total work done per unit time in a unit volume of ice subjected to simple shear. It is also true that $\frac{1}{2}\dot{\varepsilon}_{xx}\sigma'_{xx}$ is the work done by a normal stress. Thus, the total work done is $W = \frac{1}{2}\dot{\varepsilon}_{ij}\sigma'_{ij}$. Show that because $\dot{\varepsilon}_{ij} = \lambda\sigma'_{ij}$, $W = \dot{\varepsilon}_e\sigma_e$.

9.3 A laboratory ice deformation experiment is run using biaxial compression with applied stresses σ_1 and σ_2 on the faces of a cube. Stresses in the third direction are atmospheric. Strain rates are $\dot{\varepsilon}_1$ and $\dot{\varepsilon}_2$ in the σ_1 and σ_2 directions. Determine the effective stress and the effective strain rate.

9.4 An experimental system is designed to run tests in combined uniaxial compression and simple shear (see sketch below). Determine the effective stress and effective strain rate for this stress configuration.

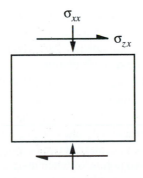

9.5 The third invariant of the stress tensor, J_3, can be interpreted in terms of the stress configuration. To do this, we define a stress configuration parameter, λ, by $\lambda = J_3^*$. Here, J_3^* is a normalized value of J_3 obtained by dividing all of the stresses by a constant factor, c, which you will derive below. The motivation for doing this is that it simplifies the expression for λ. We proceed as follows:

The octahedral shear stress is defined by: $\sigma_o^2 = \frac{1}{3}(\sigma_{ij}'\sigma_{ij}')$. Let us normalize σ_o by dividing all stresses by a constant factor, c, thus:

$$\sigma_o^{2*} = \frac{1}{3c^2}(\sigma_{ij}'\sigma_{ij}').$$

(Here, the $*$ is used to indicate the normalized value.) Let us further select c such that $\sigma_o^* = 2^{1/6}$. Obtain an expression for c in terms of the second invariant of the stress tensor, J_2.

This c is the normalization factor used in calculating J_3^*. Because the normalized stresses must retain the sign of the original stresses, use $|c|$ where necessary.

Show that:

a. The deviatoric stresses in uniaxial compression under a compressive stress, σ_3, are: $(-\sigma_3/3, -\sigma_3/3, 2\sigma_3/3)$, and (remembering that σ_3 is negative) that $\lambda = -1$ for this case.

b. The deviatoric stresses in pure shear under stresses $-\sigma_1$, σ_3, are: $(-\sigma_1, 0, \sigma_3)$, and that $\lambda = 0$ for this case.

c. For the stress configuration in Problem 9.4 above, obtain an expression for λ in terms of σ_{xx} and σ_{xz}, and evaluate this for:

$$\sigma_{xx} = -0.1 \text{ MPa}, \quad \sigma_{xz} = 0.1 \text{ MPa, and}$$
$$\sigma_{xx} = -0.1 \text{ MPa}, \quad \sigma_{xz} = 0 \text{ MPa,}$$

CHAPTER 10

10.1 Calculate and plot velocity profiles for an infinitely wide glacier that is 300 m thick with a surface slope of 0.046. Use $n = 3$ and $B = 0.141$ MPa $a^{1/3}$. Calculate one profile for $\dot{\varepsilon}_{xx} = 0.0$ and one for $\dot{\varepsilon}_{xx} = 0.1$ a^{-1}. Explain the difference between the profiles. Assume a basal velocity, u_b, of 20 m a^{-1}.

CHAPTER 11

11.1 For comparison of borehole deformation rates, with shear stress σ_{zx} and shear strain rate $\dot{\varepsilon}_{zx}$, with deformation rates in uniaxial compression, the stress in uniaxial compression must be multiplied by $1/\sqrt{3}$ and the strain rate by $\sqrt{3}/2$ (Nye, 1953). Show that this is true by calculating σ_e and $\dot{\varepsilon}_e$ for the two stress configurations.

11.2 Starting with equation (11.48), verify equation (11.49). To do this it is necessary to use only one of equations (11.48) (i.e., for $i = 1$).

11.3 Starting with $\dot{\varepsilon} = (\sigma/B)^n$ and $\dot{\varepsilon}_{ij} = \lambda \sigma_{ij}'$, show that the constants n and B in Glen's flow law can be evaluated if one can determine only one component of the stress tensor, the corresponding component of the strain rate tensor, and $\dot{\varepsilon}$ at two or more places. Explain how n and B are determined when there are only two points, or when there are more than two points.

11.4 Determine n and B in MPa $a^{1/n}$ for the following two sets of velocity derivatives.

Depth	Velocity	x	y	z
9.4 m	u	−0.00445	0.00000	−0.01417
	v	0.00000	0.00013	0.00361
	w	0.00099	−0.00040	0.00432
12.5 m	u	−0.00414	0.00000	−0.02296
	v	0.00000	0.00012	0.00393
	w	0.00099	−0.00468	0.00402

The surface slope is 0.152. Depths are in meters and velocities in m a^{-1}. The y-axis is transverse and the z-axis is normal to the surface and directed downward. Data are from the Barnes Ice Cap boreholes.

11.5 Verify that the vertically downward forces on the top of the semi-cylinder of ice in Figure 11.6 balance the vertically upward forces on the bed when the pressure in the ice is $\mathcal{P}$ and the pressure in the conduit is $\Delta\mathcal{P}$.

CHAPTER 12

12.1 We found [equation (5.7)] that one could obtain a first-order estimate of the surface velocity (assuming no sliding) from

$$u_s = \frac{2}{n+1}\left(\frac{\rho g \sin \alpha}{B}\right)^n H^{n+1}, \tag{P12.1}$$

where H is the ice thickness and α is the surface slope. By expressing this in differential form and using the approximation $\alpha \cong \sin \alpha \ (\cong \tan \alpha)$, show that

$$\frac{du_s}{u_s} = (n+1)\frac{dH}{H} + n\frac{d\alpha}{\alpha}. \tag{P12.2}$$

12.2 Using Figure 12.13b:

a. Compare the observed change in velocity of Nisqually Glacier from 1949 to 1951 with that calculated from equation (P12.2) above. Use $h \approx 80$ m.

b. Estimate the change in flux, Δq, and use equation (P12.2) to estimate the fraction of the change that is due to the change in thickness and that due to the change in surface slope, respectively. Then calculate c_0 and D_0 from their definitions, expressed in finite difference form. (Note that q, c_0, and D_0 are defined in terms of a unit width.)

12.3 Consider a perbation, $b_1 = 0.1$ m a^{-1} on Storglaciären, which is 2.8 km long and has a maximum velocity, $u_{0\max} \approx 35$ m a^{-1}. Calculate $\Delta\ell$ and the response time based only on the time needed to accumulate the additional mass. Use the model of Johannesson and others generalized to three dimensions. For this you will need to know the geometry of the glacier (Figs. 8.14 and 11.9), and you will need:

  Glacier area = 3.03×10^6 m^2
  Width at terminus, $W(\ell_o) = 400$ m
  Ablation rate at terminus = 1.3 m a^{-1}

($W(\ell_o)$ is larger than shown in Fig. 11.9 because the topography is such that the width would expand significantly as the glacier thickened here.) Assume that the average ice thickness over a cross section is $\sim\frac{1}{2}$ the thickness shown on the profile. [Remember that ΔV is not $\bar{h}\bar{W}\Delta\ell$ because $\bar{W} > W(\ell_o)$.]

References

ALLEY, R. B. 1989a. Water pressure coupling of sliding and bed deformation: I. Water system. *Journal of Glaciology*, v. 35, no. 119, pp. 108–118.

ALLEY, R. B. 1989b. Water-pressure coupling of sliding and bed deformation: II. Velocity-depth profiles. *Journal of Glaciology*, v. 35, no. 119, pp. 119–129.

ALLEY, R. B. 1991. Deforming bed origin for the southern Laurentide till sheets? *Journal of Glaciology*, v. 37, no. 125, pp. 67–76.

ALLEY, R. B. 1992. Flow-law hypotheses for ice-sheet modeling. *Journal of Glaciology*, v. 38, no. 129, pp. 245–256.

ALLEY, R. B., BLANKENSHIP, D. D., BENTLEY, C. R., and ROONEY, S. T. 1987a. Till beneath Ice Stream B 3. Till deformation: Evidence and implications. *Journal of Geophysical Research*, v. 92, no. B9, pp. 8921–8929.

ALLEY, R. B., BLANKENSHIP, D. D., ROONEY, S. T., and BENTLEY, C. R. 1987b. Till beneath Ice Stream B 4. A coupled ice-till flow model. *Journal of Geophysical Research*, v. 92, no. B9, pp. 8931–8940.

ALLEY, R. B., and WHILLANS, I. M. 1991. Changes in the West Antarctic Ice Sheet. *Science*, v. 254, no. 5034, pp. 259–263.

ALLEY, R. B., et al. 1993. Abrupt increase in Greenland snow accumulation at the end of the Younger Dryas event. *Nature*, v. 362, no. 6420, pp. 527–529.

ATKINSON, B. K. 1984. Subcritical crack growth in geological materials. *Journal of Geophysical Research*, v. 89, no. B6, pp. 4077–4144.

ATKINSON, B. K., and RAWLINGS, R. D. 1981. Acoustic emission during stress corrosion cracking in rocks. In D. W. SIMPSON and P. G. RICHARDS (Eds.), E*arthquake Prediction. An International Review* (Ewing Series, v. 4). Washington, DC: American Geophysical Union, pp. 605–619.

BAKER, R. W. 1978. The influence of ice-crystal size on creep. *Journal of Glaciology*, v. 21, no. 85, pp. 485–500.

BAKER, R. W. 1981. Textural and crystal-fabric anisotropies and the flow of ice masses. *Science*, v. 211, no. 4486, pp. 1043–1044.

BAKER, R. W. 1982. A flow equation for anisotropic ice. *Cold Regions Science and Technology*, v. 6, no. 3, pp. 141–148.

BAKER, R. W. 1987. Is the creep of ice really independent of the third deviatoric stress invariant? In: *The Physical Basis for Ice Sheet Modelling. Proceedings of the Vancouver Symposium*, 1987, International Association of Hydrological Sciences Publication 170, pp. 7–16.

BARNES, P., TABOR, D., and WALKER, J. C. F. 1971. Friction and creep of polycrystalline ice. *Proceedings of the Royal Society of London*, Series A, v. 324, no. 1557, pp. 127–155.

BENOIST, J. -P. 1979. The spectral power density and shadowing function of a glacial microrelief at the decimetre scale. *Journal of Glaciology*, v. 23, no. 89, pp. 57–66.

BENSON, C. S. 1961. Stratigraphic studies in the snow and firn of the Greenland Ice Sheet. *Folia Geographica Danica*, v. 9, pp. 13–37.

BENSON, C. S. 1962. Stratigraphic studies in the snow and firn of the Greenland Ice Sheet. *U.S. Snow, Ice, and Permafrost Research Establishment*. Research Report 70.

BIEGEL, R. L., SAMMIS, C. G., and DIETERICH, J. H. 1989. The frictional properties of simulated gouge having a fractal particle distribution. *Journal of Structural Geology*, v. 11, no. 7, pp. 827–846.

BINDSCHADLER, R. A., and SCAMBOS, T. A. 1991. Satellite-image-derived velocity field of an Antarctic ice stream. *Science*, v. 252, no. 5003, pp. 242–246.

BJÖRNSSON, H. 1992. Jökulhlaups in Iceland: Prediction, characteristics, and simulation. *Annals of Glaciology*, v. 16, pp. 95–106.

BLANKENSHIP, D. D., BENTLEY, C. R., ROONEY, S. T., and ALLEY, R. B. 1986. Seismic measurements reveal a saturated, porous layer beneath an active Antarctic ice stream. *Nature*, v. 322, no. 6074, pp. 54–57.

BODVARSSON, G. 1955. On the flow of ice sheets and glaciers. *Jökull*, v. 5, pp. 1–8.

BOULTON, G. S., and HINDMARSH, R. C. A. 1987. Sediment deformation beneath glaciers: Rheology and geological consequences. *Journal of Geophysical Research*, v. 92, no. B9, pp. 9059–9082.

BROECKER, W. S. 1994. Massive iceberg discharges as triggers for global climate change. *Nature*, v. 372, no. 6505, pp. 421–424.

BROWN, N. L., HALLET, B., and BOOTH, D. B. 1987. Rapid soft-bed sliding of the Puget glacial lobe. *Journal of Geophysical Research*, v. 92, no. B9, pp. 8985–8997.

BRUGGER, K. A. 1992. *A comparative study of the response of Rabots Glaciär and Storglaciären to recent climatic change.* Ph.D. thesis, University of Minnesota, Minneapolis, 295 pp.

BUDD, W. F. 1969. The dynamics of ice masses. *Australian National Antarctic Expeditions Scientific Reports*, Series A (IV) Glaciology. Publication No. 108.

BUDD, W. F., JENSEN, D., and RADOK, U. 1971. Derived physical characteristics of the Antarctic Ice Sheet. *Australian National Antarctic Expeditions Interim Reports*, Series A (IV) Glaciology. Publication No. 120.

BUDD, W. F., KEAGE, P. L., and BUNDY, N. A. 1979. Empirical studies of ice sliding. *Journal of Glaciology*, v. 23, no. 89, pp. 157–170.

BUDD, W. F., and JACKA, T. H. 1989. A review of ice rheology for ice sheet modelling. *Cold Regions Science and Technology*, v. 16, no. 2, pp. 107–144.

CLARK, P. U., and WALDER, J. S. 1994. Subglacial drainage, eskers, and deforming beds beneath the Laurentide and Eurasian ice sheets. *Geological Society of America Bulletin*, v. 106, no. 2, pp. 304–314.

CLARK, P. U., and HANSEL, A. R. 1989. Clast ploughing, lodgement and glacier sliding over a soft glacier bed. *Boreas*, v. 18, no. 3, pp. 201–207.

DAHL-JENSEN, D., and GUNDESTRUP, N. S. 1987. Constitutive properties of ice at Dye 3, Greenland. In: *The Physical Basis for Ice Sheet Modelling. Proceedings of the Vancouver Symposium*, 1987, International Association of Hydrological Sciences Publication 170, pp. 31–43.

DEELEY, R. M., and PARR, P. H. 1914. The Hintereis Glacier. *Philosophical Magazine*, v. 6, pp. 153–176.

DE LA CHAPELLE, S., DUVAL, P., and BAUDET, B. 1995. Compressive creep of polycrystalline ice containing a liquid phase. *Scripta Metallurgica et Materialia*, v. 33, no. 3, pp. 447–450.

DEMOREST, M. 1941. Glacier flow and its bearing on the classification of glaciers. *Geological Society of America Bulletin*, v. 52, no. 12, pp. 2024–2025.

DEMOREST, M. 1942. Glacier regimens and ice movement within glaciers. *American Journal of Science*, v. 240, no. 1, pp. 31–66.

DRAKE, L., and SHREVE, R. L. 1973. Pressure melting and regelation of ice by round wires. *Proceedings of the Royal Society of London*, Series A, v. 332, no. 1588, pp. 51–83.

DUVAL, P. 1977. The role of water content on the creep rate of polycrystalline ice. In: *Isotopes and Impurities in Snow and Ice. Proceedings of the Grenoble Symposium*, Aug.–Sept. 1975, International Association of Hydrological Sciences Publication 118, pp. 29–33.

DUVAL, P. 1978. Anelastic behavior of polycrystalline ice. *Journal of Glaciology*, v. 21, no. 85, pp. 621–628.

DUVAL, P., ASHBY, M. F., and ANDERMAN, I. 1983. Rate-controlling processes in the creep of polycrystalline ice. *Journal of Physical Chemistry*, v. 87, no. 21, pp. 4066–4074.

DUVAL, P., and CASTELNAU, O. 1995. Dynamic recrystallization of ice in polar ice sheets. *Journal de Physique*, IV, Colloque C3, supplement to *Journal de Physique*, III, v. 5, pp. C3-197–C3-205.

ENGELHARDT, H. F., HARRISON, W. D., and KAMB, B. 1978. Basal sliding and conditions at the glacier bed as revealed by bore-hole photography. *Journal of Glaciology*, v. 20, no. 84, pp. 469–508.

ENGELHARDT, H. F., HUMPHREY, N., KAMB, B., and FAHNESTOCK, M. 1990. Physical conditions at the base of a fast-moving Antarctic ice stream. *Science*, v. 248, no. 4951, pp. 57–59.

EYLES, N., SALDEN, J. A., and GILROY, S. 1982. A depositional model for stratigraphic complexes and facies superimposition in lodgement till. *Boreas*, v. 11, no. 4, pp. 317–333.

FISHER, D. A., and KOERNER, R. M. 1986. On the special rheological properties of ancient microparticle-laden Northern Hemisphere ice as derived from bore-hole and core measurements. *Journal of Glaciology*, v. 32, no. 112, pp. 501–510.

FOUNTAIN, A. G. 1989. The storage of water in, and hydraulic characteristics of, the firn of South Cascade Glacier, Washington State, U.S.A. *Annals of Glaciology*, v. 13, pp. 69–75.

FOWLER, A. C. 1987. Sliding with cavity formation. *Journal of Glaciology*, v. 33, no. 115, pp. 255–267.

GLASSTONE, S., LAIDLER, K. J., and Eyring, H. 1941. *The Theory of Rate Processes*. New York: McGraw-Hill.

GLEN, J. W. 1955. The creep of polycrystalline ice. *Proceedings of the Royal Society*, Series A, no. 1175, v. 228, pp. 519–538.

GOLD, L. W. 1958. Some observations on the dependence of strain on stress for ice. *Canadian Journal of Physics*, v. 36, no. 10, pp. 1265–1275.

GOLDTHWAIT, R. P. 1951. Development of end moraines in east-central Baffin Island. *Journal of Geology*, v. 59, no. 6, pp. 567–577.

GRIFFITH, A. A. 1924. Theory of rupture. *Proceedings of the First International Congress Applied Mechanics*, Delft, pp. 55–63.

GROVE, J. M. 1988. *The Little Ice Age*. London: Methuen.

HAEFELI, R. 1962. The ablation gradient and the retreat of a glacier tongue. *Symposium of Obergurgl*, International Association of Hydrological Sciences Publication 58, pp. 49–59.

HALLET, B. 1976a. Deposits formed by subglacial precipitation of $CaCO_3$. *Geological Society of America Bulletin*, v. 87, no. 7, pp. 1003–1015.

HALLET, B. 1976b. The effect of subglacial chemical processes on sliding. *Journal of Glaciology*, v. 17, no. 76, pp. 209–221.

HALLET, B. 1979. Subglacial regelation water film. *Journal of Glaciology*, v. 23, no. 89, pp. 321–334.

HALLET, B., LORRAIN, R. D., and SOUCHEZ, R. A. 1978. The composition of basal ice from a glacier sliding over limestones. *Geological Society of America Bulletin*, v. 89, no. 2, pp. 314–320.

HAMILTON, W. C., and IBERS, J. A. 1968. *Hydrogen Bonding in Solids: Methods of Molecular Structure Determination*. New York: W.A. Benjamin.

HARRISON, W. D. 1972. Temperature of a temperate glacier. *Journal of Glaciology*, v. 11, no. 61, pp. 15–29.

HAUSMANN, M. R. 1990. *Engineering Principles of Ground Modification*. New York: McGraw-Hill.

HAYS, J. D., IMBRIE, J., and SHACKLETON, N. S. 1976. Variations in the Earth's orbit: Pacemaker of the ice ages. *Science*, v. 194, no. 4270, pp. 1121–1132.

HOBBS, P. V. 1974. *Ice Physics*. New York: Oxford University Press.

HOCK, R., and HOOKE, R. LeB. 1993. Further tracer studies of internal drainage in the lower part of the ablation area of Storglaciären, Sweden. *Geological Society of America Bulletin*, v. 105, no. 4, pp. 537–546.

HODGE, S. M. 1974. Variations in sliding of a temperate glacier. *Journal of Glaciology*, v. 13, no. 69, pp. 349–369.

HOLMLUND, P. 1987. Mass balance of Storglaciären during the 20th century. *Geografiska Annaler*, v. 69A, no. 3–4, pp. 439–447.

HOLMLUND, P. 1988. Internal geometry and evolution of moulins, Storglaciären, Sweden. *Journal of Glaciology*, v. 34, no. 117, pp. 242–248.

HOOKE, R. LeB. 1970. Morphology of the ice-sheet margin near Thule, Greenland. *Journal of Glaciology*, v. 9, no. 57, pp. 303–324.

HOOKE, R. LeB. 1973a. Flow near the margin of the Barnes Ice Cap and the development of ice-cored moraines. *Geological Society of America Bulletin*, v. 84, no. 12, pp. 3929–3948.

HOOKE, R. LeB. 1973b. Structure and flow in the margin of Barnes Ice Cap, Baffin Island, N.W.T., Canada. *Journal of Glaciology*, v. 12, no. 66, pp. 423–438.

HOOKE, R. LeB. 1976. Pleistocene ice at the base of the Barnes Ice Cap, Baffin Island, N.W.T., Canada. *Journal of Glaciology*, v. 17, no. 75, pp. 49–60.

HOOKE, R. LeB. 1977. Basal temperatures in polar ice sheets: A qualitative review. *Quaternary Research*, v. 7, no. 1, pp. 1–13.

HOOKE, R. LeB. 1981. Flow law for polycrystalline ice in glaciers: Comparison of theoretical predictions, laboratory data, and field measurements. *Reviews of Geophysics and Space Physics*, v. 19, no. 4, pp. 664–672.

HOOKE, R. LeB. 1984. On the role of mechanical energy in maintaining subglacial conduits at atmospheric pressure. *Journal of Glaciology*, v. 30, no. 105, pp. 180–187.

HOOKE, R. LeB. 1989. Englacial and subglacial hydrology: A qualitative review. *Arctic and Alpine Research*, v. 21, no. 3, pp. 221–233.

HOOKE, R. LeB. 1991. Positive feedbacks associated with the erosion of glacial cirques and overdeepenings. *Geological Society of America Bulletin*, v. 103, no. 8, pp. 1104–1108.

HOOKE, R. LEB., CALLA, P., HOLMLUND, P., NILSSON, M., and STROEVEN, A. 1989. A three-year record of seasonal variations in surface velocity, Storglaciären, Sweden. *Journal of Glaciology*, v. 35, no. 120, pp. 235–247.

HOOKE, R. LEB., and CLAUSEN, H.B. 1982. Wisconsin and Holocene $\delta^{18}O$ variations, Barnes Ice Cap, Canada. *Geological Society of America Bulletin*, v. 93, no. 8, pp. 784–789.

HOOKE, R. LEB., and ELVERHØI, A. 1996. Sediment flux from a fjord during glacial periods, Isfjorden, Spitsbergen. *Global and Planetary Change*, v. 12, pp. 237–249.

HOOKE, R. LEB., GOULD, J. E., and BRZOZOWSKI, J. 1983. Near-surface temperatures near and below the equilibrium line on polar and subpolar glaciers. *Zeitschrift für Gletscherkunde und Glazialgeologie*, v. 19, no. 1, pp. 1–25.

HOOKE, R. LEB., and HANSON, B. H. 1986. Borehole deformation experiments, Barnes Ice Cap, Canada. *Cold Regions Science and Technology*, v. 12, no. 3, pp. 261–276.

HOOKE, R. LEB., HANSON, B., IVERSON, N. R., JANSSON, P., and FISCHER, U. H. In press. Rheology of till beneath Storglaciären, Sweden. *Journal of Glaciology*.

HOOKE, R. LEB., and HUDLESTON, P. J. 1980. Ice fabrics in a vertical flowplane, Barnes Ice Cap, Canada. *Journal of Glaciology*, v. 25, no. 92, pp. 195–214.

HOOKE, R. LEB., and HUDLESTON, P. J. 1981. Ice fabrics from a borehole at the top of the South Dome, Barnes Ice Cap, Baffin Island. *Geological Society of America Bulletin*, v. 92, no. 5, pp. 274–281.

HOOKE, R. LEB., and IVERSON, N. R. 1995. Grain-size distribution in deforming subglacial tills: Role of grain fracture. *Geology*, v. 23, no. 1, pp. 57–60.

HOOKE, R. LEB., JOHNSON, G. W., BRUGGER, K. A., HANSON, B. and HOLDSWORTH, G. 1987. Changes in mass balance, velocity, and surface profile along a flow line on Barnes Ice Cap, 1970–1984. *Canadian Journal of Earth Sciences*, v. 24, no. 8, pp. 1550–1561.

HOOKE, R. LEB., LAUMANN, T., and KOHLER, J. 1990. Subglacial water pressures and the shape of subglacial conduits. *Journal of Glaciology*, v. 36, no. 122, pp. 67–71.

HOOKE, R. LEB., and POHJOLA, V. A. 1994. Hydrology of a segment of a glacier situated in an overdeepening, Storglaciären, Sweden. *Journal of Glaciology*, v. 40, no. 134, pp. 140–148.

HOOKE, R. LEB., POHJOLA, V., JANSSON, P., and KOHLER, J. 1992. Intra-seasonal changes in deformation profiles revealed by borehole studies, Storglaciären, Sweden. *Journal of Glaciology*, v. 38, no. 130, pp. 348–358.

HULL, D. 1969. *Introduction to Dislocations*. New York: Pergamon Press.

IKEN, A. 1981. The effect of the subglacial water pressure on the sliding velocity of a glacier in an idealized numerical model. *Journal of Glaciology*, v. 27, no. 97, pp. 407–421.

IKEN, A., and BINDSCHADLER, R. A. 1986. Combined measurements of subglacial water pressure and surface velocity of Findelengletscher, Switzerland: Conclusions about the drainage system and sliding mechanism. *Journal of Glaciology*, v. 32, no. 110, pp. 101–119.

IKEN, A., and TRUFFER, M. In press. The relationship between subglacial water pressure and velocity of Findelengletscher during its advance and retreat. *Journal of Glaciology*.

IVERSON, N. 1989. *Theoretical and Experimental Analyses of Glacial Abrasion and Quarrying*. Ph.D. thesis, University of Minnesota, Minneapolis, 233 pp.

IVERSON, N. R. 1991. Potential effects of subglacial water-pressure fluctuations on quarrying. *Journal of Glaciology*, v. 37, no. 125, pp. 27–36.

IVERSON, N. R. 1993. Regelation of ice through debris at glacier beds: Implications for sediment transport. *Geology*, v. 21, no. 6, pp. 559–562.

IVERSON, N. R., HANSON, B., HOOKE, R. LEB., and JANSSON, P. 1995. Flow mechanics of glaciers on soft beds. *Science*, v. 267, no. 5194, pp. 80–81.

IVERSON, N. R., HOOYER, T.S., and HOOKE, R. LEB. 1996. A laboratory study of sediment deformation: Stress heterogeneity and grain-size evolution. *Annals of Glaciology*, v. 22, pp. 167–175.

JACKA, T. H. 1984. Laboratory studies on the relationship between ice crystal size and flow rate. *Cold Regions Science and Technology*, v. 10, no. 1, pp. 31–42.

JANSSON, E. P. 1995. Water pressure and basal sliding on Storglaciären, northern Sweden. *Journal of Glaciology*, v. 41, no. 138, pp. 232–240.

JEZEK, K. C., ALLEY, R. B., and THOMAS, R. H. 1985. Rheology of glacier ice. *Science*, v. 227, no. 4692, pp. 1335–1337.

JÓHANNESSON, T., RAYMOND, C. F., and WADDINGTON, E. 1989. Time-scale for adjustment of glaciers to changes in mass balance. *Journal of Glaciology*, v. 35, no. 121, pp. 355–369.

JÓHANNESSON, T. 1995. Written communication dated December 23, 1995.

JÓHANNESSON, T. 1996. Written communications dated November 7 and 14, 1996.

JOHNSON, A. 1960. Variation in surface elevation of the Nisqually glacier Mt. Rainier, Washington. *International Association of Scientific Hydrology Bulletin* 19, pp. 54–60.

JOHNSON, W., and MELLOR, P. B. 1962. *Plasticity for Mechanical Engineers*. New York: Van Nostrand. (There is also a 1973 edition, in which the relevant pages are 44–49.)

JOHNSEN, S. J., DANSGAARD, W., CLAUSEN, H. B., and LANGWAY, C. C., Jr. 1972. Oxygen isotope profiles through the Antarctic and Greenland ice sheets. *Nature*, v. 235, no. 5339, pp. 429–434.

JONES, S. J., and CHEW, H. A. M. 1983. Effect of sample and grain size on the compressive strength of ice. *Annals of Glaciology*, v. 4, p. 129–132.

KAMB, B. 1965. Structure of Ice VI. *Science*, v. 150, no. 3693, pp. 205–209.

KAMB, B. 1970. Sliding motion of glaciers: Theory and observation. *Reviews of Geophysics and Space Physics*, v. 8, no. 4, pp. 673–728.

KAMB, B. 1987. Glacier surge mechanism based on linked cavity configuration of the basal water conduit system. *Journal of Geophysical Research*, v. 92, no. B9, pp. 9083–9100.

KAMB, B. 1991. Rheological nonlinearity and flow instability in the deforming bed mechanism of ice stream motion. *Journal of Geophysical Research*, v. 96, no. B10, pp. 16, 585–16, 595.

KAMB, B. and LACHAPELLE, E. 1964. Direct observation of the mechanism of glacier sliding over bedrock. *Journal of Glaciology*, v. 5, no. 38, pp. 159–172.

KAMB, B., RAYMOND, C. F., HARRISON, W. D., ENGELHARDT, H., ECHELMEYER, K. A., HUMPHREY, N., BRUGMAN, M. M., and PFEFFER, T. 1985. Glacier surge mechanism: 1982–1983 surge of Variegated Glacier, Alaska. *Science*, v. 227, no. 4686, pp. 469–479.

KENDALL, K. 1978. The impossibility of comminuting small particles by compression. *Nature*, v. 272, no. 5655, pp. 710–711.

KETCHAM, W. M., and HOBBS, P. V. 1969. An experimental determination of the surface energies of ice. *Philosophical Magazine*, 8th Series, v. 19, no. 162, pp. 1161–1173.

KINOSITA, S. 1962. Transformation of snow into ice by plastic compression. *Low Temperature Science*, Series A, v. 20, pp. 131–157.

KUHN, M. 1981. *Climate and Glaciers*. International Association of Scientific Hydrology Publication 131, pp. 3–20.

KUHN, M. 1989. The response of the equilibrium line altitude to climate fluctuations: Theory and observations. In J. Oerlemans (Ed.), *Glacier Fluctuations and Climatic Change*. Dordrecht: Kluwer Academic Publishers, pp. 407–417.

LI, J., JACKA, T. H., and BUDD, W. F. 1996. Deformation rates in combined compression and shear for ice which is initially isotropic and after the development of strong anisotropy. *Annals of Glaciology*, v. 23, pp. 247–252.

LIGHTHILL, M. J., and WHITMAN, G. B. 1955. On kinematic waves: I. Flood movement in long rivers. *Proceedings of the Royal Society*, Series A, v. 229, no. 1178, pp. 281–316.

LIU, C. -H., NAGEL, S. R., SCHECTER, D. A., COPPERSMITH, S. N., MAJUMDAR, S., NARAYAN, O., and WITTEN, T. A. 1995. Force fluctuations in bead packs. *Science*, v. 269, no. 5223, pp. 513–515.

LLIBOUTRY, L. 1964. *Traité de Glaciologie* (Vol. 1). Paris: Masson.

LLIBOUTRY, L. 1968. General theory of subglacial cavitation and sliding of temperate glaciers. *Journal of Glaciology*, v. 7, no. 49, pp. 21–58.

LLIBOUTRY, L. 1970. Ice flow law from ice sheet dynamics. *Proceedings of the International Symposium on Antarctic Glaciological Exploration*, Hanover, NH, September, 3–7, 1968. International Association of Hydrological Sciences Publication 86, pp. 216–228.

LLIBOUTRY, L. 1971. Permeability, brine content, and temperature of temperate ice. *Journal of Glaciology*, v. 10, no. 58, pp. 15–30.

LLIBOUTRY, L. 1975. Loi de glissement d'un glacier sans cavitation. *Annals of Geophysics*, v. 31, no. 2, pp. 207–226.

LLIBOUTRY, L. 1983. Modifications to the theory of intraglacial waterways for the case of subglacial ones. *Journal of Glaciology*, v. 29, no. 102, pp. 216–226.

LOEWE, F. 1970. Screen temperatures and 10 m temperatures. *Journal of Glaciology*, v. 9, no. 56, pp. 263–268.

MACAYEAL, D. R. 1989. Large-scale ice flow over a viscous basal sediment: Theory and application to Ice Stream B, Antarctica. *Journal of Geophysical Research*, v. 94, no. B4, pp. 4071–4087.

MACAYEAL, D. R. 1993a. A low-order model of the Heinrich event cycle. *Paleoceanography*, v. 8, no. 6, pp. 767–773.

MACAYEAL, D. R. 1993b. Binge/purge oscillations of the Laurentide Ice Sheet as a cause of the North Atlantic's Heinrich events. *Paleoceanography*, v. 8, no. 6, pp. 775–784.

MANDL, G., DE JONG, L. N. J., and MALTHA, A. 1977. Shear zones in granular material: An experimental study of their structure and mechanical genesis. *Rock Mechanics*, v. 9, no. 2–3, pp. 95–144.

MARTINERIE, P., RAYNAUD, D., ETHERIDGE, D. M., BARNOLA, J. -M., and MAZAUDIER, D. 1992. Physical and climatic parameters which influence the air content in polar ice. *Earth and Planetary Science Letters*, v. 112, no. 1/4, pp. 1–13.

MEIER, M. F. 1961. Mass budget of South Cascade Glacier, 1957–1960. U.S. Geological Survey Professional Paper 424-B, pp. 206–211.

MEIER, M. F. 1962. Proposed definitions for glacier mass balance terms. *Journal of Glaciology*, v. 4, no. 33, pp. 252–263.

MEIER, M. F. 1965. Glaciers and climate. In H.E. WRIGHT JR. and D.G. FREY (Eds.), *The Quaternary of the United States*. Princeton, NJ: Princeton University Press, pp. 795–805.

MELLOR, M., and TESTA, R. 1969. Effect of temperature on the creep of ice. *Journal of Glaciology*, v. 8, no. 52, pp. 131–145.

MITCHELL, J. K. 1993. *Fundamentals of Soil Behavior* (2nd ed.). New York: John Wiley.

MITCHELL, J. K., CAMPANELLA, R.G., and SINGH, A. 1968. Soil creep as a rate process. *Journal of the Soil Mechanics and Foundations Division, American Society of Civil Engineers*, v. 94, no. SM1, pp. 231–253.

MOOERS, H. D. 1990. A glacial-process model: The role of spatial and temporal variations in glacier thermal regime. *Geological Society of America Bulletin*, v. 102, no. 2, pp. 243–251.

MORAN, S. R., CLAYTON, L., HOOKE, R. LeB., FENTON, M. M., and ANDRIASHEK, L. D. 1980. Glacier bed landforms of the prairie region of North America. *Journal of Glaciology*, v. 25, no. 93, pp. 457–476.

MÜLLER, F. 1962. Zonation in the accumulation area of the glaciers of Axel Heiberg Island, N.W.T., Canada. *Journal of Glaciology*, v. 4, no. 33, pp. 302–318.

NAKASE, A., and KAMEI, T. 1986. Influence of strain rate on undrained shear strength characteristics of K_o-consolidated cohesive soils. *Soils and Foundations*, v. 26, pp. 85–95.

NYE, J. F. 1951. The flow of glaciers and ice sheets as a problem in plasticity. *Proceedings of the Royal Society of London*, Series A, v. 207, no. 1091, pp. 554–572.

NYE, J. F. 1952a. Reply to Mr. Joel E. Fisher's comments. *Journal of Glaciology*, v. 2, no. 11, pp. 52–53.

NYE, J. F. 1952b. Mechanics of glacier flow. *Journal of Glaciology*, v. 2, no. 12, pp. 82–93.

NYE, J. F. 1953. The flow law of ice from measurements in glacier tunnels, laboratory experiments, and the Jungfraufirn borehole experiment. *Proceedings of the Royal Society of London*, Series A, v. 219, no. 1139, pp. 477–489.

NYE, J. F. 1957. The distribution of stress and velocity in glaciers and ice sheets. *Proceedings of the Royal Society of London*, Series A, v. 239, no. 1216, pp. 113–133.

NYE, J. F. 1960. The response of glaciers and ice sheets to seasonal and climatic changes. *Proceedings of the Royal Society*, Series A, v. 256, no. 1287, pp. 559–584.

NYE, J. F. 1963a. On the theory of the advance and retreat of glaciers. *Geophysical Journal of the Royal Astronomical Society*, v. 7, no. 4, pp. 432–456.

NYE, J. F. 1963b. The response of glaciers to changes in the rate of nourishment and wastage. *Proceedings of the Royal Society of London*, Series A, v. 257, no. 1360, pp. 87–112.

NYE, J. F. 1965a. The flow of a glacier in a channel of rectangular, elliptic, or parabolic cross section. *Journal of Glaciology*, v. 5, no. 41, pp. 661–690.

NYE, J. F. 1965b. A numerical method for inferring the budget history of a glacier from its advance and retreat. *Journal of Glaciology*, v. 5, no. 41, pp. 589–607.

NYE, J. F. 1969. The calculation of sliding of ice over a wavy surface using a Newtonian viscous approximation. *Proceedings of the Royal Society of London*, Series A, v. 311, no. 1506, pp. 445–467.

NYE, J. F. 1973. The motion of ice past obstacles. In E. WHALLEY, S.J. JONES and L.W. GOLD (Eds.), *The Physics and Chemistry of Ice*. Ottawa: Royal Society of Canada, pp. 387–394.

NYE, J. F., and MAE, S. 1972. The effect of non-hydrostatic stress on intergranular water veins and lenses in ice. *Journal of Glaciology*, v. 11, no. 61, pp. 81–101.

NYE, J. F., and FRANK, F. C. 1973. Hydrology of intergranular veins in a temperate glacier. *IUGG-AIHS Symposium on the Hydrology of Glaciers, Cambridge, September 7–13, 1969*. International Association of Scientific Hydrology Publication 95, pp. 157–161.

PARKER, G. 1979. Hydraulic geometry of active gravel rivers. *Journal of the Hydraulics Division, American Society of Civil Engineers*, v. 105, no. HY9, pp. 1185–1201.

PATERSON, W. S. B. 1971. Temperature measurements in Athabasca Glacier, Alberta, Canada. *Journal of Glaciology*, v. 10, no. 60, pp. 339–349.

PATERSON, W. S. B. 1977. Secondary and tertiary creep of glacier ice as measured by borehole closure rates. *Reviews of Geophysics and Space Physics*, v. 15, no. 1, pp. 47–55.

PATERSON, W. S. B. 1994. *Physics of Glaciers* (3rd ed.). New York: Pergamon Press.

PHILBERTH, K., and FEDERER, B. 1971. On the temperature profile and age profile in the central part of cold ice sheets. *Journal of Glaciology*, v. 10, no. 58, pp. 3–14.

PIMIENTA, P., and DUVAL, P. 1987. Rate-controlling processes in the creep of polar glacier ice. *Journal de Physique*, v. 48, Colloque C1, Supplement to no. 3, pp. C1-243–C1-248.

RASMUSSEN, E. M. 1984. El Niño: The ocean/atmosphere connection. *Oceanus*, v. 27, no. 2, pp. 5–12.

RATCLIFFE, E. H. 1962. Thermal conductivity of ice. *Philosophical Magazine*, 8th Series, v. 7, pp. 1197–1203.

RAYMOND, C. F. 1971. Flow in a transverse section of Athabasca Glacier, Alberta, Canada. *Journal of Glaciology*, v. 10, no. 58, pp. 55–84.

RAYMOND, C. F. 1973. Inversion of flow measurements for stress and rheological parameters in a valley glacier. *Journal of Glaciology*, v. 12, no. 64, pp. 19–44.

RAYMOND, C. F., and HARRISON, W. D. 1975. Some observations on the behavior of liquid and gas phases in temperate glacier ice. *Journal of Glaciology*, v. 14, no. 71, pp. 213–234.

RAYMOND, C. F., and HARRISON, W. D. 1988. Evolution of Variegated Glacier, U.S.A., prior to its surge. *Journal of Glaciology*, v. 34, no. 117, pp. 154–165.

RAYNAUD, D., JOUZEL, J., BARNOLA, J. -M., CHAPPELLAZ, J., DELMAS, R. J., and LORIUS, C. 1993. The ice core record of greenhouse gases. *Science*, v. 259, no. 5097, pp. 926–934.

RAYNAUD, D., and WHILLANS, I. M. 1982. Air content of the Byrd core and past changes in the West Antarctic Ice Sheet. *Annals of Glaciology*, v. 3, pp. 269–273.

RIGSBY, G. P. 1958. Effect of hydrostatic pressure on velocity of shear deformation of single ice crystals. *Journal of Glaciology*, v. 3, no. 24, pp. 273–278.

ROBIN, G. DeQ. 1955. Ice movement and temperature distribution in glaciers and ice sheets. *Journal of Glaciology*, v. 2, no. 18, pp. 523–532.

ROBIN, G. DeQ. 1970. Stability of ice sheets as deduced from deep temperature gradients. *International Symposium on Antarctic Glaciological Exploration (ISAGE)*, Hanover, NH, September 3–7, 1968. International Association of Hydrological Sciences Publication 86, pp. 141–151.

ROBIN, G. DeQ. 1976. Is the basal ice of a temperate glacier at the pressure melting point? *Journal of Glaciology*, v. 16, no. 74, pp. 183–195.

RÖTHLISBERGER, H. 1972. Water pressure in intra- and subglacial channels. *Journal of Glaciology*, v. 11, no. 62, pp. 177–204.

RÖTHLISBERGER, H., and IKEN, A. 1981. Plucking as an effect of water-pressure variations at the glacier bed. *Annals of Glaciology*, v. 2, pp. 57–62.

RUSSELL-HEAD, D. S., and BUDD, W. F. 1979. Ice-sheet flow properties derived from bore-hole shear measurements combined with ice-core studies. *Journal of Glaciology*, v. 24, no. 90, pp. 117–130.

SAMMIS, C. G., KING, G., and BIEGEL, R. 1987. The kinematics of gouge deformation. *Pure and Applied Geophysics*, v. 125, no. 5, pp. 777–812.

SEABERG, S. Z., SEABERG, J. Z., HOOKE, R. LeB., and WIBERG, D. W. 1988. Character of the englacial and subglacial drainage system in the lower part of the ablation area of Storglaciären, Sweden, as revealed by dye-trace studies. *Journal of Glaciology*, v. 34, no. 117, pp. 217–227.

SEGALL, P. 1984. Rate-dependent extensional deformation resulting from crack growth in rock. *Journal of Geophysical Research*, v. 89, no. B6, pp. 4185–4195.

SHABTAIE, S., and BENTLEY, C. R. 1987, West Antarctic ice streams draining into the Ross Ice Shelf: Configuration and mass balance. *Journal of Geophysical Research*, v. 92, no. B2, pp. 1311–1336.

SHABTAIE, S., and BENTLEY, C. R. 1988. Ice-thickness map of the West Antarctic ice streams by radar sounding. *Annals of Glaciology*, v. 11, pp. 126–136.

SHARP, M. 1982. Modification of clasts in lodgement tills by glacial erosion. *Journal of Glaciology*, v. 28, no. 100, pp. 475–481.

SHREVE, R. L. 1972. Movement of water in glaciers. *Journal of Glaciology*, v. 11, no. 62, pp. 205–214.

SHREVE, R. L. 1985a. Esker characteristics in terms of glacier physics, Katahdin Esker system, Maine. *Geological Society of America Bulletin*, v. 96, no. 5, pp. 639–646.

SHREVE, R. L. 1985b. Late Wisconsin ice-surface profile calculated from esker paths and types, Katahdin esker system, Maine. *Quaternary Research*, v. 23, no. 1, pp. 27–37.

SHREVE, R. L., and SHARP, R. P. 1970. Internal deformation and thermal anomalies in lower Blue Glacier, Mount Olympus, Washington, USA. *Journal of Glaciology*, v. 9, no. 55, pp. 65–86.

SHUMSKII, P. A. 1964. *Principles of Structural Glaciology*. New York: Dover.

SKEMPTON, A. W. 1985. Residual strength of clays in landslides, folded strata, and the laboratory. *Géotechnique*, v. 25, no. 1, pp. 3–18.

SOMMERFELD, R., and LaCHAPELLE, E. 1970. The classification of snow metamorphism. *Journal of Glaciology*, v. 9, no. 55, pp. 3–17.

TAYLOR, L. D. 1963. Structure and fabric on the Burroughs Glacier, south-east Alaska. *Journal of Glaciology*, v. 4, no. 36, pp. 731–752.

THOMAS, R. H. 1973a. The creep of ice shelves: Theory. *Journal of Glaciology*, v. 12, no. 64, pp. 45–53.

THOMAS, R. H. 1973b. The creep of ice shelves: Interpretation of observed behavior. *Journal of Glaciology*, v. 12, no. 64, pp. 55–70.

THOMPSON, L. G., MOSLEY-THOMPSON, E., DANSGAARD, W., and GROOTES, P.M. 1986. The Little Ice Age as recorded in the stratigraphy of the tropical Quelccaya Ice Cap. *Science*, v. 234, no. 4774, pp. 361–364.

TRESCA, M. H. 1864. Mémoire sur l'écoulement des corps solides soumis à de fortes pressions. *Comptes Rendus des Séances de l'Academie des Sciences, Paris*. v. 59, pp. 754–758.

VALLON, M., PETIT, J -R., and FABRE, B. 1976. Study of an ice core to the bedrock in the accumulation zone of an alpine glacier. *Journal of Glaciology*, v. 17, no. 75, pp. 13–28.

VAN BEAVER, H. G. 1971. The significance of the distribution of clasts within the Great Pond esker and adjacent till. Master's thesis, University of Maine, Orono, 61 p.

VAN DER VEEN, C. J., and WHILLANS, I. M. 1989. Force budget: I. Theory and numerical methods. *Journal of Glaciology*, v. 35, no. 119, pp. 53–60.

VAN DE WAL, R. S. W., and OERLEMANS, J. 1995. Response of valley glaciers to climatic change and kinematic waves: A study with a numerical ice flow model. *Journal of Glaciology*, v. 41, no. 137, pp. 142–152.

VON MISES, R. 1913. Mechanik der festen Körper im plastisch-deformablen Zustand. *Nachrichten von der Königlichen Gesellschaft der Wissenschaften zu Göttingen, Mathematisch-physikalische Klasse*, pp. 582–592.

WALDER, J. S. 1982. Stability of sheet flow of water beneath temperate glaciers and implications for glacier surging. *Journal of Glaciology*, v. 28, no. 99, pp. 273–293.

WALDER, J. S., and FOWLER, A. 1994. Channelized subglacial drainage over a deformable bed. *Journal of Glaciology*, v. 40, no. 134, pp. 3–15.

WALTERS, R., and MEIER, M. F. 1989. Variability of glacier mass balances in western North America. *American Geophysical Union, Geophysical Monograph* 55, pp. 365–374.

WEERTMAN, J. 1957a. On sliding of glaciers. *Journal of Glaciology*, v. 3, no. 21, pp. 33–38.

WEERTMAN, J. 1957b. Deformation of floating ice shelves. *Journal of Glaciology*, v. 3, no. 21, pp. 38–42.

WEERTMAN, J. 1964. Glacier sliding. *Journal of Glaciology*, v. 5, no. 39, pp. 287–303.

WEERTMAN, J. 1972. General theory of water flow at the base of a glacier or ice sheet. *Reviews of Geophysics and Space Physics*, v. 10, no. 1, pp. 287–333.

WEERTMAN, J. 1983. Creep deformation of ice. *Annual Reviews of Earth and Planetary Science*, v. 11, pp. 215–240.

WEERTMAN, J., and BIRCHFIELD, G.E. 1983. Stability of sheet water flow under a glacier. *Journal of Glaciology*, v. 29, no. 103, pp. 374–382.

WHILLANS, I. M. 1977. The equation of continuity and its application to the ice sheet near "Byrd" Station, Antarctica. *Journal of Glaciology*, v. 18, no. 80, pp. 359–371.

WHILLANS, I. M., and van der Veen, C. J. 1993. New and improved determinations of velocity of Ice Streams B and C, West Antarctica. *Journal of Glaciology*, v. 39, no. 133, pp. 483–490.

WHILLANS, I. M., and TSENG, Y. -H. 1995. Automatic tracking of crevasses on satellite images. *Cold Regions Science and Technology*, v. 23, no. 2, pp. 201–214.

YARNAL, B. 1984. Relationships between synoptic-scale atmospheric circulation and glacier mass balance in south-western Canada during the International Hydrological Decade, 1965–74. *Journal of Glaciology*, v. 30, no. 105, pp. 188–198.

Index